JPL Publication 96-2

Venus Gravity Handbook

Alexander S. Konopliv
William L. Sjogren

January 1996

National Aeronautics and
Space Administration

Jet Propulsion Laboratory
California Institute of Technology
Pasadena, California

The research described in this publication was carried out by the Jet Propulsion Laboratory, California Institute of Technology, under a contract with the National Aeronautics and Space Administration.

Reference herein to any specific commercial product, process, or service by trade name, trademark, manufacturer, or otherwise does not constitute or imply its endorsement by the United States Government or the Jet Propulsion Laboratory, California Institute of Technology.

Abstract

This report documents the Venus gravity methods and results to date (model MGNP90LSAAP). It is called a handbook in that it contains many useful plots (such as geometry and orbit behavior) that are useful in evaluating the tracking data. We discuss the models that are used in processing the Doppler data and the estimation method for determining the gravity field. With Pioneer Venus Orbiter and Magellan tracking data, the Venus gravity field was determined complete to degree and order 90 with the use of the JPL Cray T3D Supercomputer. The gravity field shows unprecedented high correlation with topography and resolution of features to the 200km resolution. In the procedure for solving the gravity field, other information is gained as well, and, for example, we discuss results for the Venus ephemeris, Love number, pole orientation of Venus, and atmospheric densities. Of significance is the Love number solution which indicates a liquid core for Venus. The ephemeris of Venus is determined to an accuracy of 0.02 mm/s (tens of meters in position), and the rotation period to 243.0194±0.0002 days.

Acknowledgements

We thank Chuck Yoder for his input on the Love number determination, Myles Standish for work with the Venus ephemeris, Nicole Rappaport for the Venus topography model, Connie Dang and Herb Royden for their special efforts in delivering the ionospheric and tropospheric calibrations used in this work, Fred Krogh for help with the numerical integration, and Algis Kucinskas for discussions on the geophysics of Venus. The supercomputing development was funded by a 1995 JPL Director's Discretionary Fund proposal. Special thanks are due Chuck Lawson and Jon Giorgini for their efforts in developing the gravity software for the Cray. The JPL MGSO provided funds to finish and document the task (this report). The Cray Supercomputer used in this investigation was provided by funding from the NASA Offices of Mission to Planet Earth, Aeronautics, and Space Science.

Table of Contents

Abstract iii

Acknowledgements iv

Table of Contents v

List of Figures vii

List of Tables ix

1. Introduction 1
2. Spacecraft Tracking 2
3. Models 5
   - a. Observable 6
   - b. Geometric Models 7
   - c. Dynamic Models 10
4. Estimation Procedure 16
   - a. Determination of Unconstrained Gravity Field 16
   - b. Gravity A Priori 17
5. Gravity Results 23
   - a. Global Gravity Model 23
   - b. Principal Axes 48
   - c. Love Number 50
6. Venus Constants 51
   - a. Venus Ephemeris 51
   - b. Venus Pole and Rotation 54

7. Solutions for Auxiliary Forces 55
    a. Atmospheric Drag 55
    b. Solar Pressure.. 59

8. Summary 60

References 61

Appendix A: PVO Low-Altitude Periapse Information A-1

Appendix B: PVO High-Altitude Periapse Information B-1

Appendix C: Magellan Cycle 4 (Elliptical) Information C-1

Appendix D: Magellan Cycle 5 and 6 (Circular) Information D-1

Appendix E: Magellan X-Band Orbit Information E-1

Appendix F: PVO and Magellan Data Arcs F-1

Appendix G: MGNP90LSAAP Gravity Coefficients to Degree and Order 40 ... G-1

Appendix H: Correlations Between Estimated Parameters H-1

Appendix I: Regional Gravity Maps for MGNP90LSAAP........................... I-1

List of Figures

1. Space91 and Texas UT1 Daily Differences 9
2. Space91 and Texas X-Pole Daily Differences 9
3. Space91 and Texas Y-Pole Daily Differences 9
4. Daily Solar Flux for PVO and Magellan Time Frame 10
5. Radial Acceleration Profile for High-Altitude PVO 15
6. Radial Acceleration Profile for Low-Altitude PVO 15
7. Radial Acceleration Profile for Magellan Cycle 4 15
8. Radial Acceleration Profile for Magellan Cycle 5&6 15
9. Acceleration Profiles for Venus Unconstrained Surface Gravity 19
10. Degree Strength for MGNP90LSAAP 20
11. Acceleration Profiles for Venus Constrained Surface Gravity 22
12. Gravity Correlation with Topography of Surface Constraint Minus Kaula Constraint 22
13. Vertical Gravity of Venus at the Surface (mgals) 25
14. Geoid of Venus (meters) 26
15. Vertical Gravity Uncertainty at the Surface (mgals) 27
16. Venus Geoid Uncertainty (meters) 28
17. Vertical Gravity Error to the Degree Strength 30
18. Unconstrained Vertical Gravity (mgals), Degree=90 31
19. Unconstrained Vertical Gravity (mgals), Degree=75 32
20. Unconstrained Vertical Gravity (mgals), Degree=60 33
21. Unconstrained Vertical Gravity (mgals), Degree=40 34
22. RMS Magnitude Gravity Spectrum 37
23. Correlation of Gravity with Topography and Error Bars 37
24. Comparison of Correlation with Topography for Gravity Solutions 37

25.	Correlation with Topography for N=2 to 10	38
26.	Correlation with Topography for N=11 to 20	39
27.	Correlation with Topography for N=21 to 30	40
28.	Correlation with Topography for N=61 to 90	41
29.	MGNP90LSAAP Admittance and Theoretical Compensation Depths	42
30.	Apparent Depth of Compensation for N=41 to 60	44
31.	Bouguer Gravity for MGNP90LSAAP	45
32.	Isostatic Perturbations for MGNP90LSAAP	46
33.	Isostatic Anomaly Map for Mead Crater	47
34.	Gravity Coefficient Correlation Matrix	48
35.	Ephemeris Solution for PVO wrt DE403	52
36.	Ephemeris Solution for Magellan Cycle 4 wrt DE403	52
37.	Ephemeris Solution for Magellan Cycle 5&6 wrt DE403	52
38.	Venus wrt Earth Radial Position Error	54
39.	PVO Low-Orbit Density Solution at 140 km Altitude	56
40.	Difference of PVO Density Solutions and VIRA Values at Spacecraft Altitude	56
41.	PVO Low-Orbit Atmospheric Lift Solution	56
42.	PVO Low-Orbit Solar Radiation Pressure (GR) Solution	57
43.	PVO Low-Orbit Solar Radiation Pressure (GX) Solution	57
44.	PVO Low-Orbit Solar Radiation Pressure (GY) Solution	57
45.	PVO High-Orbit Solar Radiation Pressure (GR) Solution	58
46.	PVO High-Orbit Solar Radiation Pressure (GX) Solution	58
47.	PVO High-Orbit Solar Radiation Pressure (GY) Solution	58

List of Tables

1. Tracking Station Cylindrical Coordinates 7
2. DSN Plate Motion Velocities in cm/year 7
3. VIRA Atmosphere Model 13
4. Gravitational Constant Times the Mass of Venus (GM in km^3/s^2) 24
5. Gravity Peaks for Venusian Features of Interest 29
6. Comparisons of Spherical Harmonics with Line-of-Sight Reductions (mgals) 35
7. Principal Axes for Venus in Degrees 49
8. Normalized Second Degree Coefficients with Formal Uncertainties ($\times 10^{10}$) 50
9. Love Number Solutions 51
10. Venus and Earth Ephemeris Solutions ($\times 10^9$) 53
11. Venus Pole and Rotation Rate Solutions 55

1. Introduction

Two spacecraft orbiters, Magellan and Pioneer Venus Orbiter (Pioneer 12 or PVO), provide nearly a global gravity data set for Venus. Together they have been used to solve for a 90th degree and order spherical harmonic gravity field of Venus as well as the GM of Venus (gravitational constant times the mass), ephemeris of Venus with respect to the Earth, tide or Love number (k_2) of Venus, pole and rotation rate of Venus, and other addition models such as atmospheric densities and Venus albedo variations. This handbook documents the data used in the reduction and presents the results to date.

Prior to Magellan and PVO, the Mariner 2, 5, and 10 flybys of Venus provided mass estimates and an upper bound on J_2 (Anderson and Efron, 1969 and Howard et al, 1974) showing that the oblateness for Venus was several orders of magnitude smaller than the Earth's oblateness. The Russian Venera 9 and 10 spacecraft provided a more accurate estimate of $J_2 = 4.0 \pm 1.5 \times 10^{-6}$ (Akim et al, 1978). The initial spherical harmonic determinations of the Venus gravity field from PVO were low degree and order solutions by Ananda et al (1980) to degree and order six and Williams et al (1983) to degree and order seven. Mottinger et al (1985) extended the harmonic solution to degree and order ten by using only high-altitude periapse (about 1000 km) data from PVO, and Bills et al (1987) solved for a degree and order 18 field by combining the high-altitude data with low-altitude data arcs. In addition to harmonic analyses, others such as Phillips et al (1979), Sjogren et al (1980, 1983, 1984), Reasenberg et al (1981, 1982), and more recently Reasenberg and Goldberg (1992) have solved for high resolution surface mass distributions either regionally or globally.

With the arrival of faster computers with more memory and disk space, the spherical harmonic solutions have shown a drastic increase in resolution. In support of the Magellan Navigation effort, McNamee et al (1992) reprocessed the low-altitude PVO data to produce a 21st degree and order model. Nerem (1991) with the additional high-altitude PVO data set produced the Preliminary Goddard Venus Model 1, a 36th degree and order field, and ushered in dramatic increases in resolution for harmonic fields. Konopliv (1992) followed with a 42nd degree and order field and Nerem et al (1993) with a 50th degree and order field, all based upon PVO low- and high-altitude data sets.

With Magellan now in orbit, McNamee et al (1993) produced another 21st degree and order field incorporating Magellan high-altitude (periapse altitude of 250 km) data with PVO. After September 1992, Magellan began to be tracked during periapse (altitude of 170 km). Prior to this, the high gain antenna was pointed toward Venus to acquire Synthetic Aperture Radar (SAR) images and no Doppler tracking was obtained within 30 minutes of periapse. Konopliv et al (1993a) produced a 60th degree and order model by combining the PVO data with four months (or about one-half of longitude coverage) of Magellan data. After Magellan successfully aerobraked into a near circular orbit in August of 1993, Konopliv and Sjogren (1994a) produced another 60th degree and order model incorporating much of the near circular orbit data, and Konopliv et al (1994b) produced a 75th degree and order model with all the Magellan gravity data. This report presents the results for a 90th degree and order gravity solution (named MGNP90LSAAP). This model is the highest resolution spherical harmonic model for a planet, including the Earth, that is based upon spacecraft tracking data. The current solutions for the Earth gravity field extend to degree and order 70 for solutions based upon spacecraft tracking data only (Nerem et al, 1994) and to degree and order 360 with surface measurements included (Rapp et al, 1991). Nerem et al (1995) summarize the history of gravity determination.

In addition to the spherical harmonic gravity field, line-of-sight (LOS) accelerations with respect to a spherical harmonic gravity field (degree and order 75, JPL solution MGNP75ISAAP, for the pre-aerobraking data and degree and order 40, JPL gravity solution MGN40E, for the post-aerobraking data) have been produced for the Magellan Doppler residuals. The procedure is to solve for the spacecraft state and other spacecraft specific dynamical models using the base spherical harmonic model. The residuals then contain the remaining gravity signature in the data that is not incorporated into the harmonic field. Orbits are processed one at a time, and so this eliminates long-term modeling errors (and long-term gravity information). The pre-aerobraking LOS data were processed by Barriot and Balmino (1994) to produce detailed gravity maps over Eistla Regio and the western portion of Aphrodite Terra. Kaula (1995), Smrekar (1994), and others have produced regional gravity determinations from the LOS data as well.

2. Spacecraft Tracking

The gravity measurements used for Venus gravity field determination are two-way coherent Doppler tracking of the PVO and Magellan spacecraft acquired at the Deep Space Network (DSN) complexes at Goldstone, California; Madrid, Spain; and Canberra, Australia. The PVO spacecraft operated at S-band with the DSN stations transmitting at 2.11 GHz. The transponder on board the spacecraft then multiplied the frequency by 240/221 to obtain a downlink frequency of 2.30 GHz. For PVO, only two-way data were acquired, i.e., the receiving DSN station is the same as the transmitting station. The spacecraft transponder also provided a minimal amount of X-band downlink data (S-band and X-band were received simultaneously at the ground station), but these data were not processed. The Magellan spacecraft had an S-band transponder with an X to S-band uplink converter and S to X-band downlink converter. The resulting system had either an X-band or S-band uplink and an S-band and/or X-band downlink. The X-band uplink and X-band downlink (8.43 GHz) provided the high resolution gravity data because of reduced charged particle effects on the X-band signal. All two-way Magellan data were processed beginning with the gravity data of cycle four. Three-way data (i.e., the receiving DSN complex is different from the transmitting complex) were also taken for generation of differenced Doppler data in support of navigation, but these data were not incorporated into the gravity solution.

PVO was inserted into orbit about Venus on December 4, 1978 and provided several years of low-altitude periapse data (150 to 170 km) until July of 1980 when no maneuvers were performed to maintain a low periapse altitude. By November of 1980, periapse altitude reached 400 km due to solar perturbations on the orbit. The low-altitude data set is continuous from December 9, 1978 to December 4, 1980 except for a data gap from July 12, 1979 to December 18, 1979, during which superior conjunction occurred and periapse was occulted, and so these data will not add much to the gravity solution. Tracking data were acquired during this time, but these data have not been recovered from archive tapes and may or may not exist. Due to the superior resolution of the Magellan data, the PVO data were compressed substantially to 60 seconds for the interval within thirty minutes each side of periapse, 300 seconds for the next two hours around periapse, and 600 seconds near apoapse. The spacecraft velocity at periapse is 9 km/s and 60 second samples decrease the periapse resolution in the PVO data, but this is easily recovered with the Magellan data and the three to four longitudinal coverages of the PVO data. This compression time still retains the long term gravity, rotational, tidal, and ephemeris information and greatly reduces computer time required for filtering. Future solutions may reintroduce a higher number of samples at periapse. The original PVO data, as processed by Konopliv et al (1993a) and others have 5 second samples at periapse. The total number of Doppler observations processed for the low-altitude PVO data is 170,000.

The time histories of the low-altitude orbit semi-major axis, eccentricity, inclination, latitude at periapse, longitude at periapse, altitude at periapse, plane-of-sky inclination, one-way light time, Sun-Earth-Venus angle, Earth-Venus-Probe at periapse angle, and local solar time at periapse are given in Appendix A. PVO had a highly eccentric (e=0.8) orbit with a period of 24 hours and was nearly polar with an inclination of 106°. Because of the high eccentricity, the altitude climbed to over 1000 km for the high latitude regions (>60°N, <30°S) as shown by the altitude profile in Appendix A. The approximately weekly maneuvers to lower periapse altitude are clearly evident in the semi-major axis, eccentricity, and periapse altitude plots. The geometry of the orbit provides good gravity data since the plane-of-sky inclination (angle between line-of-sight and orbit plane normal) rarely fell below 20° to near face-on geometries. Initially, periapse was occulted and became visible in March of 1979. Also, as periapse rose at the end of the low-altitude data set, periapse became occulted again (as shown by the Earth-Venus-Probe at periapse angle plot in Appendix A). Since the rotational period of Venus is 243 days, the low-altitude data provide several longitudinal coverages of Venus with the high resolution data contained within a narrow latitude band around periapse. The low-altitude data set, with the exception of the first three months of data to March 1979, is identical to the data set used by McNamee et al (1993) and Nerem et al (1993).

The second PVO data set extends from November 6, 1981 to September 7, 1982. During this time, periapse increased from 980 km to 1340 km due to the solar perturbation. At the beginning of this time span, periapse was just coming out of occultation, and near the end of the high-altitude data, periapse again entered occultation in August of 1982. After 1982, periapse altitude increased unimpeded until its maximum of about 2500 km in 1986 and then it decreased until 1992. At that time, periapse raise maneuvers were performed to keep PVO from burning up in the atmosphere. However, with propellant exhausted, PVO entered too deep into the atmosphere and the DSN lost signal on Oct. 8, 1992. This data set, due to sparseness of tracking, is not included in this gravity solution. Again, future studies may include this tracking. The high-altitude PVO was tracked continuously for three days once a week for this part of the mission (Mottinger et al, 1985), and amounts to 34,000 observations with the same compression scheme as the low-altitude PVO data. This data set is identical to the high-altitude data set used by Mottinger et al (1985) and is included in the solution by Nerem et al (1993).

Appendix B shows the high PVO orbit time histories of the same variables as the low-altitude orbit (Appendix A). The nearly conservative behavior of the orbit (due to minimal atmospheric drag) is shown by the periodic motion of the osculating semi-major axis. Since drag is small, the solar pressure force and Venus albedo forces have the major effect on the Venus orbit. For this reason, the Earth-Venus-Sun angle is also given in

Appendix B to show the angle of the PVO antenna with respect to the Venus-Sun line. With ten months of data, the high-altitude orbit provides more than one full longitudinal coverage of Venus. However, the first four months of the high-altitude data were in a nearly face-on geometry.

The Magellan spacecraft was inserted into orbit about Venus on August 10, 1990. The first three cycles (one cycle is one Venus rotation period of 243 days) were dedicated to SAR imaging of the Venus surface. This required the high gain antenna to be pointed to the Venus surface within about thirty minutes of periapse passage. Thus only high-altitude tracking (>2500 km) was obtained when the high gain antenna was returned to point at Earth. These data still, however, have some long term information on the gravity field and it was used by McNamee et al (1993). The altitude of periapse for the first three cycles is 250 km and is higher than the gravity cycle four. Except for the periapse altitude, the orbit shape for cycle four is identical to the previous cycles and should contain all the gravity information that is in the previous cycles and more. In addition for the first three cycles, the modeling of the solar pressure force for a rotating antenna through periapse passage is complicated, and solar pressure is a significant force for the higher periapse due to diminished drag. For these reasons, the first three cycles of data are not included in the gravity solution.

Cycle four began on September 15, 1992 and continued to May 24, 1993. During the complete cycle, there were no periapse altitude adjustments and the altitude varied between 185 and 165 km. Magellan was tracked through periapse with a two-second sample time, but for the spherical harmonic gravity solutions, the sample time was compressed to 10 seconds. With a periapse velocity of 8.5 km/s, 10 second samples provide two or three samples per half wavelength of a 90th degree and order field. For higher solutions, we may need to increase the number of samples near periapse. There are 770,000 10-second observations (both X and S-band) for cycle four. The time histories for Magellan cycle four are given in Appendix C. The location of periapse on the nightside and dayside (see local-solar-time plot in Appendix C) is apparent in the semi-major axis decay, with a greater drag on the dayside. Initially for cycle four, periapse is just coming out of occultation and hence a near face-on condition (<20° plane-of-sky inclination) and shows degraded gravity information. At the end of the cycle, the geometry again returns to a near face-on condition, but full longitudinal coverage was obtained with periapse tracking. However, we've noticed degraded gravity information for orbits within 20° of the face-on geometry. Due to smaller eccentricity (e=0.4), the Magellan cycle 4 data are much more sensitive at the higher latitudes than the PVO data (compare altitude vs latitude plots from the appendices), but still lack the high resolution gravity information for the higher latitudes.

At the end of May, 1993, Magellan periapse was lowered deep into the atmosphere to begin aerobraking. Over the next several months to early August, the atmospheric drag on the spacecraft changed the orbit to nearly circular to provide much lower altitude gravity tracking at the higher latitudes. From August 6, 1993 (August 17 for beginning of X-band) to October 10, 1994, Magellan was tracked in this nearly circular orbit with apoapse altitude varying from 600 km to 350 km and periapse altitude from 155 km to 220 km. Appendix D gives the time histories for the post-aerobraking orbit. This includes cycle five and part of cycle six until the Magellan spacecraft was "windmilled" into the atmosphere of Venus and lost signal. Even if the spacecraft had not been deliberately terminated, Magellan would have been lost within several weeks due to degradation of the solar arrays and loss of power. It would have never been able to fill in some tracking data gaps that remain in the gravity field. The plane-of-sky inclination versus longitude in Appendix D clearly displays the coverage of the post-aerobraking data. Initially, periapse is occulted and apoapse is tracked from longitude 100°W to 60°E, then periapse becomes visible from 220°E to 90°E except for a gap due to superior conjunction. Apoapse tracking then resumes from 110°W to 130°E, and finally periapse is tracked from 60°W to 10°E for the conclusion of the mission. As a result, there is a gap between 140°E and 220°E where there is no direct low-altitude tracking for the high latitudes. This data gap especially shows up in the southern hemisphere. During cycles five and six, there are many maneuvers to adjust periapse and apoapse altitude and these are visible in the semi-major axis plot and others. The time of the maneuvers is noted on the semi-major axis plot. The same compression time of 10 seconds is used for the post-aerobraking data and amounts to 1,230,000 observations. The velocity of the Magellan spacecraft in the nearly circular orbit is about 7 km/s, providing a sample every 70 km.

The PVO and Magellan data were processed in many arcs (a data time span which is dynamically continuous) where for each arc the initial spacecraft state and other parameters are estimated (see Appendix F). The PVO data arcs were chosen to be as long as practical given the imperfect knowledge of the spacecraft non-gravitational accelerations. The arc lengths generally were shorter in regions where uncertainties in the non-gravitational accelerations (primarily those due to atmospheric drag) were highest. In addition, the data arcs did not include any propulsive maneuvers which, as mentioned above, occurred regularly on PVO from 1978 through 1982 due to solar gravitational perturbations. The arc lengths for the low-altitude PVO tracking data varied from a minimum of one day to a maximum of ten days, with six days as the typical length. The high-altitude PVO data arcs were three days in length to match the tracking schedule. For Magellan, the data arc lengths for pre- and post-aerobraking were generally one day in length with a maximum length of two days. The long term information in the gravity field can be enhanced by increasing the arc lengths if careful attention is given to the nongravitational forces acting on the spacecraft. This will be the focus of future work.

The Magellan orbits with X-band tracking provide the highest resolution gravity information for Venus. Appendix E is a summary of all the Magellan X-band tracking and lists the orbit number, tracking data file name, number of Doppler points, tracking station number, and the time for the first observation of the orbit. The name of the tracking file also indicates whether the sample time is 10 seconds or 2 seconds. The LOS accelerations for all these orbits, with respect to the gravity fields mentioned above, were delivered to the Geoscience Node, Planetary Data System (PDS), at Washington University in St. Louis (Simpson, 1995a, gives the format of the delivery). For the LOS data, the 2-second samples were used if available; this amounted to delivery of almost 6 million observations for about 4,600 orbits.

3. Models

In this section, we discuss the observable and the geometric and dynamic models that affect the processing of the Doppler observable. The geometric models include Earth platform parameters such as station positions, precession, nutation, etc., and media calibrations for the observable due to the troposphere and ionosphere of the Earth. The dynamical models affect the spacecraft motion and are included in the numerical integration of the equations of motion for the spacecraft. In addition, any dynamic parameters that are estimated require that the partial of the spacecraft state with respect to the dynamical parameter be integrated. The parameters for the planetary ephemeris are both geometric and dynamic, but the change in force on the spacecraft due to a change in Earth position relative to Venus is negligible.

3a. Observable

The DSN station transmits either an S-band or X-band signal to the spacecraft and the spacecraft multiplies the frequency by a factor depending on the band of the uplink and downlink. For two-way S-band, the multiplier is 240/221 and for two-way X-band the factor is 880/749 (Moyer, 1987, documents the formulation for the X-Band Observable). The signal is then received at the DSN and the phase difference between the received and transmitted signal is measured with a Doppler cycle counter. The Doppler counter outputs the phase difference at integral times (in cycles and fraction of the last cycle). The Doppler observable is the average change in phase over the count or sample time (the phase difference divided by the count time), and is thus a differenced range measurement and not an instantaneous line-of-sight velocity (see Moyer, 1971).

The transmitting frequency can either be constant or ramped (linear with time). The majority of the PVO data have constant transmitting frequencies and the Magellan data were ramped until November 19, 1993 and mostly constant thereafter. From August 1993 to November 1993 for Magellan, there were some instances of incorrect reporting of the ramp rates in the last significant digit. There is no listing of these ramp errors but the error can be noticed in the residuals because of the discontinuity it causes in the observable on the order of 1 mm/s. If this discontinuity does not match with the momentum wheel desaturations, then it is probably a ramp error and it can be removed by increasing or decreasing the ramp rate by 0.000005 Hz/sec. In addition for Magellan, there are several occurrences in cycle 5 (mostly December 1993 and January 1994) where the incorrect reference frequency is reported, and this required the estimation of a Doppler bias on the order of 1000 mm/s. The reference frequency was not chosen as an integral number and thus was truncated to fit within the file format. Again for Magellan, there are also instances where one-way Doppler data were reported as two-way Doppler data, but these are easily identified as blunder points. The Doppler data file formats (ODF, Orbit Data File, and ATDF, Archival Tracking Data File) are available from the Geosciences Node, Planetary Data System (PDS), at Washington University in St. Louis, Missouri or from parts of a JPL DSN document that were contributed by Goltz (1988a and 1988b).

The major contribution to the observable is the relative velocity of Venus with respect to the Earth which varies between ±13 km/s. Since the noise of the Magellan X-band observable is less than 0.1 mm/s, this is a measure of the Venus-relative-to-Earth ephemeris in the tenth significant digit. The next major contributor to the observable is the orbit velocity of the spacecraft about Venus which varies between 1 and 9 km/s. The last major contributor to the observable is the motion of the Earth station with a velocity less than 0.5 km/s.

Table 1. Tracking Station Cylindrical Coordinates.

DSN Station	Longitude (deg)	Z height (km)	Spin radius (km)
11	243.150581357	3673.7640164	5206.3398806
12	243.194514844	3665.6309210	5212.0544322
13	243.205115543	3660.9568700	5215.4841386
14	243.110465567	3677.0522364	5203.9968496
15	243.112808584	3676.6699703	5204.2342885
16	243.126353832	3669.3861410	5209.3695231
42	148.981263719	-3674.5822320	5205.3523453
43	148.981263716	-3674.7487034	5205.2514376
44	148.977789576	-3691.3475374	5193.9817946
45	148.977682009	-3674.3815553	5205.4946063
46	148.983078061	-3674.9756700	5205.0754067
61-34m	355.750974037	4114.8843050	4862.6104361
61-26m	355.750974037	4114.8823280	4862.6081161
62	355.632163870	4116.9054796	4860.8179889
63	355.751987690	4115.1089216	4862.4509999
65	355.748578096	4114.7486263	4862.7173018
66	355.748529707	4114.9997980	4862.5302573

Table 2. DSN Plate Motion Velocities in cm/year.

Complex	East	North	Vertical
Goldstone (10)	-1.98	-0.57	-0.01
Canberra (40)	1.97	5.06	0.01
Madrid (60)	2.11	2.55	0.11

3b. Geometric Models

The planetary ephemeris is given by the state of the art JPL model DE403 as described by Standish et al (1995), and is a significant improvement over JPL model DE200 (Standish, 1990). The accuracy of DE403 is almost an order of magnitude better then DE200 for the Venus line-of-sight velocity with respect to the Earth. DE403 is oriented with respect to the International Earth Rotation Service (IERS) coordinate frame and is consistent with the IERS station coordinates that are used in this gravity solution. The initial state of the Earth and Venus are also estimated using the Set III variables of Brouwer and Clemence (1961).

The station locations are given by JPL VLBI solution "SSC(JPL) 91 R 01" which was submitted to the 1990 IERS Annual Report (Steppe et al, 1991). The station positions are listed in Table 1. They are in the IERS frame with respect to a 1988.0 epoch and include vertical corrections when used for Doppler measurements instead of VLBI measurements (7 cm for the 70m antennas and 1.5 cm for the 34m HEF X-band stations, Folkner, 1992b). The coordinates are for the geocentric metric and are scaled to the barycentric metric by the Orbit Determination Program (ODP, Moyer, 1990); this scaling has a 0.01 mm/s signature in the Doppler. The relative position uncertainty between stations is about 2-3 cm per component with a geocentric uncertainty of about 10cm

(Folkner, 1992a). The plate motion model is given by ITRF93 of IERS (Boucher et al, 1994) and is listed in Table 2. The combined uncertainties in the station positions and plate motion model result in errors in the Doppler observable of 0.01 mm/s or less for both the PVO and Magellan Doppler data. The PVO spacecraft was tracked by DSN stations 11, 12, 14, 42, 43, 44, 61, 62, and 63, and Magellan by stations 12, 14, 15, 42, 43, 45, 61, 63, and 65. The positions of stations 11, 44, and 62 are given by Moyer (1989) with respect to the 70m stations (14,43,63) from intercomplex survey data (to an accuracy of about 10cm). However, all the tracking stations for Magellan are given by Steppe et al (1991). Station 61 was upgraded from a 26-meter antenna to a 34-meter antenna during August 9, 1979 to March 31, 1980 (Moyer, 1983). The station platform was raised exactly 10 feet and this is reflected in the station coordinates given in Table 1. (Station 12 was also upgraded and raised to a 34m antenna from June 1, 1978 to December 1, 1978, but this is prior to the orbit insertion for PVO). The two-way X-band data are transmitted and received by only the 34m HEF stations 15, 45, and 65. The station coordinates are corrected by the ODP for solid earth tides (0.015 mm/s amplitude in Doppler) and pole tide (<0.001 mm/s).

The Earth orientation parameters (UT1-TAI, TAI-UTC, and polar motion) are given by the University of Texas, Center for Space Research solution "EOP(CSR) 95 L 01" submitted to IERS (1994 IERS Annual Report) for Magellan data and JPL solution "EOP(JPL) 91 C 01" submitted to IERS (Gross, 1992, also called Space91) for PVO data. For Magellan, the IERS solutions for UT1 and polar motion are better than 0.01 millisec and 0.1 milliarcsec, and the solution uncertainties for the PVO time frame are at least an order of magnitude greater than these values (1994 IERS Annual Report). Figures 1, 2, and 3 show the differences between the UT1 and polar motion solutions of U. of Texas and Space91 for the PVO time frame. The Doppler signatures from these Earth orientation errors are about 0.02 mm/s for PVO and negligible for Magellan. A UT1 bias is estimated for each PVO arc with an a priori of 1 millisec, but the resulting estimate sigmas are not significantly improved. The precession and nutation models are given by the 1976 IAU precession (Lieske et al, 1977) and 1980 IAU nutation models (Seidelmann, 1982). Corrections to the IAU precession and nutation models have been applied using the terms from Folkner (1994a). The resulting error in the precession and nutation for the PVO and Magellan time frame (1979-1994) is less than 10 nrad (6 cm).

The DSN stations record the Doppler observation in Station Time (ST). The differences between ST and UTC at each station complex are available for the Magellan data but not the PVO data. The time differences for Magellan are less than one microsecond and contribute only a small signature to the Doppler residual (<0.01 mm/s). The conversion from UTC to TAI is then given by "EOP(CSR) 95 L 01" or "EOP(JPL) 91 C 01" and the ET-TAI calculation is outlined by Moyer (1981). The relativistic delay of the signal due to the gravitational mass of the Sun is given by Moyer (1977) as a function of the relativity parameter γ. This effect on the Doppler observable is on the order of 1 mm/s for the Sun and is negligible for the relativistic terms of the planets (< 0.002 mm/s).

A seasonal troposphere model for each DSN complex in the form of a Fourier series (see Estefan and Sovers, 1994) is applied to the Magellan and PVO observables. The errors in the seasonal model can be as large as 0.1 mm/s for the low elevation data. Future gravity solutions will include a daily tropospheric correction to the seasonal model for the Magellan Doppler data only (supplied by the JPL Tracking System Analytical Calibration or

Figure 1: Space91 and Texas UT1 Daily Differences

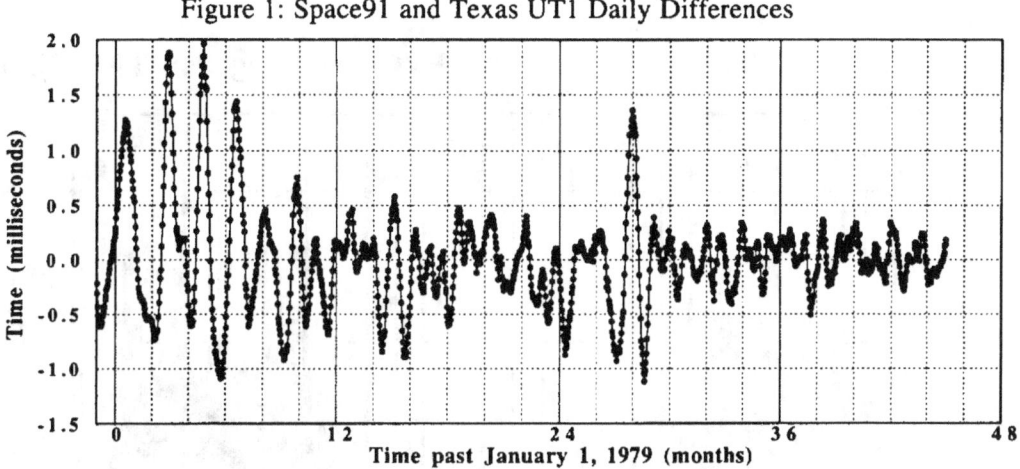

Figure 2: Space91 and Texas X-Pole Daily Differences

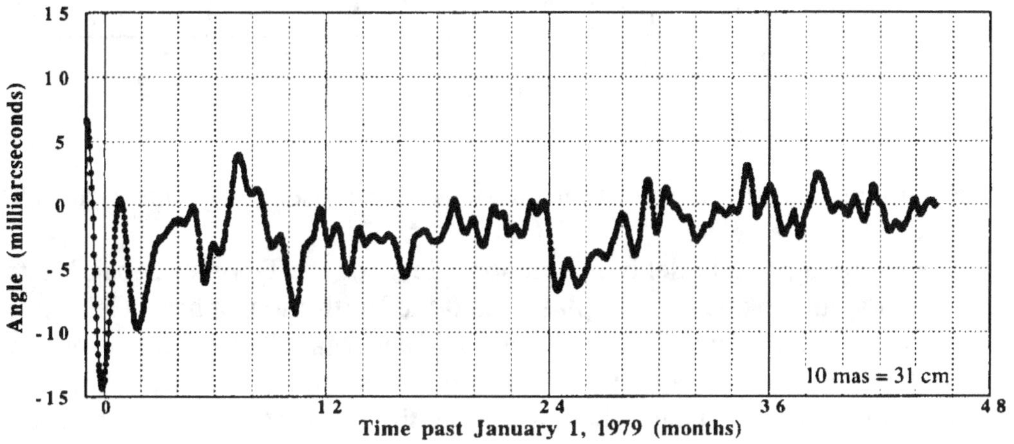

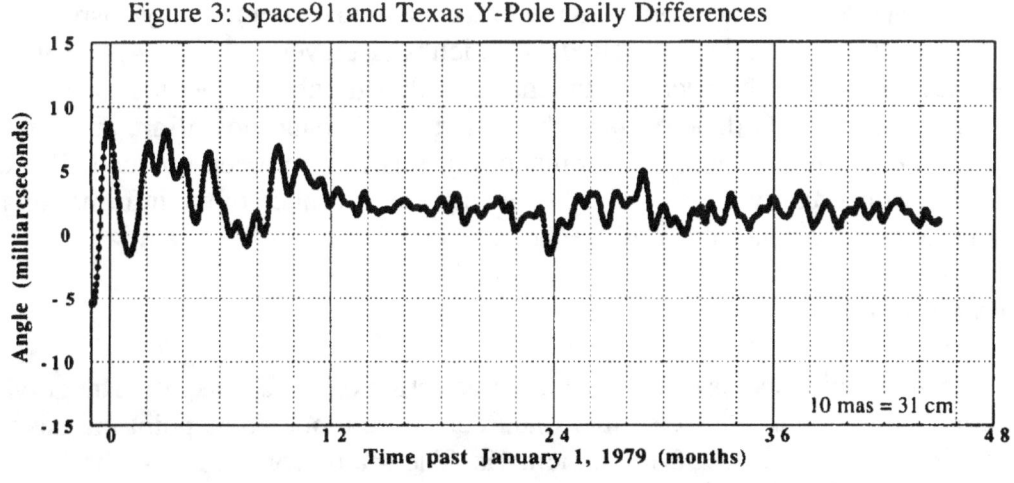

Figure 3: Space91 and Texas Y-Pole Daily Differences

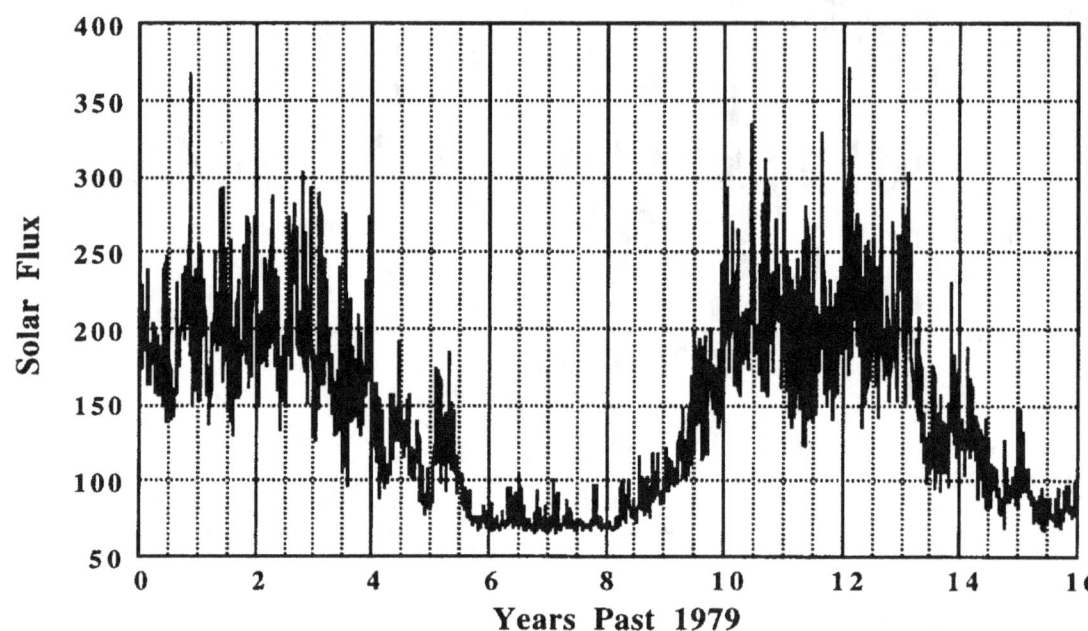

Figure 4: Daily Solar Flux for PVO and Magellan Time Frame

TSAC group and based upon weather data). At the present, any observations with an elevation below 10° are deleted.

The charged particle delay is a major effect for the Doppler S-band data (on the order of 1 mm/s) and is due to solar plasma and the Earth's ionosphere. Calibrations from the Earth's ionosphere were used for both PVO and Magellan (again provided by JPL TSAC) and are accurate to about 10% (Royden, private communication, JPL, 1995). However for the older PVO data, there are some time spans where Faraday rotation data could not be obtained. These times are the first three months of PVO data where no calibrations were used for Madrid and for March 1982 to June 1982 for Canberra which were filled in with an empirical seasonal (Bent et al, 1976) model prediction accurate to about 30%. The Magellan calibrations are based upon GPS data. The ionosphere delay is approximately proportional to the solar flux which is displayed in Figure 4, and all the PVO data are near maximum solar activity and the Magellan gravity data are near solar minimum. Some very limited PVO data exists with S-band and X-band downlink, and ionospheric delays for the downlink could be computed, but this has not been pursued. The charged particle delay uncertainty for the PVO data is the limit for much of the information that can be obtained from the data, such as the ephemeris of Venus.

3c. Dynamic Models

The dynamic models consist of all forces acting on the spacecraft. The gravitational attraction of the Sun and planets other than Venus are modeled as point masses with the positions given by the JPL DE403 ephemeris. The major contributor is the Sun with an acceleration of 1×10^{-9} km/s^2. The relativistic acceleration on the spacecraft (Moyer, 1971) is largest due to Venus and is about 1×10^{-9} km/s^2 at periapse with the Sun contribution four orders of magnitude smaller. The force is predominantly in the radial direction and inversely proportional to the distance squared (equivalent to a change in GM).

The gravitational potential of Venus is modeled by a spherical harmonic expansion with normalized coefficients ($\overline{C}_{nm}$, $\overline{S}_{nm}$) and is given by

$$U = \frac{GM}{r} + \frac{GM}{r} \sum_{n=2}^{\infty} \sum_{m=0}^{n} \left(\frac{a_e}{r}\right)^n \overline{P}_{nm}(\sin \phi) [\overline{C}_{nm}\cos m\lambda + \overline{S}_{nm}\sin m\lambda] \quad (1)$$

where n is the degree and m is the order, $\overline{P}_{nm}$ are the fully normalized associated Legendre polynomials, a_e is the reference radius of Venus (6051.0 km for our models), ϕ is the latitude, and λ is the longitude. The normalized coefficients are related to the unnormalized by (see Kaula, 1966)

$$(\overline{C}_{nm} ; \overline{S}_{nm}) = \left[\frac{(n+m)!}{(2-\delta_{0m})(2n+1)(n-m)!}\right]^{1/2} (C_{nm} ; S_{nm}) \quad (2)$$

where δ_{0m} is the Kronecker delta function and $\overline{C}_{n0} = -\overline{J}_n$. The harmonic coefficients of degree one are fixed to zero since the origin of the coordinate system is chosen to be the center of mass of the body. The body-fixed coordinate system is nominally given by the 1991 IAU values (Davies et al, 1992a, 1992b) for Venus pole position and rotation rate. The pole and rate are fixed to the 1991 IAU values for the final gravity solution delivered to the scientific community. However, we also present below our solutions for the pole and rate.

The normalized gravity coefficients are estimated complete to degree and order 90 (8277 coefficients). However, the gravity analysis for Venus is not complete with a 90th degree and order solution. The Magellan Doppler tracking data contain information about the Venus gravity field to about harmonic 120 for much of the planetary surface with the equatorial band resolution perhaps to degree 180 (Kaula, 1995). Determination of a higher degree and order gravity field for Venus will be a part of future work. A 120th degree and order field (15,000 parameters) is easily within reach on the JPL Cray T3D Supercomputer. The nominal gravity field for this 90th degree solution (MGNP90LSAAP) is the previous 90th degree solution (or iteration, MGNP90KSAAP).

Also modeled and estimated is the tidal effect on Venus due to the Sun. This force causes a time varying component of the second degree harmonics as a function of the body-fixed position of the Sun. The relationships for corrections to the normalized coefficients are (see McCarthy et al, 1989)

$$\Delta\overline{C}_{20} = \frac{1}{\sqrt{5}} k_2 \frac{GM_s R^3}{GM_v r_s^3} \left[\frac{3}{2} \sin^2\phi_s - \frac{1}{2}\right]$$

$$\Delta\overline{C}_{21} = \sqrt{\frac{3}{5}} k_2 \frac{GM_s R^3}{GM_v r_s^3} \sin \phi_s \cos \phi_s \cos \lambda_s$$

$$\Delta \overline{S}_{21} = \sqrt{\frac{3}{5}}\, k_2\, \frac{GM_s R^3}{GM_v r_s^3}\, \sin\phi_s \cos\phi_s \sin\lambda_s$$

$$\Delta \overline{C}_{22} = \sqrt{\frac{3}{20}}\, k_2\, \frac{GM_s R^3}{GM_v r_s^3}\, \cos^2\phi_s \cos 2\lambda_s$$

$$\Delta \overline{S}_{22} = \sqrt{\frac{3}{20}}\, k_2\, \frac{GM_s R^3}{GM_v r_s^3}\, \cos^2\phi_s \sin 2\lambda_s$$

(3)

where ϕ_s is the latitude of the Sun, λ_s is the longitude of the Sun, and M_v is the mass of Venus. The Love number k_2 is estimated. Since the latitude of the Sun is only a few degrees, the only significant time varying term occurs for the $\overline{C}_{22}$ and $\overline{S}_{22}$ coefficients. The expected amplitude for the normalized coefficients for a Love number of 0.25 is 7.1×10^{-9}; the other periodic terms have amplitudes two orders of magnitude smaller. If the Doppler tracking data determine the $\overline{C}_{22}$ and $\overline{S}_{22}$ coefficients to this level over various solar longitudes then the tidal effect of Venus can be determined. The solution for $\overline{J}_2$ given in this report has the permanent (or constant) part of the tide removed from it. To get the total $\overline{J}_2$, one must add 4.1×10^{-9}.

A major dynamic effect on the spacecraft is atmospheric drag with alongtrack accelerations for PVO reaching 1×10^{-6} km/s^2 for the daytime atmosphere at 150 km altitude. The density profile for the Venus atmosphere is given by the Venus International Reference Atmosphere (VIRA) model. It is a multilayered exponential model with density values at 5-km intervals in altitude and profiles given at different local solar times (Keating et al, 1985). The local solar time (LST) is defined by the direction of Venus rotation which is retrograde (the morning terminator=6am), and is the angle between the longitude of periapse and the longitude of the Sun. The 23 total atmospheric layers extend from 140 km to 250 km altitude and the VIRA model is symmetric about noon and midnight LST. The atmospheric drag on the orbit is estimated for every periapse passage for both Magellan and PVO. The exponential scale-height values for each layer are held fixed and the density at the lowest layer of 140 km is estimated (thus changing the density at each layer). For periapse altitudes above 250 km (including the 1000-km altitude PVO data), a single-layered atmosphere is used with scale-height values remaining a function of LST. Table 3 gives the scale height for each layer of the VIRA model and the base density at 140 km. For a given spacecraft LST at periapse, the atmospheric densities are determined by linearly interpolating the corresponding densities from Table 3.

For the PVO low-altitude data, a lift to drag coefficient is estimated for every data arc. The PVO spacecraft is basically a cylinder (with the antenna being a smaller component) and with the axis of symmetry pointed to the north ecliptic. The cylinder has a radius of 50 inches and a height of 48 inches. Since PVO approaches periapse from the northern latitudes, the angle of attack (the angle between the symmetry axis and the velocity vector) is about 22° and so the flat plate at the base of the cylinder is the major contributor to the lift force. For free molecular flow, the coefficient of lift to drag, $C_{l/d}$, should be in the direction away from Venus (this is a negative $C_{l/d}$ as given by the ODP software).

Table 3. VIRA Atmosphere Model.

Altitude (km)	Midnight	5	6	7	8	10	Noon
			Local Solar Time				
			Base Density, 10^{-12} (gm/cm^3)				
140	0.31	0.48	1.15	2.52	3.58	3.09	6.13
			Scale Heights (km)				
140	4.1	4.4	4.4	4.8	5.0	6.6	5.3
145	5.1	5.4	5.2	5.5	5.7	6.6	5.8
150	6.1	6.5	6.1	6.3	6.5	6.6	6.3
155	6.9	7.7	7.3	7.2	7.4	7.5	7.1
160	7.4	8.7	8.6	8.3	8.4	8.4	8.0
165	7.7	9.7	9.8	9.5	9.5	9.5	8.9
170	8.1	10.4	11.0	10.7	10.6	10.7	10.0
175	8.5	11.0	11.9	11.8	11.8	11.8	11.1
180	9.0	11.9	12.7	13.0	13.1	12.8	12.2
185	9.7	12.2	13.2	13.9	13.8	13.8	13.4
190	10.9	13.1	13.9	14.6	14.8	14.9	13.9
195	12.7	14.2	14.1	15.2	15.6	15.4	14.9
200	15.4	15.5	14.5	15.7	16.2	16.0	15.5
205	19.3	17.0	14.9	16.1	16.7	16.6	15.9
210	24.3	19.1	15.4	16.4	17.0	16.8	16.5
215	30.7	21.3	15.6	16.7	17.5	17.2	16.7
220	36.5	24.9	16.4	17.0	17.6	17.4	16.9
225	43.6	27.3	16.7	17.2	17.9	17.7	17.4
230	50.6	31.0	17.8	17.2	18.1	17.9	17.5
235	55.2	34.9	18.6	17.6	18.3	18.1	17.7
240	59.5	38.4	19.7	17.6	18.3	18.1	17.8
245	62.1	41.5	21.1	18.5	18.8	18.3	18.1
250	62.1	41.5	21.1	18.5	18.8	18.3	18.1

The solar radiation pressure force for both PVO (1.3×10^{-10} km/s^2) and Magellan (2.7×10^{-10} km/s^2) is accounted for in each arc by estimating one coefficient in each of three orthogonal directions (Sun-spacecraft line, ecliptic north direction, and in the ecliptic normal to the Sun-spacecraft direction). The PVO spacecraft is a simple spinning cylinder with a smaller antenna continuously pointed to Earth. The coefficients will change with time due to the change in antenna position. This is noticeable in the high-altitude PVO data. The Magellan spacecraft, however, goes through many orientation changes to either heat up or cool down ("hide") the spacecraft. These changes are modeled by estimating an acceleration vector for each hide, and may be as short as 10 minutes or as long as two hours. The spacecraft orientation is changed with the momentum wheels and there is no thrusting. The acceleration thus absorbs the change in the solar pressure force on the spacecraft due to the new orientation. Also, the Magellan spacecraft goes through an orientation change near apoapse for star calibration. In that case, the solar pressure force change is modeled with a small (a priori of 0.3 mm/s) delta velocity estimation. With the current arc lengths of one day for Magellan, these small solar pressure effects (hides, star calibrations) are small and have negligible effects on the gravity estimation. However, with longer data arcs, these effects may become important. The hide history for Magellan is listed in Appendix F.

The radiation force from Venus albedo is a ring model (Knocke and Ries, 1987) where a simple bus model is used for Magellan and a cylindrical model is used for PVO. The albedo force is basically undetermined for Magellan and the low-altitude PVO data due to the atmospheric drag. For the high-altitude PVO, the albedo force is significant. The mean albedo value is 0.76 (Taylor et al, 1983) and variations in albedo are allowed for by estimating a scale factor for the albedo for each data arc. The albedo force is approximately 1×10^{-10} km/s^2, and is about four times smaller than the tidal force (k_2). The radial components of the tide, albedo, and drag forces for PVO and Magellan are displayed in Figures 5, 6, 7, and 8. The examples are for orbits with a local solar time near noon. The drag force for the high-altitude PVO orbits is several orders of magnitude smaller than the albedo and tidal forces, but reaches 1×10^{-8} km/s^2 in the radial direction for the low-altitude PVO orbits. In solutions for these forces, the tide generally correlates with the initial spacecraft state and solar pressure and albedo with the drag due to a nonconservative part of the force (the albedo force has a mean acceleration in the velocity direction). The variational equation for the semi-major axis (a) is (Battin, 1987)

$$\frac{da}{dt} = \frac{2va^2}{GM_v} a_{dt} \qquad (4)$$

where v is the spacecraft velocity, a_{dt} is the acceleration along the velocity direction. Even though the tide force is also larger than albedo in the along-track direction, the average rate of change of the semi-major axis (the integral over one orbit in equation (4)) for the albedo force is an order of magnitude greater (1 m/day vs 0.1 m/day) than the tide or drag. The formal uncertainty in the semi-major axis for the high-altitude 3-day PVO arcs is about 2 meters. Hence, the albedo is fairly well determined for the PVO high-altitude orbits but the tidal effect is better determined for the low-altitude orbits. The drag force is substantially lower for Magellan than PVO due to higher altitudes and lower spacecraft velocity through the atmosphere. The drag force for the near circular orbit is even lower still for the same reasons. The tidal force can, in part, be better separated from the other forces because on the nightside of Venus the tidal force remains the same, the albedo force vanishes, and the drag force diminishes by an order of magnitude.

The Magellan spacecraft attitude was maintained with the use of momentum wheels, and they were also used to change the orientation of the spacecraft. These changes did not impart any force on the spacecraft. However, due to atmospheric drag torques on the spacecraft, the momentum wheels had to be despun or desaturated every orbit or about 8 times per day for cycle four and 15 times per day for cycles five and six. Prior to the gravity cycles, the desaturations occurred about three times per day. The desaturations imparted an incremental velocity to the spacecraft of about 1 mm/s and required estimation of the three components of the velocity vector. These desaturations greatly reduced the long term gravity information in the Magellan data.

All the above forces on the spacecraft are included in the numerical integration of the spacecraft state ($d^2\mathbf{r}/dt^2$ = forces) in rectangular coordinates of the Earth mean equator of J2000 coordinate system. The integrator is a multistep Adams type predictor-corrector that varies the order to obtain the largest possible step size. We use an absolute integration

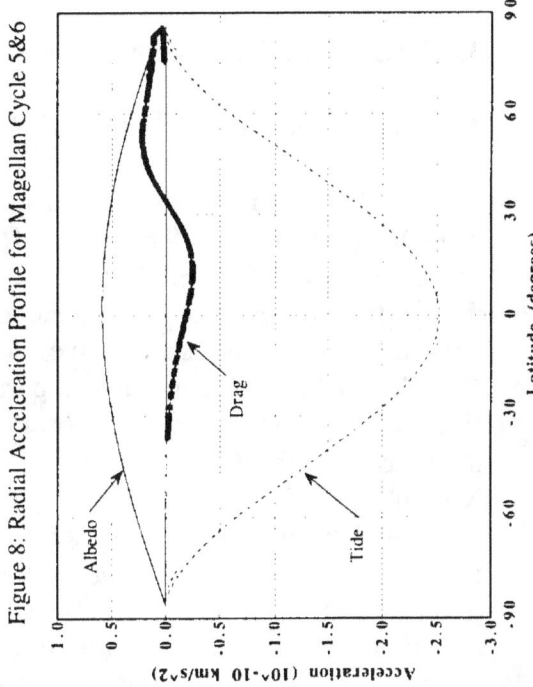

Figure 6: Radial Acceleration Profile for Low-Altitude PVO

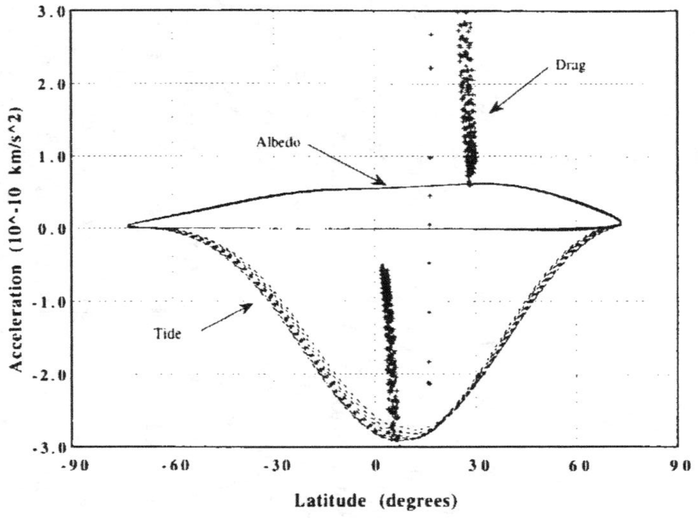

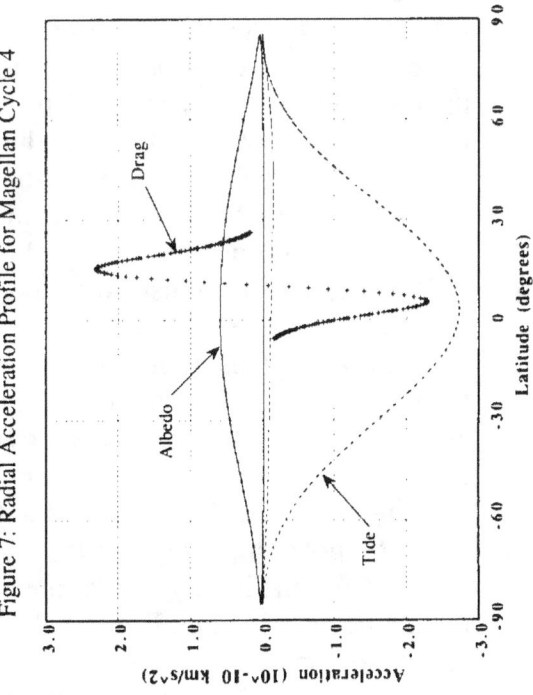

Figure 8: Radial Acceleration Profile for Magellan Cycle 5&6

Figure 5: Radial Acceleration Profile for High-Altitude PVO

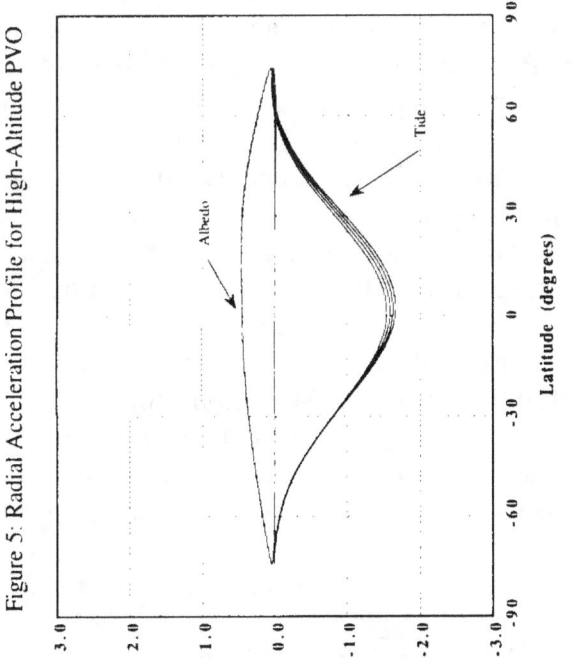

Figure 7: Radial Acceleration Profile for Magellan Cycle 4

tolerance of 2×10^{-11}, and this results in numerical noise for the Doppler observable of less than 0.01 mm/s.

4. Estimation Procedure

4a. Determination of Unconstrained Gravity Field

The JPL gravity estimation software is based upon the Orbit Determination Program or ODP (see Moyer, 1971); the software set used at JPL for navigation of all planetary spacecraft. This FORTRAN code was used for the numerical integration and processing of the Doppler observable to produce an observation equation (using all the geometric and dynamic models discussed above). The ODP was modified by the authors for use on the JPL Cray T3D supercomputer, a parallel computer with 256 DEC alpha processors. The filter which then processes the observation equations, however, was written specifically for the Cray T3D (by C.L. Lawson at JPL, for more specific details of the T3D software see Konopliv, 1995b). The use of the Cray has dramatically reduced the amount of computer time from years to days to generate the gravity solution. The spacecraft state and other parameters are estimated using a weighted least-squares filter based upon the square root of the information matrix (see Lawson and Hanson, 1995; Bierman, 1977). The parameters that are estimated consist of arc-dependent or local variables (spacecraft state, atmospheric densities, etc.) that are determined separately for each data arc (i.e., the arcs listed in Appendix F) and global variables (harmonic coefficients, etc.) that are common to all the data arcs.

Initially, we converge the data arcs by estimating only the local variables using the nominal values for the global variables. The observations of each arc are weighted according to data root mean square (rms) of that arc with a separate rms for each tracking station pass and with the rms including corrections for the count times of the observations. The actual data weight used is the rms multiplied by a factor of two with an additional correction factor for the observation elevation. Since the PVO and Magellan orbits are nearly polar, the groundtracks converge near the pole and the observations become more dense. For this reason, the observation sigma is adjusted for latitude ϕ ($\sigma_{new} = \sigma_{old} * \cos^{-1/2}\phi$).

The Doppler data (i.e. in the frequency domain, f) are treated as white noise, and theoretically, the rms of the data is inversely proportional to the square root of the sample time. However, due to the charged particle effects of the solar plasma, the noise follows a $f^{-8/3}$ power spectrum instead of the white f^{-2} spectrum (Folkner, 1994b). This is true for count times greater than about one minute; for count times shorter than one minute, the noise is white. This implies that there is more noise in the data for the longer count times than what is modeled. As a result, the uncertainties from the estimation process are too small (or optimistic) for the shorter wavelengths of the gravity field and too large (or pessimistic) for the longer wavelengths. At this time we have not devised a method to account for this noise spectrum. For the $f^{-8/3}$ spectrum, the rms is inversely proportional to the sixth root of the sample time.

Once the local variables are converged, the global parameters are determined with a technique described by Ellis (1980) that merges only the global parameter portion of the square root information (or SRIF) arrays from all the arcs, but is equivalent to solving for the global parameters plus local parameters of all arcs. For each data arc, the local variables estimated are the spacecraft state, three solar pressure coefficients, a factor for the Venus albedo, the base density for each periapse passage through the atmosphere, the lift-to-drag coefficient for the low-altitude PVO orbits, velocity vector increments for the momentum wheel desaturations and star calibrations of Magellan, acceleration vectors for the hides of Magellan, and a UT1 bias for the PVO arcs. The a priori uncertainties for the spacecraft state are large (20 km). The a priori base density uncertainties for the PVO orbits are large but are more tightly constrained for Magellan ($1 \times 10^{-12} \pm 1 \times 10^{-12}$ gm/cm^3). Future work will constrain the Magellan base densities more closely to the VIRA model. The a priori on the Magellan desaturations are 5×10^{-6} km/s and for the star calibrations are 3×10^{-7} km/s. The hides are constrained to 10^{-10} km/s^2.

The following global parameters are estimated: the normalized spherical harmonic coefficients ($\overline{C}_{nm}, \overline{S}_{nm}$) of the gravity field complete to degree and order 90, the gravitational constant times the mass of Venus (GM), the ephemerides of Earth and Venus (12 parameters), the Love number for Venus, the right ascension and declination of the Venus spin axis or pole, and the rotation rate for Venus. An SRIF array for the global parameters is obtained for three data sets (PVO, Magellan cycle four, and Magellan cycles five and six) where there is no constraint on the global variables. Combinations of these SRIF arrays together with a priori constraints on the global parameters are used to find solutions for the global variables. The a priori constraint on the gravity field is discussed in the next section.

4b. Gravity A Priori

Once all the global information is packed from all the data arcs, the gravity field is constrained with an a priori. The common method is to constrain each harmonic coefficient toward zero with an uncertainty given by the Kaula rule (Kaula, 1966) for that particular planet (used, for example, in Konopliv et al. 1993a, Konopliv et al. 1993b, Nerem et al. 1993, McNamee et al. 1993, Smith et al., 1993, Lemoine, 1992, Lemoine et al 1995). The Kaula rule used for Venus is $1.2 \times 10^{-5} / n^2$ where n is the degree of the coefficient. The second a priori constraint method is a spatial constraint and is also outlined for Mars in Konopliv and Sjogren (1995a).

The a priori constraint applied for this gravity field evaluates the radial acceleration and its uncertainty on the reference sphere (i.e., $r = a_e$). At that surface, the radial acceleration (a_n) from all coefficients of degree n is given by

$$a_n = \frac{GM}{a_e^2} (n+1) \sum_{m=0}^{n} \overline{P}_{nm}(\sin \phi) (\overline{C}_{nm} \cos m\lambda + \overline{S}_{nm} \sin m\lambda) \tag{5}$$

To create a profile of acceleration contributions versus degree, the rms of the acceleration a_n is obtained over the sphere. The mean of the square of the acceleration $(a_n)_{ms}$ of equation (5) is given by

$$(a_n)_{ms} = [\frac{GM}{a_e^2}(n+1)]^2 \frac{1}{4\pi} \times$$
$$\int_0^{2\pi} \int_{-\pi/2}^{\pi/2} [\sum_{m=0}^{n} \overline{P}_{nm}(\sin \phi)(\overline{C}_{nm} \cos m\lambda + \overline{S}_{nm} \sin m\lambda)]^2 \cos \phi \, d\phi \, d\lambda$$

Since the spherical harmonics are orthogonal, we obtain

$$(a_n)_{ms} = \left[\frac{GM}{a_e^2}(n+1)\right]^2 \sum_{m=0}^{n} (\overline{C}_{nm}^2 + \overline{S}_{nm}^2)$$

As a good approximation, the rms magnitude spectrum of the gravity coefficients follows the Kaula rule and is given by

$$\left[\frac{\sum_{m=0}^{n}(\overline{C}_{nm}^2 + \overline{S}_{nm}^2)}{2n+1}\right]^{1/2} = K/n^2$$

where K is the constant for the particular planet (1.2 x 10^{-5} for Venus). The expected acceleration profile is then given by (for n >> 1)

$$(a_n)_{rms} = \frac{GM}{a_e^2} K\sqrt{2/n} \tag{6}$$

which for Venus is

$$(a_n)_{rms} = 15/\sqrt{n} \text{ milligals} \tag{7}$$

This is the expected "signal" for the acceleration at each point on the surface of the reference sphere. The signal could also be determined empirically by taking the rms of a given gravity field over different regions. However, for this work, only one signal profile is used for all latitudes and longitudes.

The next task is to map the acceleration uncertainty at the surface into an uncertainty or "noise" profile showing the error in acceleration versus harmonic degree. The acceleration uncertainty from the summed contributions of all coefficients from degree 2 to n, $\sigma(a_{2,n})$, is given by

$$\sigma(a_{2,n}) = \frac{\partial a_{2,n}^T}{\partial G_{2,n}} \mathbf{P}_{noap(2,n)} \frac{\partial a_{2,n}}{\partial G_{2,n}}$$

where $G_{2,n}$ is the vector of all normalized gravity coefficients from degree 2 to n and $\mathbf{P}_{noap(2,n)}$ is the corresponding covariance. The covariance of the coefficients from degree 2 to n is the covariance as if the higher degree coefficients (>n) are not estimated. Hence, it is a truncation, or submatrix, of the full 90th degree and order covariance without any constraint applied to the gravity field. The partial of the acceleration with respect to the coefficients of degree n and order m are functions of latitude and longitude and are given by

$$\frac{\partial a_{2,n}}{\partial \overline{C}_{nm}} = \frac{GM}{a_e^2}(n+1)\overline{P}_{nm}(\sin\phi)\cos m\lambda \tag{8}$$

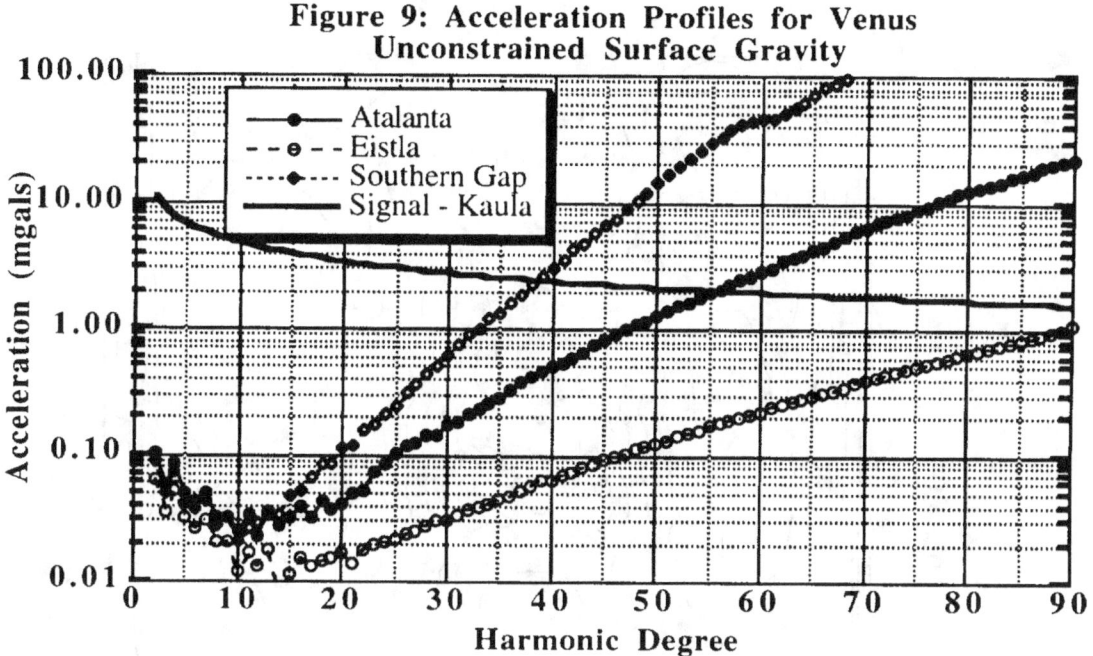

Figure 9: Acceleration Profiles for Venus Unconstrained Surface Gravity

$$\frac{\partial a_{2,n}}{\partial \bar{S}_{nm}} = \frac{GM}{a_c^2} (n + 1) \bar{P}_{nm}(\sin \phi) \sin m\lambda$$

The uncertainty for the coefficients of degree n, $\sigma(a_n)$ is then given by the difference of the sum total error to degree n and the sum total error to degree n-1 as

$$\sigma(a_n) = \sigma(a_{2,n}) - \sigma(a_{2,n-1}) \tag{9}$$

Figure 9 shows the expected acceleration profile from the Kaula rule of equation (6) and the unconstrained acceleration uncertainty profile as given by equation (9) for Atalanta, the periapse region for Magellan cycle 4 (e.g. eastern Eistla Regio), and the gap in Magellan cycle 5 data in the southern hemisphere (160°E to 220°E, 30°S to 80°S). The crossing point of the Kaula signal with the acceleration uncertainty is called the degree strength of the gravity field for that particular latitude and longitude. For degrees greater than the degree strength, the "noise" in the data exceeds the "signal." Based upon the Kaula rule, the degree strengths for Atalanta, Eistla, and the southern gap are 55, 90, and 38, respectively. Figure 10 displays the degree strength on a global scale. The maximum degree strength is greater than harmonic degree 90 near the low-altitude periapse locations.

The basic idea of the gravity constraint method is to constrain the "noise" of the gravity field to zero with some uncertainty when the "noise" exceeds the "signal." The acceleration at the surface from all harmonic coefficients greater than or equal to the degree strength is constrained to zero with an uncertainty approximately equal to the expected signal at the degree strength. This amounts to generating observations over the entire

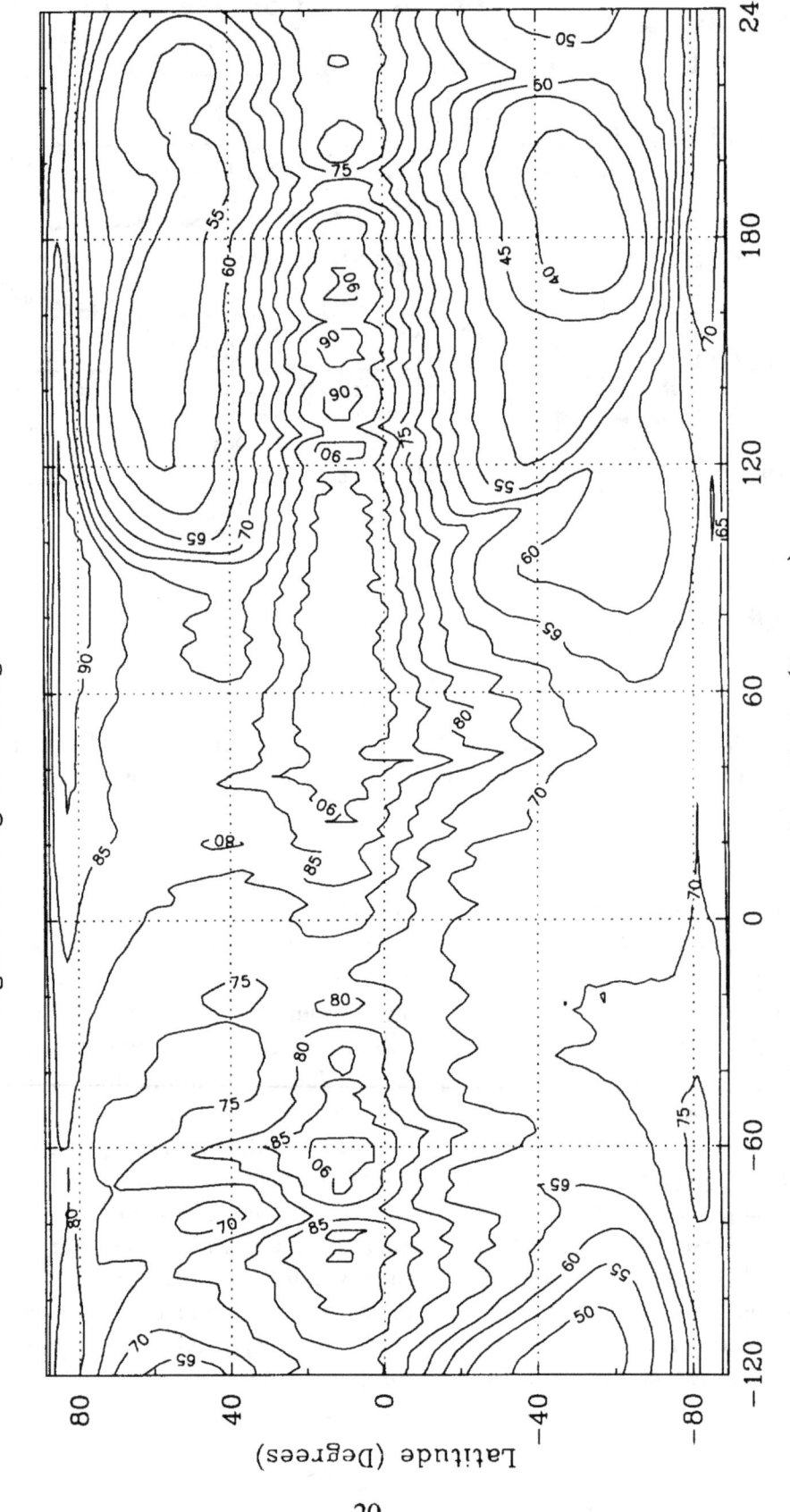

Figure 10: Degree Strength for MGNP90LSAAP

surface of the sphere based upon the degree strength at each latitude and longitude. An observation ($a_{D,90}$) for degree strength D is

$$a_{D,90} = \frac{GM}{a_e^2} \sum_{n=D}^{90} \sum_{m=0}^{n} (n+1) \bar{P}_{nm}(\sin \phi) (\bar{C}_{nm}\cos m\lambda + \bar{S}_{nm}\sin m\lambda) \qquad (10)$$

and the linearized observation equation is given by (Bierman, 1977)

$$z_i = A_i x + v_i$$

where z_i is the difference between the observed value (zero in this case) and the nominal value of the observation (the accumulated acceleration at the surface for degrees D to 90 from the a priori gravity model as given by equation (10)), A_i is the row vector of observation partials (the partial of the observation with respect to all the parameters being estimated), x is the vector of estimated parameters (differences in the gravity coefficients from the nominal gravity model), and v_i is the observation error. The partials A_i to construct the observation equation are

$$A_i = \frac{\partial a_{D,90}}{\partial G}$$

where G is the vector of all gravity coefficients. The elements of A_i for coefficients with degrees less than the degree strength D are zero and, otherwise, are again given by the partials of equation (8).

The observations are then merged with the unconstrained gravity SRIF array using Householder transformations. In normal form, the constrained gravity estimate x is written as

$$x = \left[P_{noap}^{-1} + A^T W A\right]^{-1} \left[P_{noap}^{-1} x_{noap} + A^T W z\right] \qquad (11)$$

where P_{noap} is the unconstrained covariance of the gravity coefficients, A is the matrix of observation partials with each row an observation, W is the diagonal weight matrix, x_{noap} is the unconstrained gravity estimate, and z is the vector of linearized observations. The new constrained covariance P is then

$$P = \left[P_{noap}^{-1} + A^T W A\right]^{-1} \qquad (12)$$

The observations should be spaced such that at least three observations are generated over the shortest harmonic wavelength. The weight used for an observation is then proportional to the area between observations and is approximately equal to the signal at the degree strength (i.e., 10 to 20 milligals). The observations are globally spaced on a rectangular grid of latitude and longitude with a spacing of two degrees. To obtain truer peak values, there is no constraint around Maxwell Montes, Maat Mons, and Eistla and Beta Regio.

The result on the acceleration uncertainty profiles from applying the a priori observations on the gravity field is displayed in Figure 11 for the MGNP90LSAAP gravity

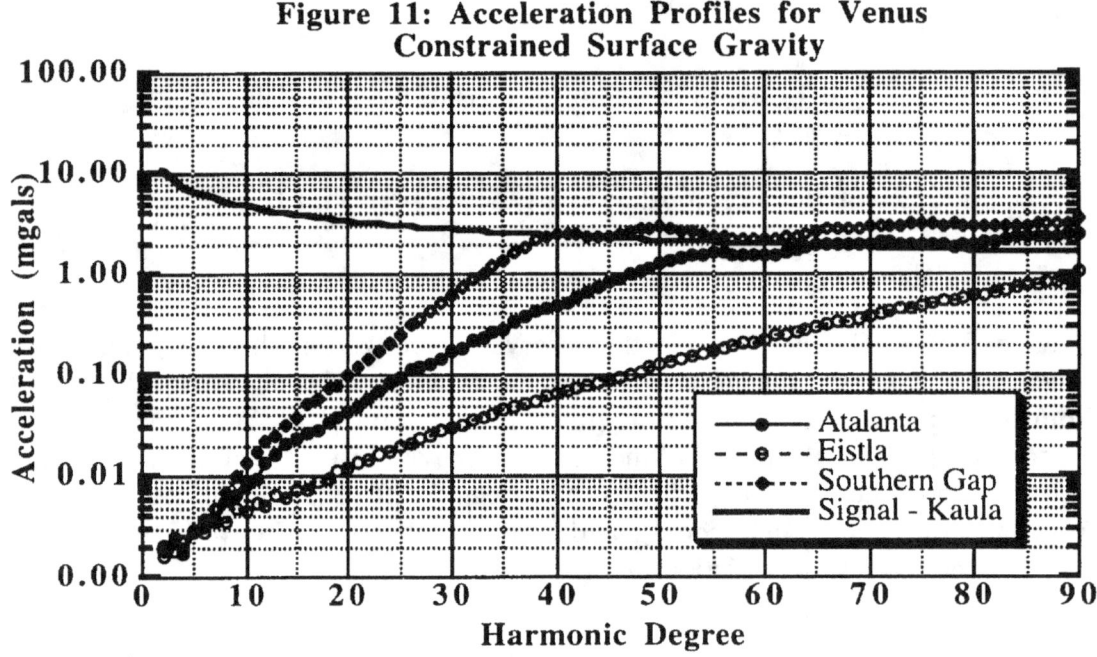

field. The result for Atalanta and the Southern gap is an approximate Kaula constraint on the "noise" part of the spectrum. The uncertainties in the lower degree harmonics (up to degree 10) are one to two orders of magnitude greater for the unconstrained case (Figure 9)

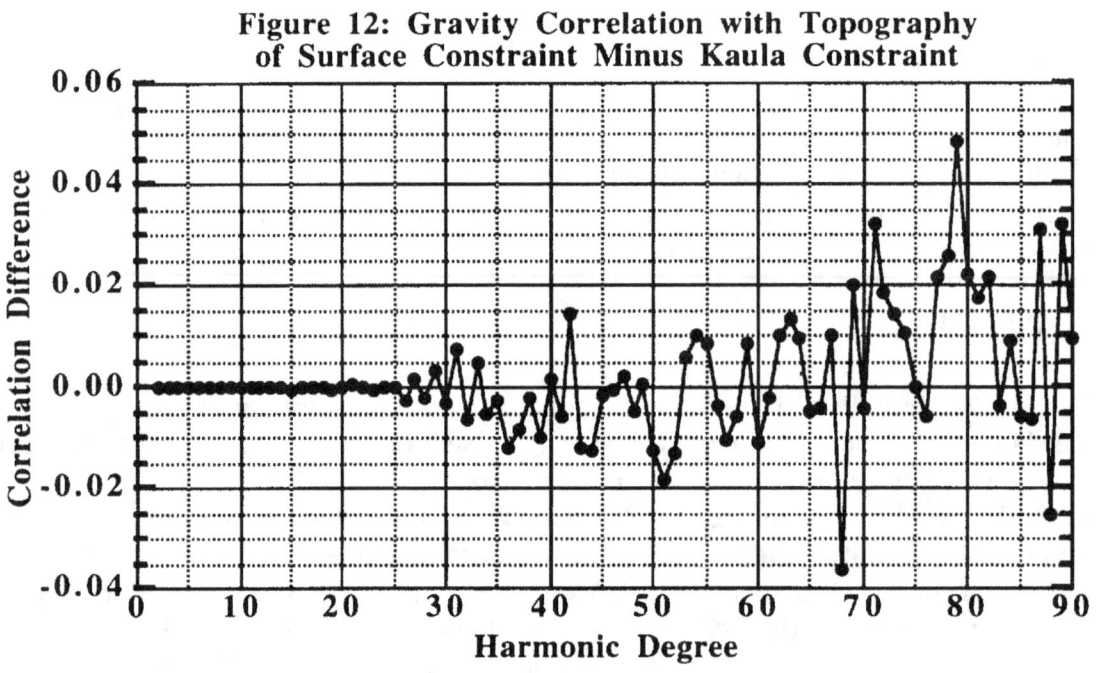

versus the constrained (Figure 11). This is due to the lack of any constraint on the estimated parameters that are not gravity harmonics (Venus ephemeris, pole, and Love number) and is not due to the constraint on the higher degree coefficients.

The main advantage of using this spatial constraint instead of a straight Kaula rule on the spectrum appears to be better determination of peak amplitudes. Since the well determined degrees are not constrained directly (only somewhat through correlations), the amplitudes (and coefficients) for those degrees are not biased toward zero. It is also flexible in allowing relaxation of selected regions for any reason, such as incorrect data weighting or a region exhibiting greater signal than the power rule. The amplitudes are reduced by about 5 to 10% when a Kaula power law is applied versus the spatial constraint (e.g., 10 milligals for Maxwell Montes and Gula Mons, and 35 milligals for Maat Mons). The correlations with topography for the two different constraint methods are very similar. The correlations from the Kaula constraint are generally slightly higher for the medium wavelength harmonics and slightly lower for the higher frequencies (see Figure 12) with the sum of correlations over the degrees in favor of the surface constraint. In general, there are only slight differences between the methods since Venus does not have strong local deviations from Kaula power spectrum. For Mars, the differences are more pronounced for the Tharsis region (Konopliv and Sjogren, 1995a).

5. Gravity Results

5a. Global Gravity Model

The normalized coefficients of the nominal gravity solution (MGNP90LSAAP, SAAP = Surface Acceleration A Priori) are given in Appendix G up to degree and order 40. A file containing the complete field to degree and order 90 can be requested from the authors at ask@krait.jpl.nasa.gov or obtained from the Geoscience Node of the PDS at Washington University in St. Louis, Missouri (in the format specified by Simpson, 1993a). Also available from the Data Node are the covariance of the gravity harmonics (a 275 MB binary file, Simpson, 1993b), vertical surface gravity and error map (Simpson, 1995b), geoid and error map (Simpson, 1995b), and Bouguer map (Simpson, 1995b).

The GM solutions for Venus are given in Table 4 for different combinations of data. The 40th degree solutions are generated by fixing the coefficients from degree 41 to 90 at the values of the nominal gravity field (MGNP90KSAAP). The 40th degree solutions allow the Venus pole and rotation rate to move (i.e., they are estimated) and the 90th degree solutions hold the pole and rate to the nominal 1991 IAU values. The variations in the GM solution are generally one formal sigma with almost a two-sigma variation for the Magellan Cycle 4 data. The best GM solution with a realistic error is 324858.601 ± 0.014 km^3/s^2 (2 x formal uncertainty). This solution agrees with our previous solutions within about two formal uncertainties. The ionosphere calibrations play a major role in the determination of the Venus GM from the PVO data, and their neglect may be a reason for unrealistically large values of previous solutions (our PVO solution without ionosphere calibrations jumps to about 324858.65).

Table 4. Gravitational Constant Times the Mass of Venus (GM in km^3/s^2).

Data Combination	Constraints	Solution	Formal Uncertainty
PVO	40, K, lp	324858.576	0.017
MGN Cycle 4	40, K, lp	324858.654	0.027
MGN Cycle 5	40, K, lp	324858.600	0.011
PVO+ MGN Cycle 4	40, K, lp	324858.589	0.011
ALL	40, K, lp	324858.600	0.007
PVO	90, K, lp	324858.581	0.018
MGN Cycle 4	90, K, lp	324858.649	0.028
MGN Cycle 5	90, K, lp	324858.605	0.011
PVO+ MGN Cycle 4	90, K, lp	324858.587	0.011
ALL	90, K, lp	324858.601	0.007
ALL (MGNP90LSAAP)	90, S, fp	324858.601	0.007

90, 40 = degree and order of solution, K = Kaula rule, S = Spatial (SAAP)
fp = fixed IAU pole, lp = loose pole (estimated pole and rotation rate for Venus)

Previous Solutions:

Sjogren et al, 1990 (PVO Only)	324858.60	0.05
Nerem et al, 1993 (PVO Only)	324858.64	0.01
McNamee et al, 1993	324858.681	0.03
Konopliv et al 1994a, MGNP60FSAAP	324858.628	0.016
Konopliv et al 1994b, MGNP75ISAAP	324858.589	0.006

The vertical (or radial) gravity at the reference surface (a sphere of 6051 km) is displayed in Figure 13 with contours every 20 milligals (10^{-8} km/s^2). The radial gravity is given by the partial of the potential, equation (1), in the radial direction without the central mass term but including J_2. A sphere is used as the reference surface for Venus since J_2 is small for Venus and comparable in size to the other spherical harmonics. The geoid or equipotential surface of equation (1) is given in Figure 14 with contours every 10 meters. The rotational contribution to the potential is neglected for Venus and the geoid is iterated to convergence. The difference between one and two iterations on the geoid or the error in Bruns' formula (Heiskanen and Moritz, 1967) is at most 3 cm in Atla and Beta Regio out of a 100+ meter geoid value.

The uncertainties in the vertical gravity and geoid are given in Figures 15 and 16. The uncertainties in the surface acceleration or potential are given by the errors up to the resolution or degree strength of the data plus the error for omission of terms beyond the resolution. From Figure 11, the uncertainties from the higher order terms (> degree strength) generally follow the Kaula power rule. So Figures 15 and 16 provide realistic errors for terms up to degree and order 90. Figure 17 shows the error in the vertical gravity up to the degree strength (i.e., the unconstrained covariance P_{noap} is truncated at the degree strength for that given latitude and longitude). The error in the gravity field that the data can

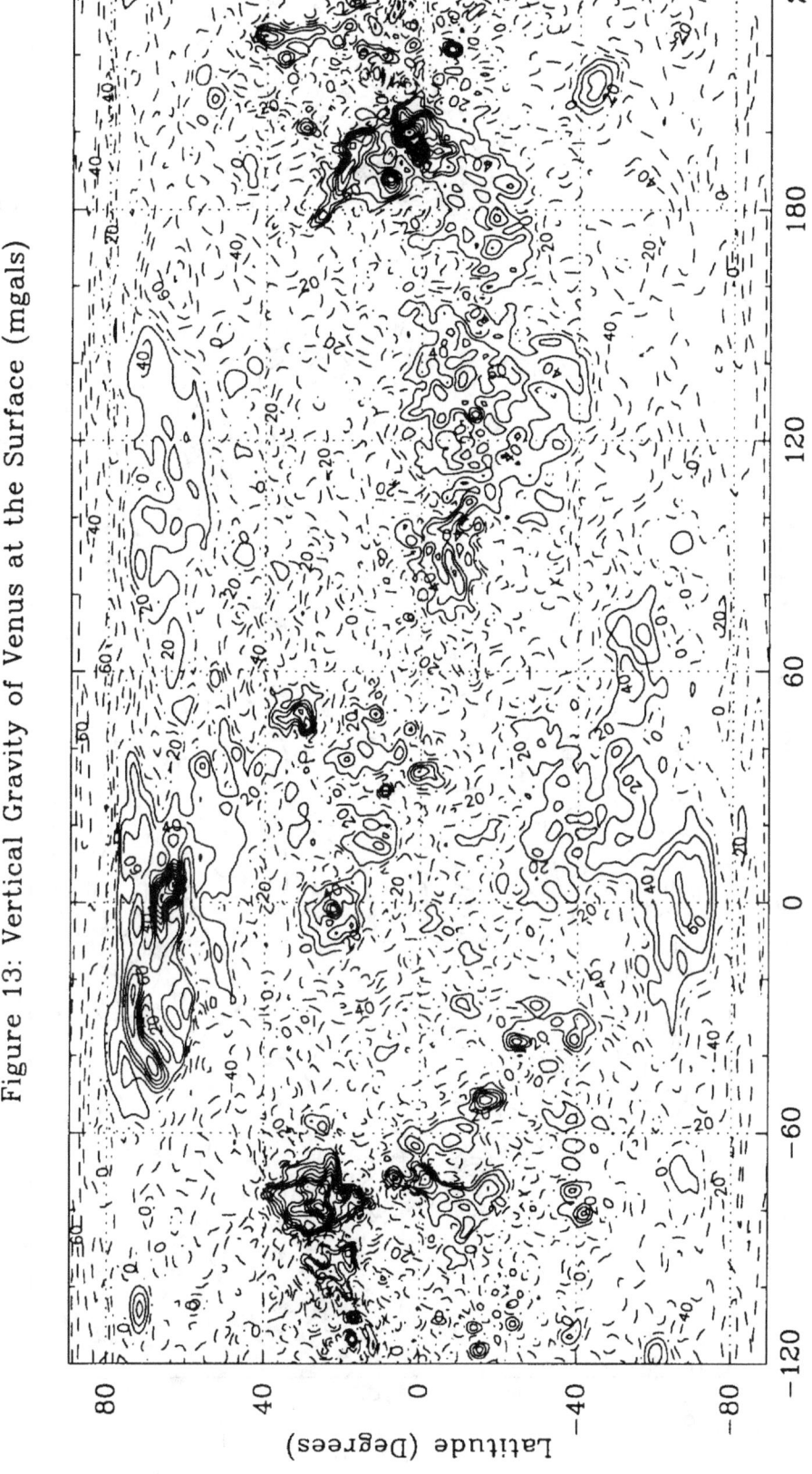

Figure 13: Vertical Gravity of Venus at the Surface (mgals)

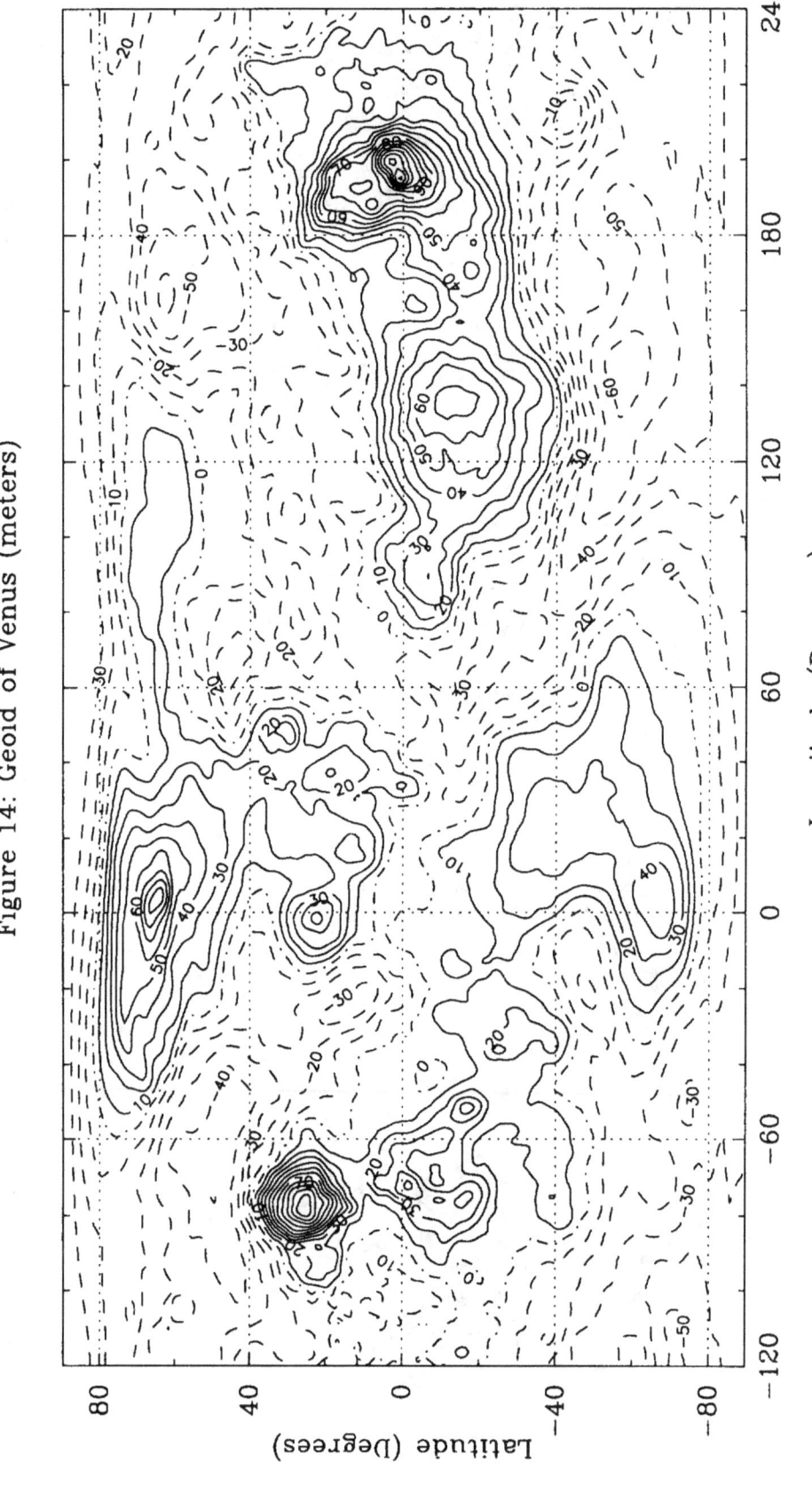

Figure 14: Geoid of Venus (meters)

Figure 15: Vertical Gravity Uncertainty at the Surface (mgals)

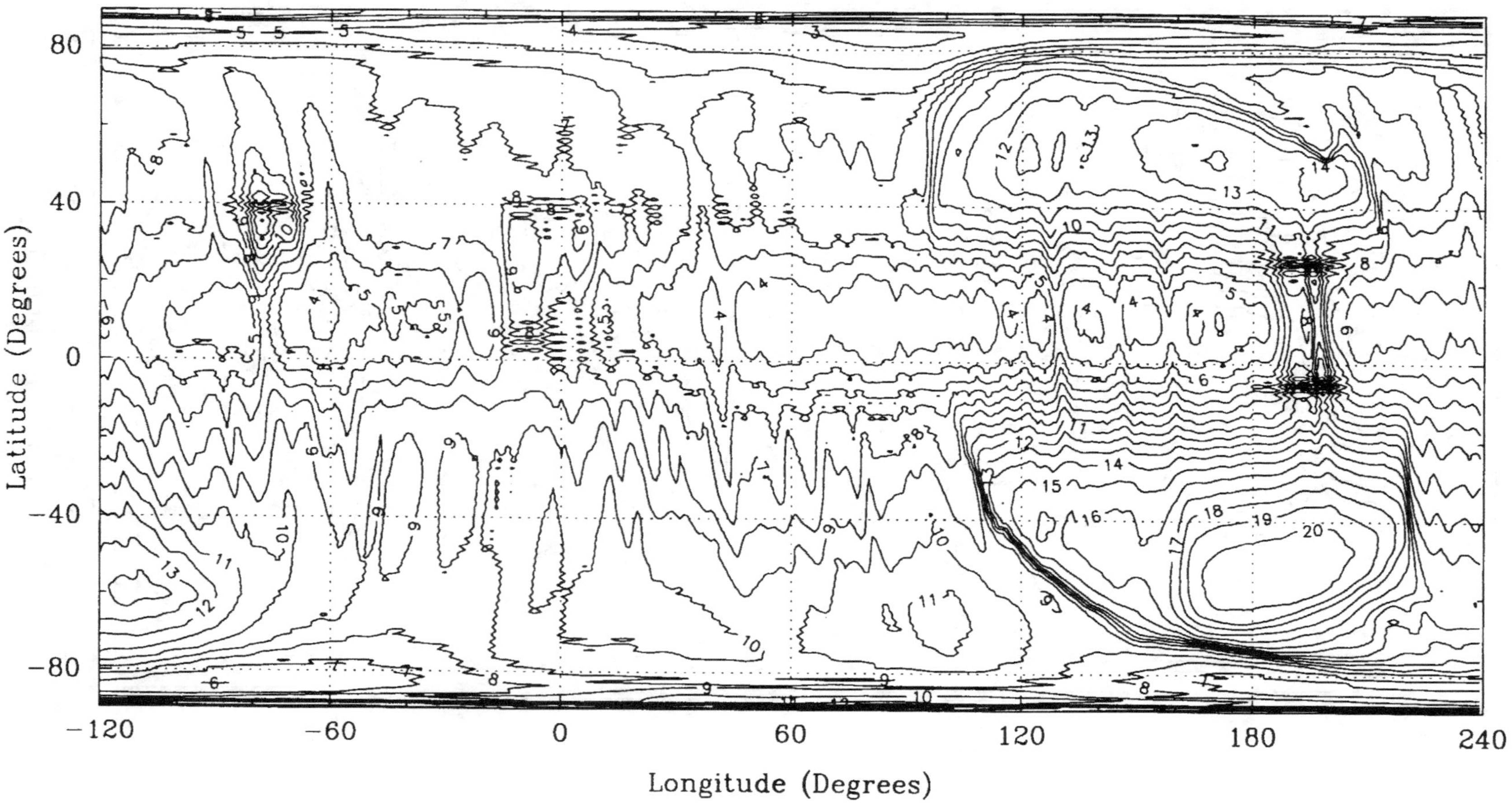

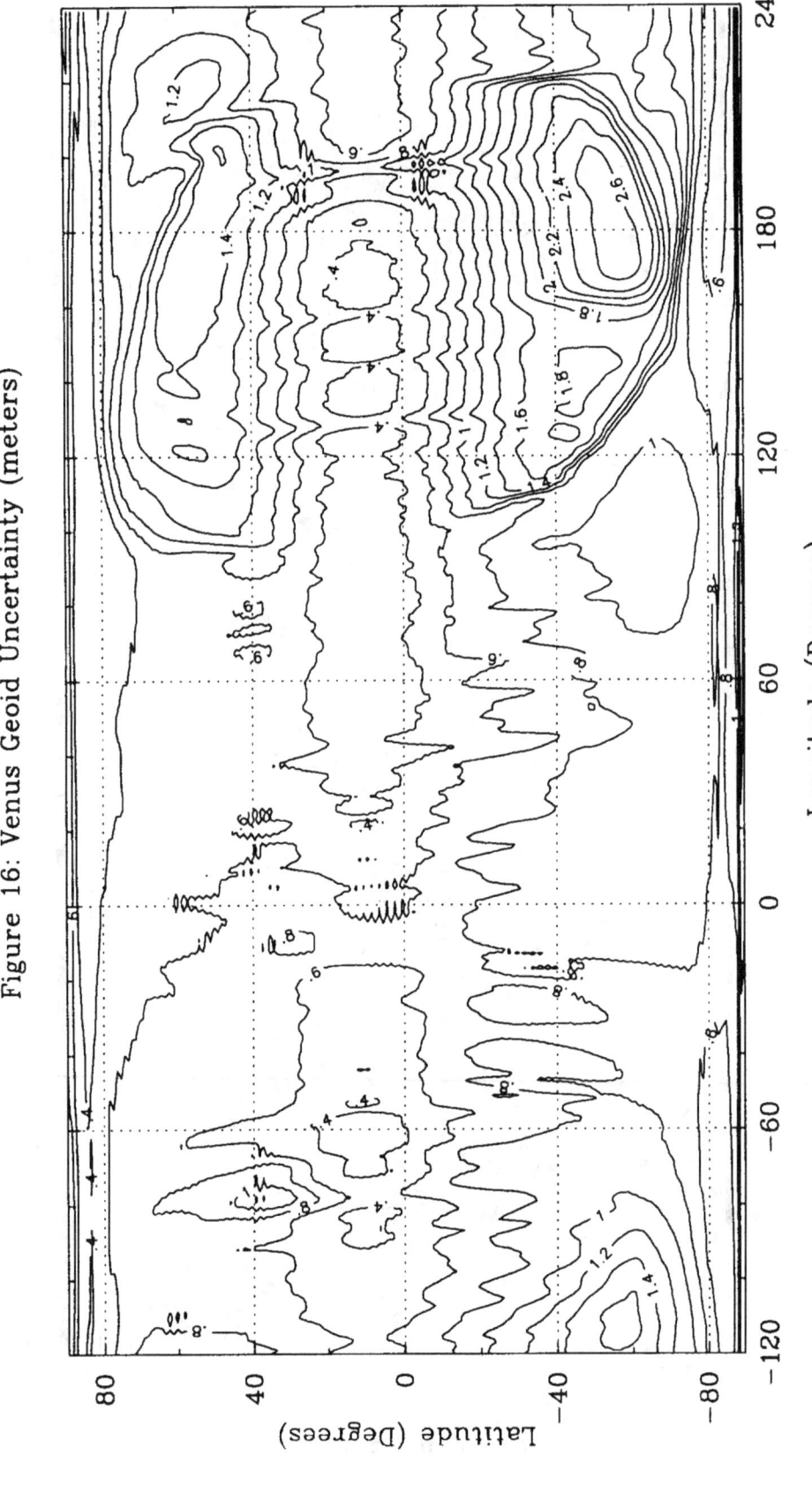

Figure 16: Venus Geoid Uncertainty (meters)

Table 5. Gravity Peaks for Venusian Features of Interest

Feature	Lon	Lat	MGNP90LSAAP	MGNP75ISAAP	MGNP60FSAAP
Maxwell	4.5	63.5	244.68	220.65	184.30
Akna	-42.5	68.5	115.17	99.57	75.52
Freya	-23.5	73.5	126.34	123.98	105.51
Bell	46.0	29.0	126.25	116.40	102.88
Beta	-79.0	25.5	234.32	231.82	211.87
Gula	-2.0	22.0	138.27	121.77	99.90
Maat	195.0	1.0	356.41	308.46	228.63
Ozza	200.0	3.5	245.52	250.75	224.63
Nokomis	190.0	19.5	132.89	136.44	124.42
Sapas	188.0	8.5	157.54	135.76	126.92
Atalanta	164.5	62.5	-84.44	-83.96	-78.34
Mead	57.2	12.6	-49.67	-39.71	-29.85

sense is globally uniform and equals about 4 to 5 milligals (the error for the sensed geoid is about 0.4 meters). After Magellan was aerobraked into a near circular orbit, there was no direct observation of the gravity field from about 160°E to 220°E, and these gaps are apparent in the uncertainty maps. The total errors (Figures 15 and 16) are largest for the southern gap in the Magellan post-aerobraking data since there the resolution is poorest and we have the greatest error from omission of higher degree terms. The largest vertical gravity and geoid errors are 20 milligals and 2.6 meters. Also visible in the error maps (and degree strength map - Figure 10), is the apoapse tracking in the cycle 5 Magellan data from about 100°E to 160°E. The face-on orbit geometry of Magellan cycle 4 data and the decrease in resolution is evident in Figure 10 near longitudes 0° and 240°E.

The degree strength map (Figure 10) can be verified by plotting the unconstrained solutions for different degree and order. The 90th degree and order solution without any gravity constraint is displayed in Figure 18 with the unconstrained solutions for degrees 75, 60, and 40 shown in Figures 19, 20, and 21, respectively. Again as in Table 4, the lower degree solutions are found by fixing the higher degree terms to the nominal gravity field (i.e. a truncation of the square root information array). The higher order coefficients (the last 5 to 10 degrees) show aliasing from signatures in the residuals due to terms with degree greater than 90. This shows up as noise in Figure 18. The 75th degree truncated solution, however, does not exhibit aliasing and is much cleaner than the unconstrained solution from MGNP75ISAAP. In general, the unconstrained solutions match the resolution map. The gravity field is completely determined to about degree and order 40 since no noise is visible in Figure 21.

The peak values of the vertical gravity for areas of interest are given in Table 5 for MGNP90LSAAP and the previous 75th degree and order solution (MGNP75ISAAP, Konopliv and Sjogren, 1994b) and 60th degree and order solution (MGNP60FSAAP, Konopliv and Sjogren, 1994a). All are maximum peak values except for the gravity lows of Atalanta and Mead Crater. The strongest gravity feature on Venus is Maat Mons, which will continue to increase in amplitude with the increasing higher degree and order gravity solutions (the next solution will be to degree and order 120). The 90th degree postfit Doppler residuals still show substantial systematic trends from the gravity for the Atla and Beta Regios. The peaks of Bell Regio show noticeable increase and also better alignment

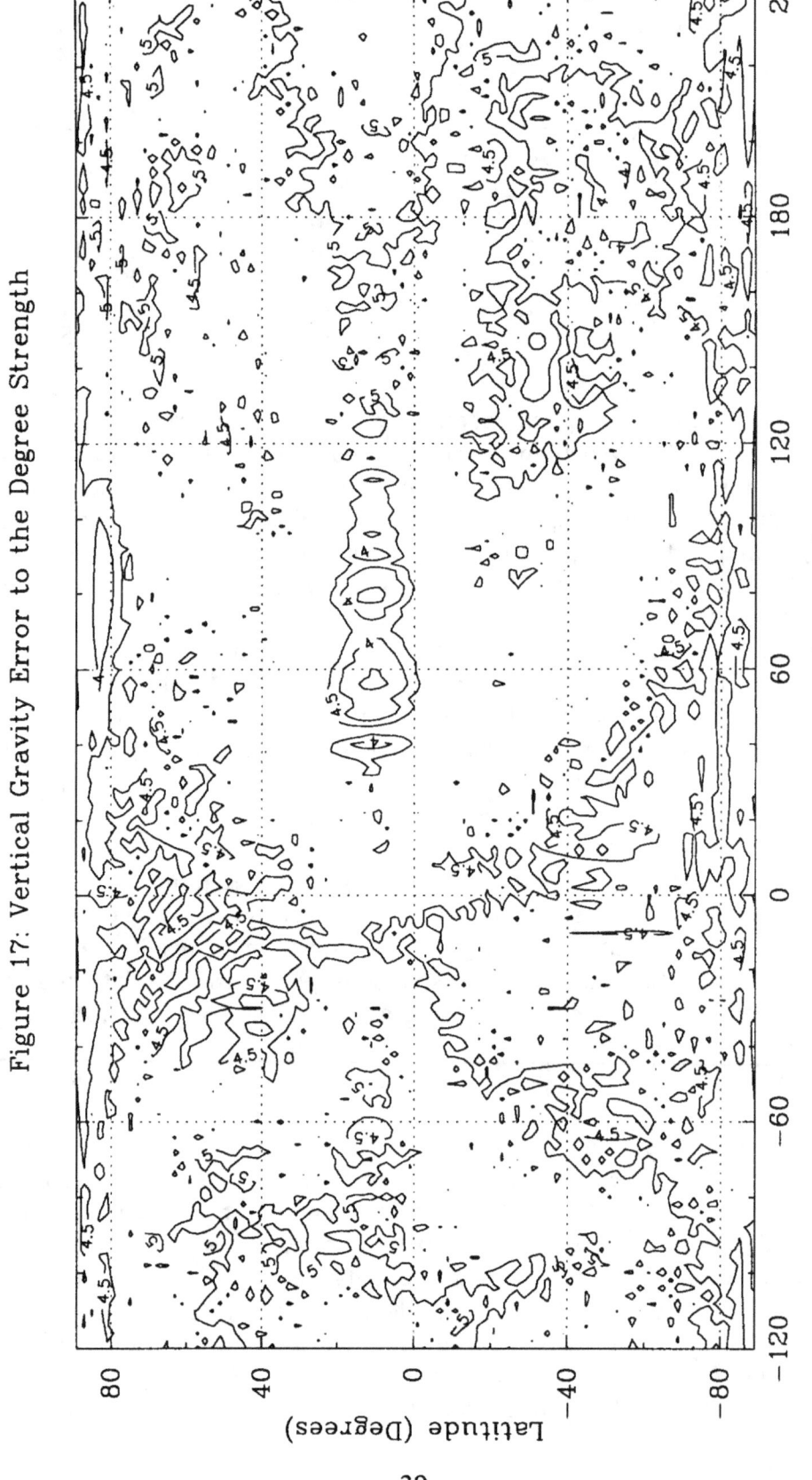

Figure 17: Vertical Gravity Error to the Degree Strength

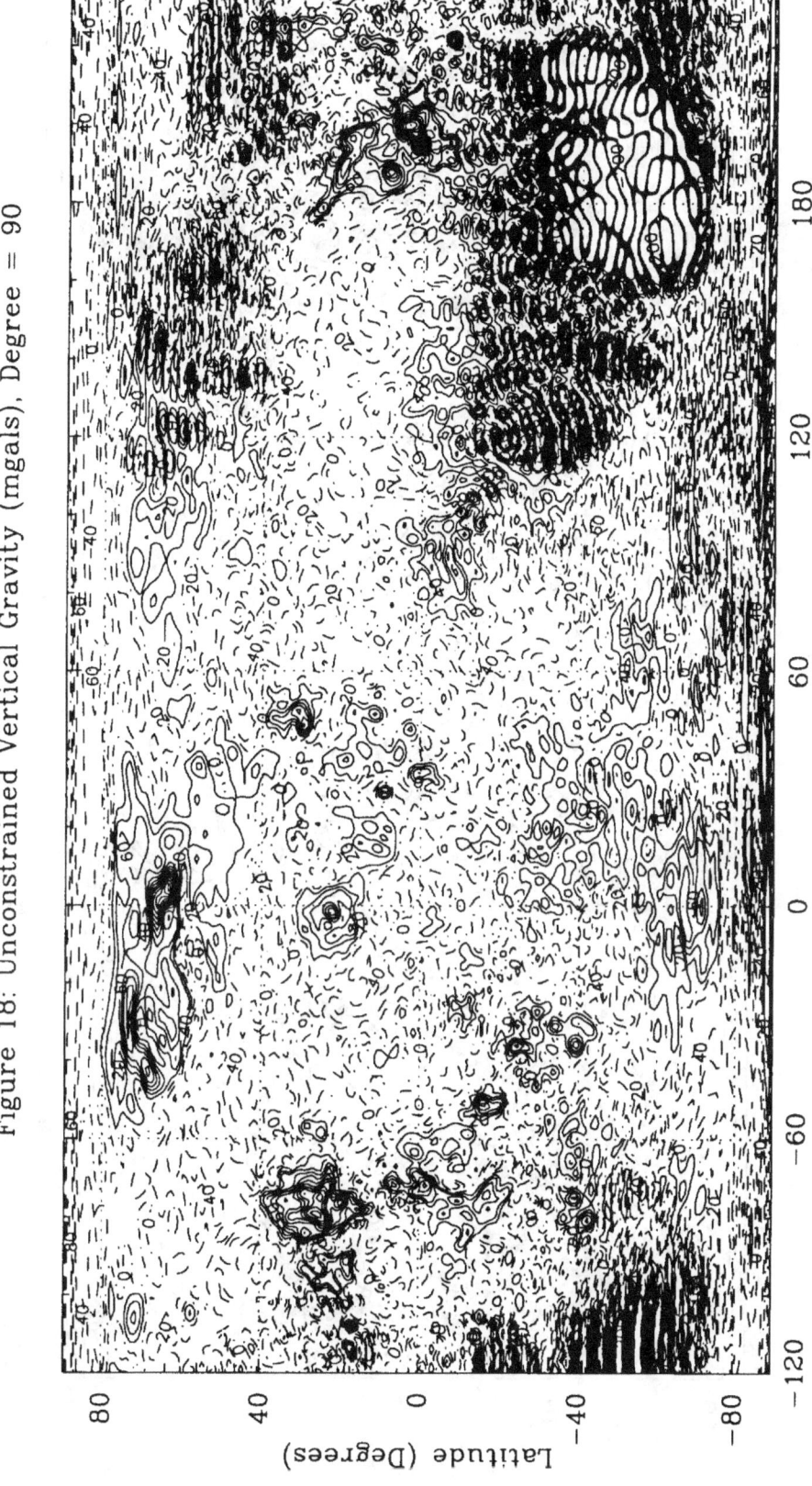

Figure 18: Unconstrained Vertical Gravity (mgals), Degree = 90

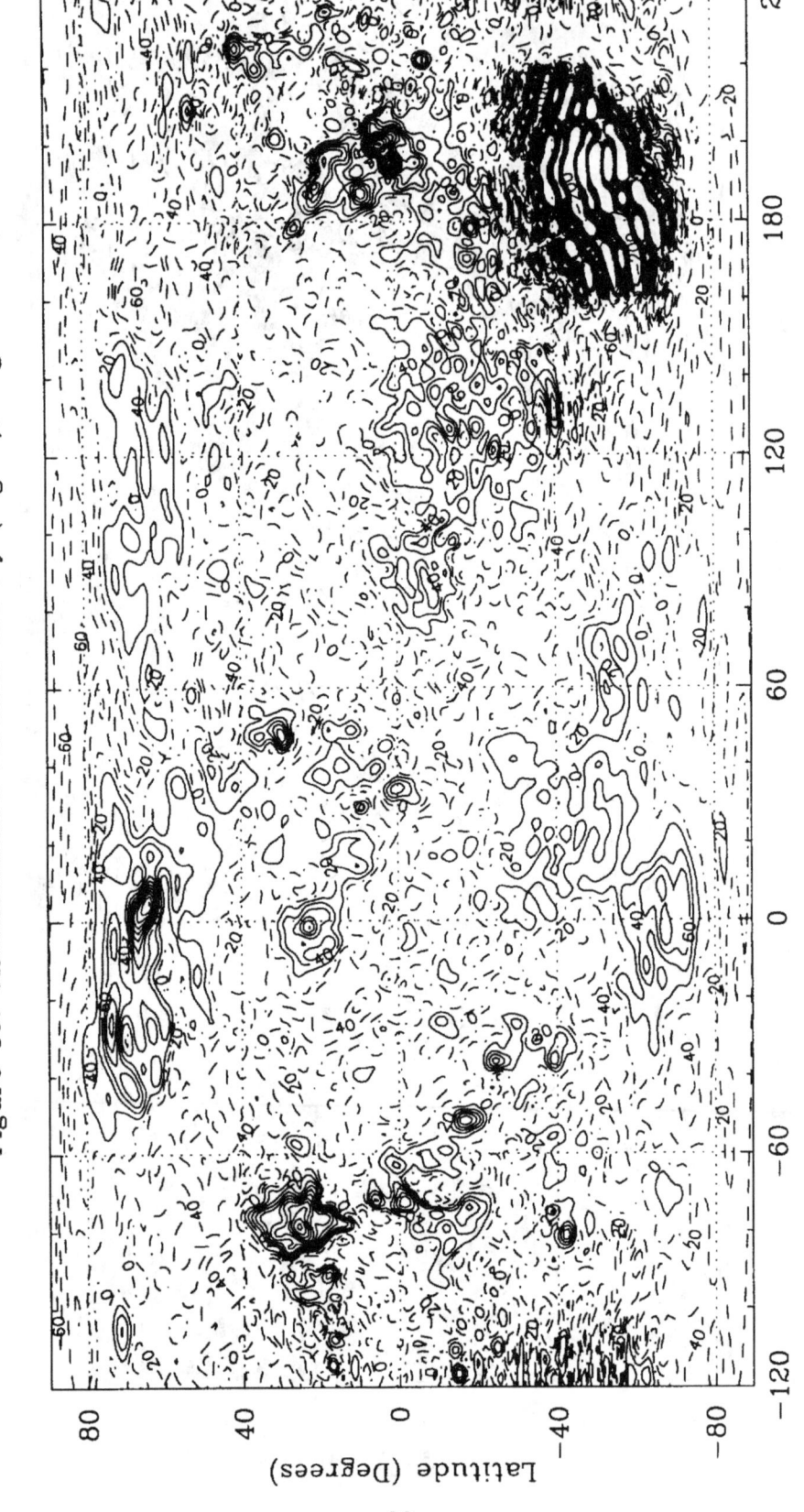

Figure 19: Unconstrained Vertical Gravity (mgals), Degree = 75

Figure 20: Unconstrained Vertical Gravity (mgals), Degree = 60

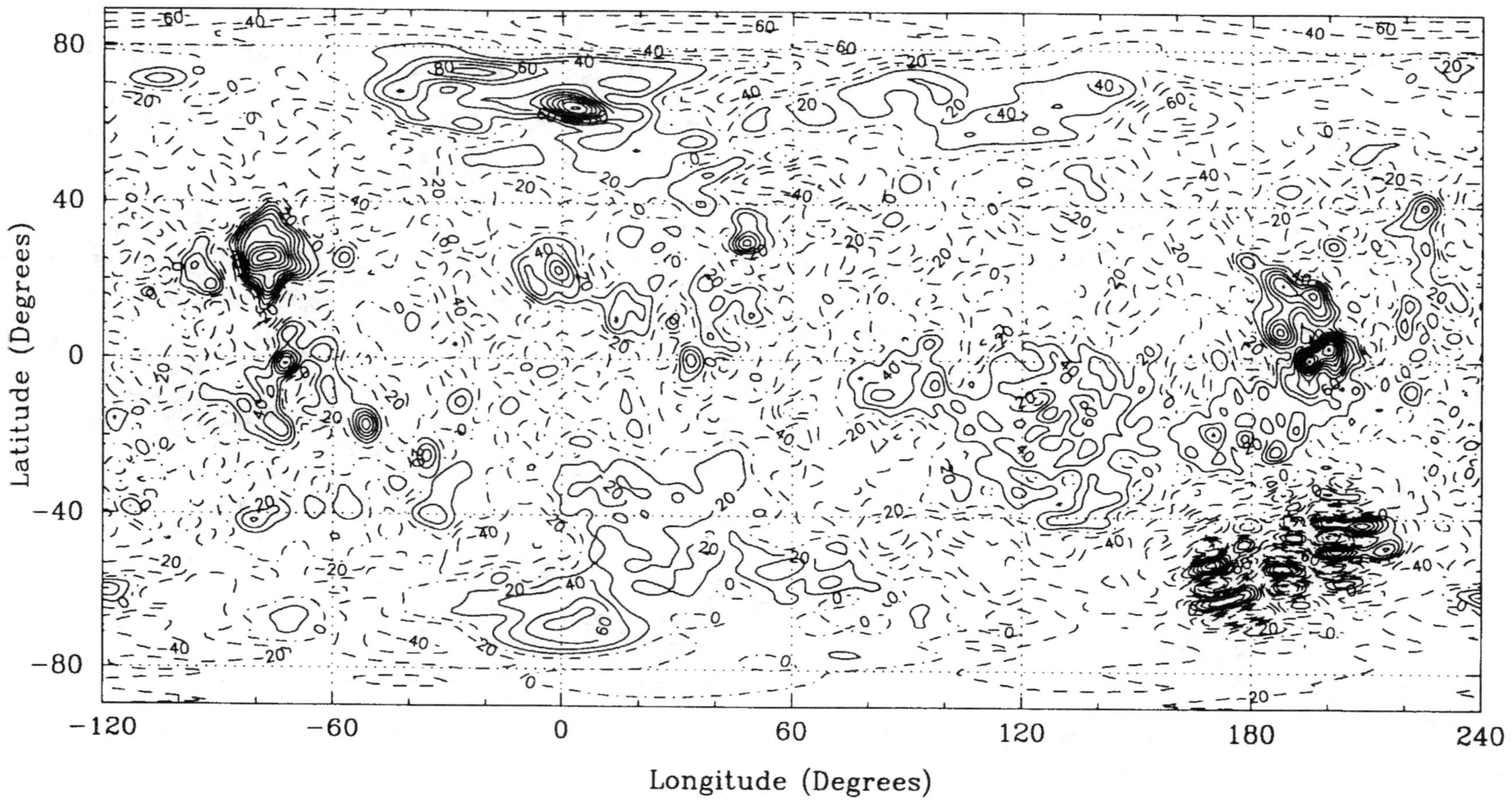

Figure 21: Unconstrained Vertical Gravity (mgals), Degree = 40

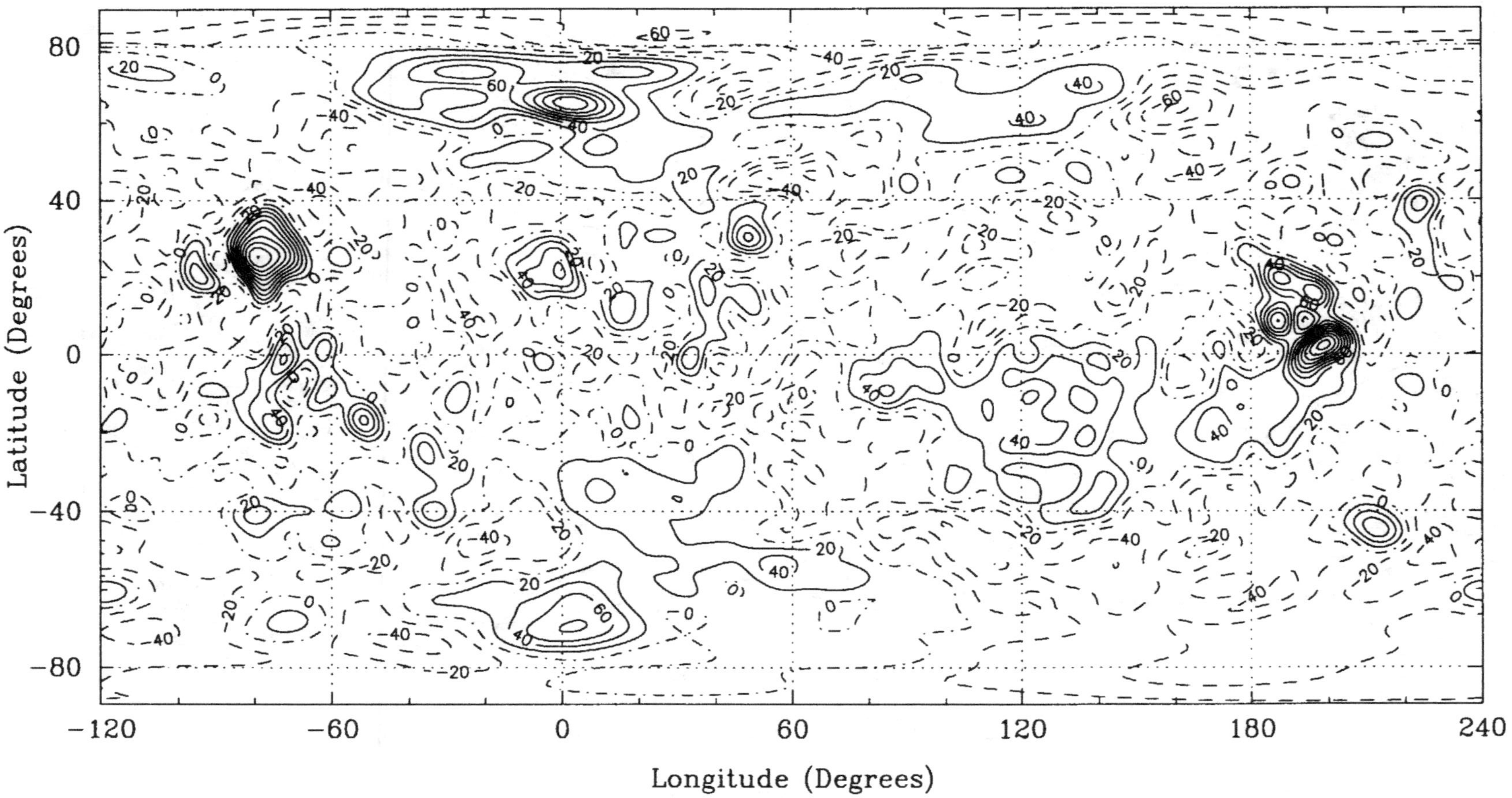

Table 6. Comparisons of Spherical Harmonics with Line-of-Sight Reductions (mgals)

Feature		Reference Altitude (km)			
Beta	Surface	187	200	250	Comments
Konopliv	234	131	128	114	
Kaula (1995)	240				2% high
McKenzie (1995)		90			31% low
Sjogren (1983)			73		43% low
Esposito (1982)			135		5% high
Smrekar (1994)				85	25% low
Maxwell	Surface	323			
Konopliv	245	68			
Kaula (1995)	200				18% low
McKenzie (1995)		39			42% low
Maat	260				
Konopliv	106				
Sjogren (1983)	64				40% low
Smrekar (1994)	75				29% low
Gula	Surface	180	202		
Konopliv	138	61	57		
Sjogren (1983)		38			34% low
Barriot (1994)	110				20% low
McKenzie (1995)			40		30% low
Smrekar (1994)			50		12% low

with the topographic highs. Appendix I contains plots of vertical gravity for different regions of Venus for MGNP90LSAAP.

A comment, which was mentioned by Sjogren (1984) when discussing his analysis of Ishtar, was that there was a need to have a positive mass placed at 51.5° N so that the acceleration profile could be fit. This gravity anomaly is now definitely revealed (see Appendix I).

A comparison of what other analysts have obtained for some of these features is shown in Table 6. They have used the line-of-sight accelerations derived from Doppler residuals and produced local estimates (all except Esposito et al, 1982, who used the raw Doppler observations and surface mass disks to estimate the Beta gravity anomaly). Most line-of-sight estimates were obtained at different reference altitudes and therefore the harmonic estimates were evaluated at those altitudes to make comparisons valid. Except for the Beta estimate of Kaula (1995, 240 versus 234 mgals), all estimates are lower than the harmonic estimates by considerable amounts. This is rather surprising since the Doppler residuals should contain the very highest resolution of the data. On the other hand, the harmonics at degree 90 leave almost no systematic signature in their residuals. An explanation for this variance may be due to the model fitting to the LOS data. The experimenters must decide on optimum block sizes or mass distributions. The data are then smoothed to avoid singularities at the surface which may reduce the amplitudes. Also, there may be amplitude reductions as a result of a larger than needed spline interval for determination of the accelerations from the Doppler residuals.

Mead was the subject of previous investigation (Banerdt et al, 1994) with a 60th degree and order gravity field. The amplitude for Mead with that model was -30 mgals with the gravity anomaly being slightly offset by about one degree from the center of the crater. The higher resolution 90th degree models show almost perfect alignment with the crater and substantially increased amplitudes, and show further confirmation that Mead is indeed mostly uncompensated (even less than the 30% maximum reported by Banerdt et al, 1994). For terms up to 90th degree, the gravity signature at the surface from uncompensated topography of Mead only and not the surrounding topography is -25.1 mgals (for up to degrees 60 and 75, the amplitudes are -9.8 and -16.7 mgals, respectively). The spherical harmonic topography model was determined by Nicole Rappaport with data from Bob Grimm and zeros out all topography except Mead. The model was determined to degree and order 120 but is truncated for comparison with the gravity. The corresponding gravity signature from Mead is also at least -25 mgals with respect to the surrounding gravity and maybe even 5 to 10 mgals larger. From degree 60 to 90, the uncompensated gravity from topography increased by 15.3 mgals and the observed gravity increases by 19.8 mgals (i.e. from Table 5). With a crust thickness of 25 and 50 km and 30% compensated, the gravity minimum from topography is -19.3 and -20.7 mgals, respectively. For the high degree terms, the uncertainty in the gravity is better than the 4 mgal formal error, and, hence, we can now say Mead is even less than 30% compensated.

The rms magnitude of the gravity spectrum for degree n (M_n) is given by $M_n = G_n/(2n+1)^{1/2}$ where G_n is the magnitude of all the gravity coefficients of degree n (given by the vector $\mathbf{G_n}$). The spectrum for MGNP90LSAAP along with the uncertainties and Kaula power rule for Venus is shown in Figure 22. Also shown is the spectrum from the unconstrained gravity solution (90lnoap or no a priori). In this case there is the same constraint on the nongravity parameters (ephemeris, pole, etc.) as MGNP90LSAAP and one can notice that the constraint on the higher degree coefficients does not affect the formal uncertainties on the lower degree coefficients (at least < degree 10, note Figures 9 and 11 show the effect of not constraining pole, etc.). From Figure 22, one can say that the gravity field is determined over the entire surface to about degree and order 40 (the crossing point of the unconstrained sigma and the Kaula power rule).

For Venus, probably the best test for evaluation of the gravity field has been the correlation with topography. As more tracking data were added to the solution and as modeling improved, the correlation with topography continued to show higher values. Figure 23 shows the correlation for MGNP90LSAAP along with the error bars. With $\mathbf{T_n}$ being the vector of all topography coefficients for degree n, the correlation for degree n is given by $\gamma_n = (\mathbf{G_n} \cdot \mathbf{T_n})/(G_n T_n)$. The topography coefficients are given by the 360th degree and order model of Rappaport and Plaut (1994). The correlation error bars in Figure 23 for each degree ($\sigma\gamma_n$) are contributions from the gravity covariance and are given by $\sigma\gamma_n^2 = \mathbf{A_n}^T \mathbf{P_n} \mathbf{A_n}$ where the matrix $\mathbf{P_n}$ is the sub-covariance with the degree n terms ($\mathbf{G_n}$) only of the full covariance matrix of MGNP90LSAAP. The vector $\mathbf{A_n}$ is the partial of the correlation for degree n (γ_n) with respect to the gravity coefficients of degree n, $\mathbf{A_n} = (\mathbf{T_n}/G_n T_n) - \gamma_n(\mathbf{G_n}/G_n^2)$. As mentioned in Konopliv et al (1994b), the uncertainties to

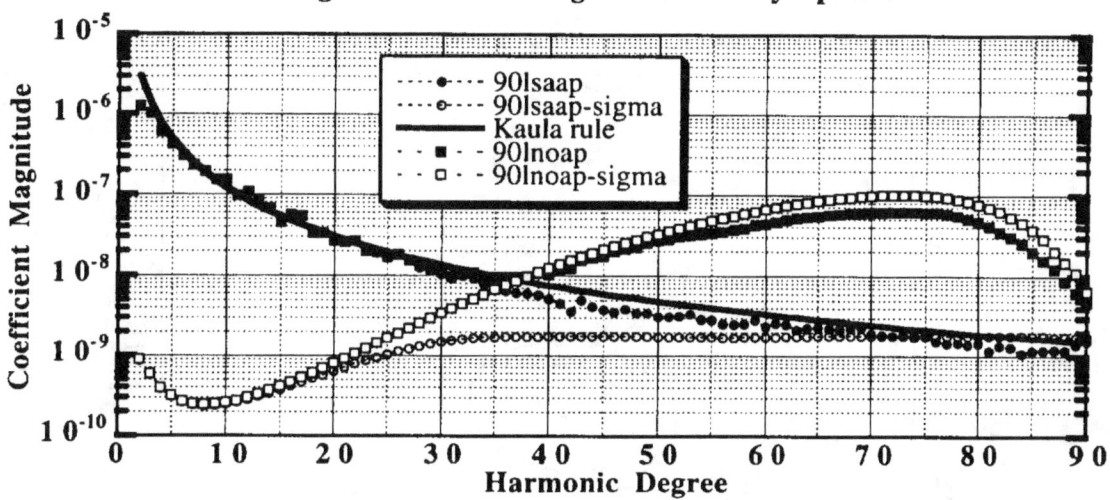

Figure 22: RMS Magnitude Gravity Spectrum

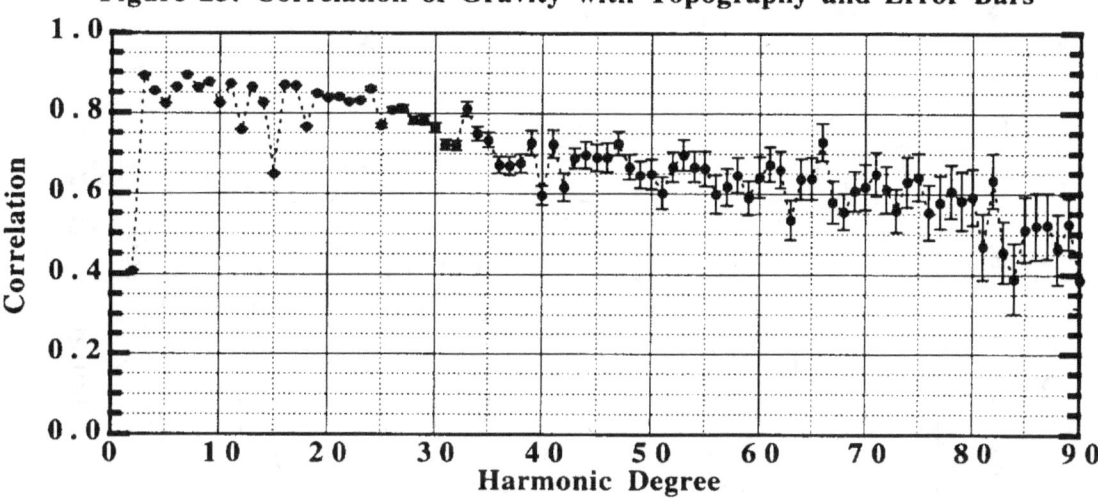

Figure 23: Correlation of Gravity with Topography and Error Bars

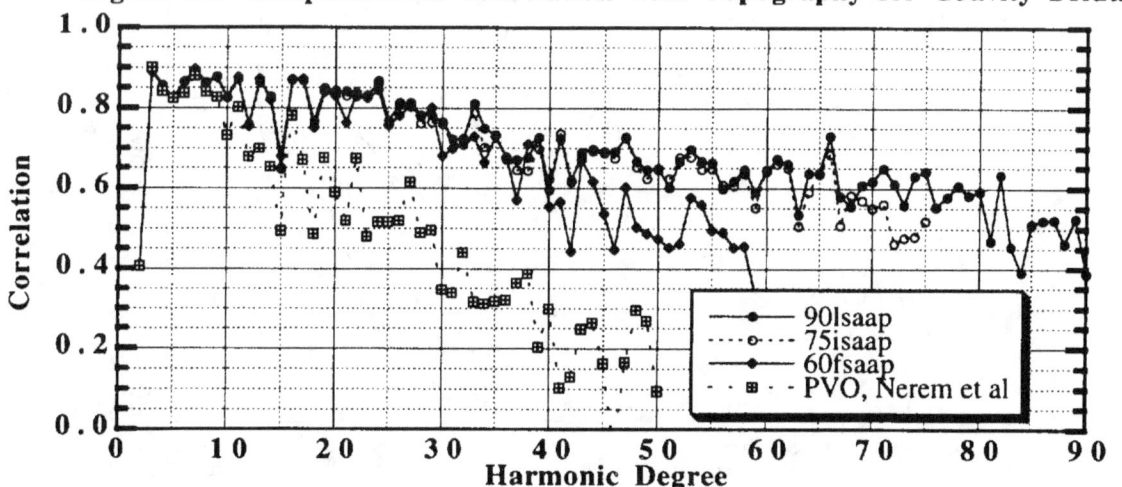

Figure 24: Comparison of Correlation with Topography for Gravity Solutions

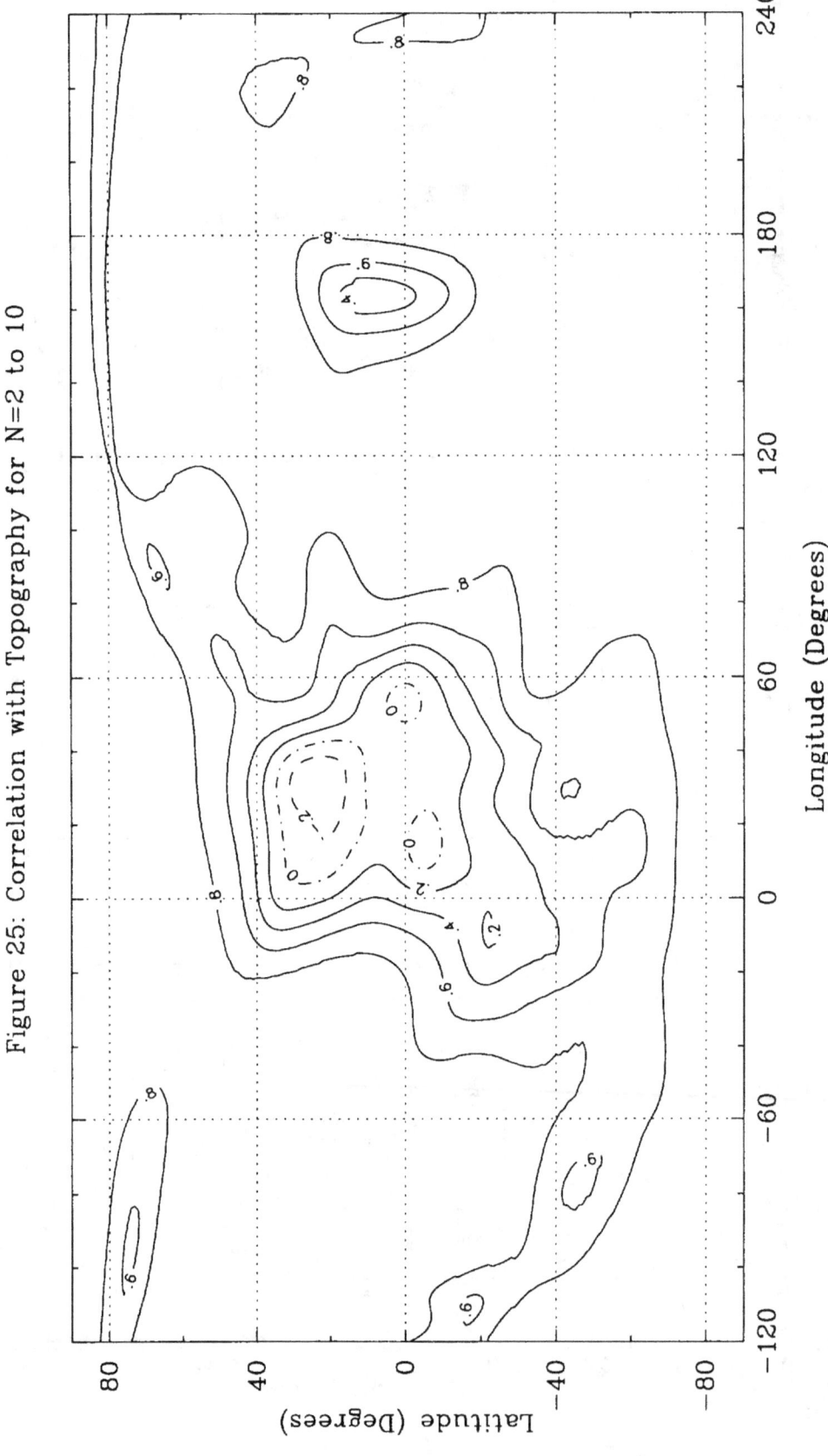

Figure 25: Correlation with Topography for N=2 to 10

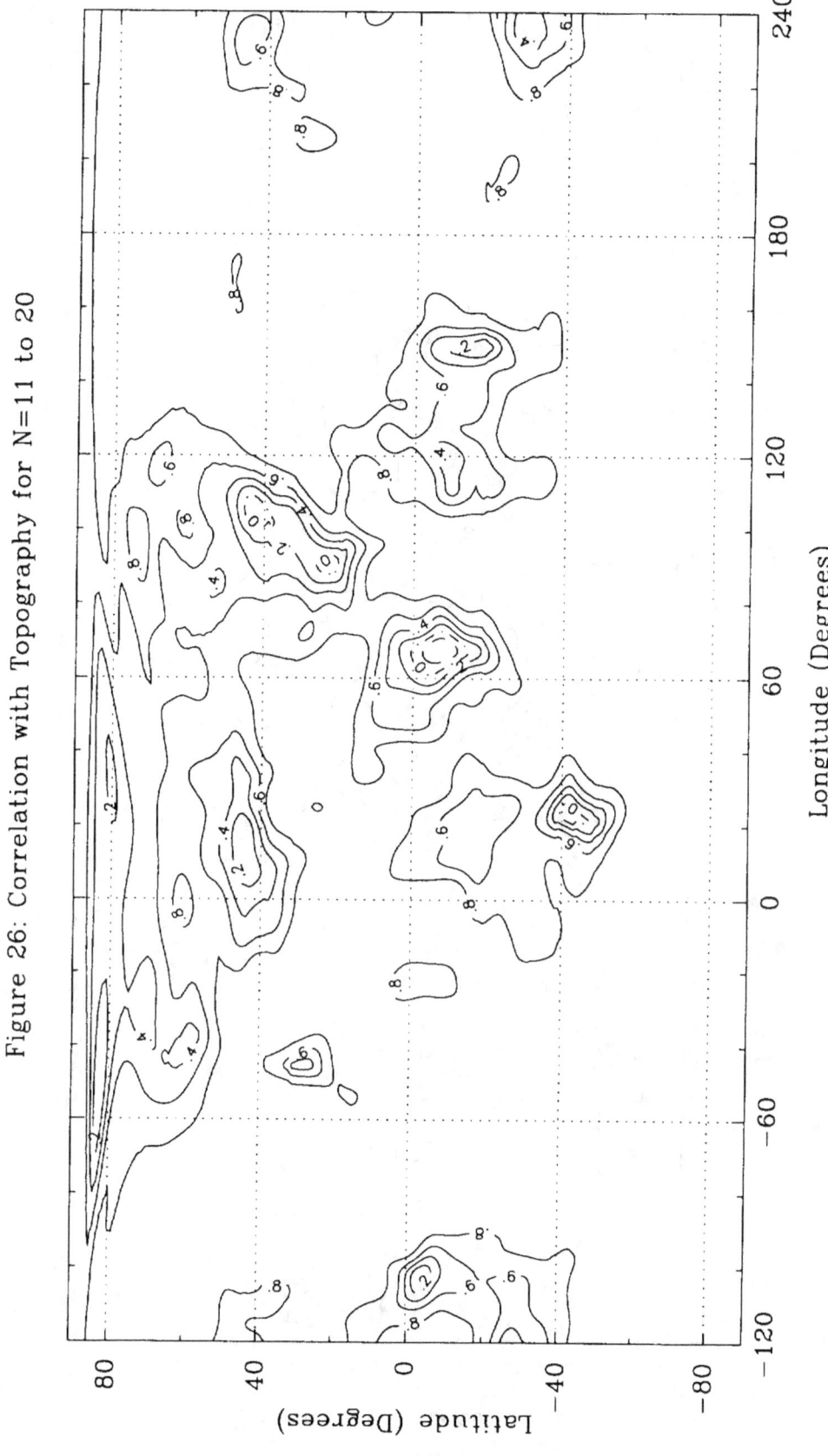

Figure 26: Correlation with Topography for N=11 to 20

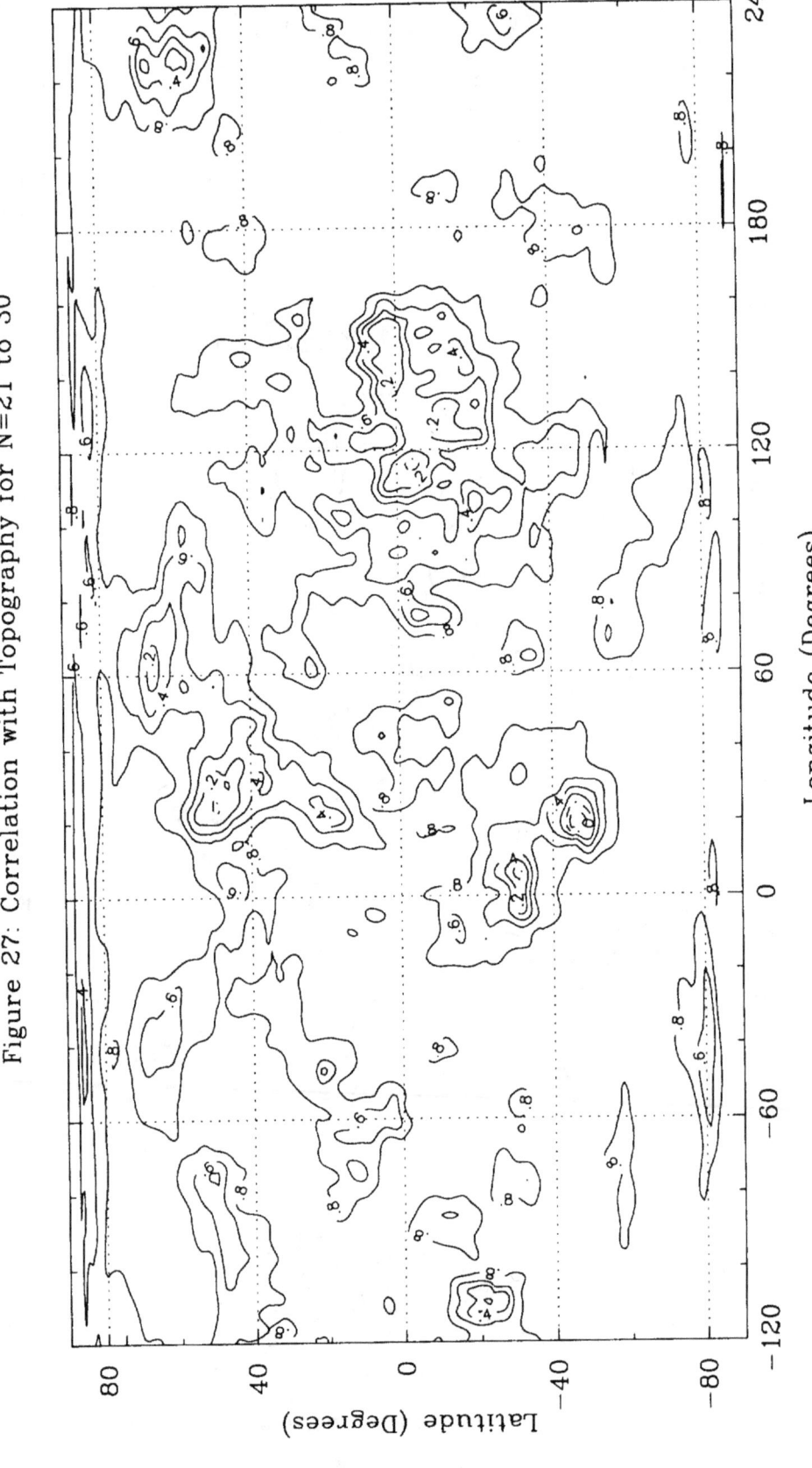

Figure 27: Correlation with Topography for N=21 to 30

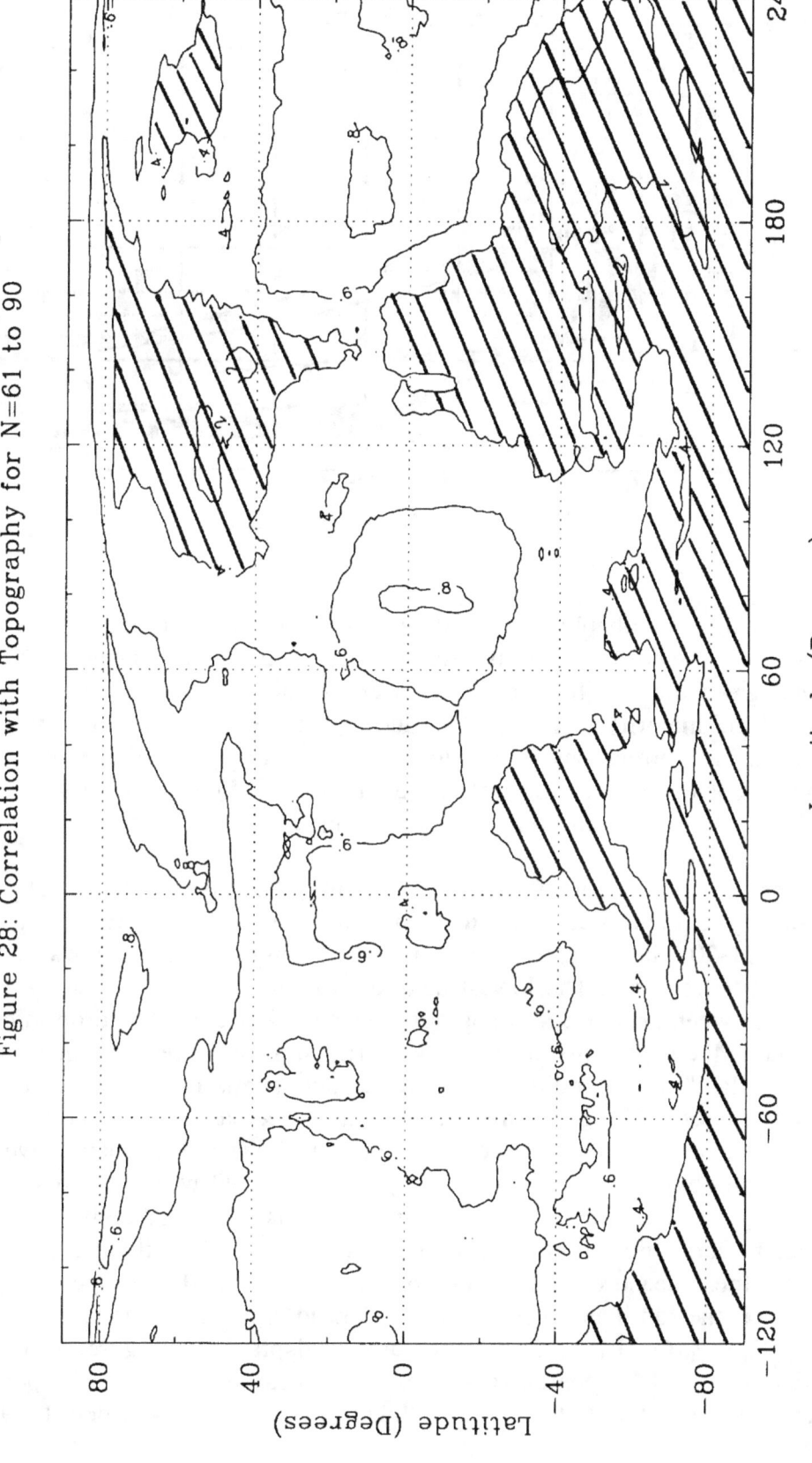

Figure 28: Correlation with Topography for N=61 to 90

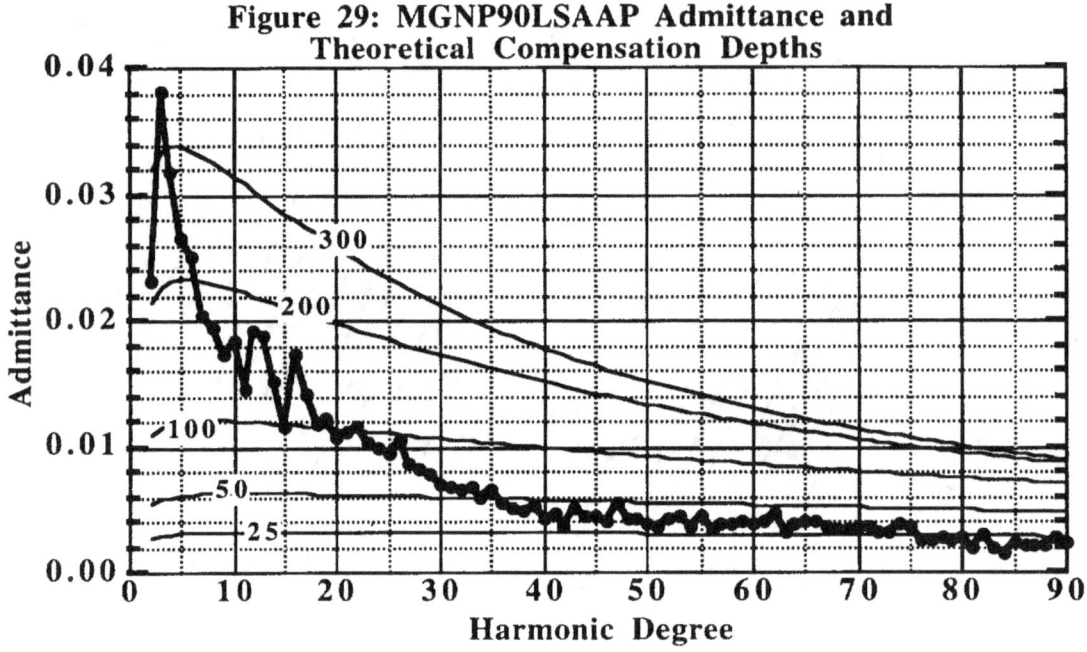

Figure 29: MGNP90LSAAP Admittance and Theoretical Compensation Depths

about degree 30 are probably dominated by uncertainties in the topography, and for degree greater than 30, the errors are mostly from gravity. Since the errors neglect the correlations between degrees, the error bars are probably optimistic. The dip in correlation at degree 15 is real and the majority of it (80%) is due to the poor correlation between the zonal coefficients. The correlations for previous solutions are presented in Figure 24. Note the increase in correlation for the last ten degrees (65-75) for MGNP90LSAAP versus MGNP75ISAAP due to the removal of aliasing in these terms for the higher degree solution.

The correlation of gravity with topography can also be plotted spatially for different ranges of degree. Figures 25, 26, and 27 show the lower degree harmonic ranges of 2-10, 11-20, and 21-30. The correlations are calculated using a sliding window equal in size to the shortest wavelength. The lowest degree harmonics (Figure 25) show a substantial mismatch between gravity and topography over Eistla Regio. The drop in correlation is mainly due to the degree two through five harmonics. In the next harmonic range, Ishtar Terra, the "claw" of Aphrodite, Tellus Regio, and Thetis Regio appear. For the higher degree harmonics (>60), one can notice a general decrease in correlations for the areas where the data are weak (see Figure 10). Figure 28 shows the correlation averaged over a 30 degree window for terms between degrees 61 and 90 with the cross-hatched areas representing correlations below 0.4. This indicates that if we had global uniform tracking coverage, the correlations for the higher degrees in Figure 23 would be greater than 0.6.

Another measure of the geophysical processes that shape the gravity and topography is the admittance between them. The admittance function for each degree n is given by $F_n = \gamma_n(G_n/T_n)$ or $G_n \cdot T_n/T_n^2$ and is displayed in Figure 29 along with the theoretical curves for apparent depth of compensation for Airy compensation. The admittance at degree n is related to the depth D by (Turcotte and McAdoo, 1979)

$$F_n = \frac{3\rho_s}{(2n+1)\bar{\rho}} \left[1 - (1 - \frac{D}{a_e})^n\right] \qquad (13)$$

where surface and mean densities are 2.9 and 5.248 gm/cm^3, respectively. The admittance, like the correlations above, can be calculated spatially for different spectral windows. Simons et al (1994) have plotted the admittance on a global scale for several different spectral windows and inferred internal processes that match the admittance. Here, we calculate the admittance for the harmonic range 40 to 60. For this range, we expect the combined effect of dynamic or thermal support for the long wavelengths and the elastic support for the short wavelengths to be minimized. With the admittance we calculate the depth of compensation on a global scale. The result is displayed in Figure 30 with averages over a 30 degree window. Assuming Airy compensation for wavelengths of degree 40 to 60, the crustal thickness is 20 to 30 km for the Ishtar and western Aphrodite regions and thicker for areas such as Atla and Beta.

The Bouguer and isostatic anomalies are also used for geophysical interpretation and are given in Figures 31 and 32. The Bouguer acceleration is the difference between the vertical gravity acceleration at the surface and the theoretical vertical acceleration from uncompensated topography. In this case, the spherical harmonic topography coefficients A_{nm}, B_{nm} are related to the theoretical gravity coefficients C^t_{nm}, S^t_{nm} for use in the potential (equation 1) by

$$C^t_{nm}, S^t_{nm} = \frac{3\rho_s}{(2n+1)\bar{\rho}} A_{nm}, B_{nm}$$

The large negative Bouguer anomalies for Aphrodite and Ishtar Terra are clearly evident in Figure 31. The isostatic anomaly evaluates the nonlinear difference between the gravity and topography. Given the gravity coefficients of degree n ($\mathbf{G_n}$) and the topography coefficients ($\mathbf{T_n}$), the isostatic coefficients are given by $\mathbf{I_n} = \mathbf{G_n} - F_n \mathbf{T_n}$ where F_n is the admittance. If the supporting mechanism (dynamic, isostasy) is linear, then the isostatic anomalies show the difference from the global average compensating mechanism. If for a positive anomaly, the isostatic anomaly is negative, then the feature has more compensation than the global average. If the gravity and isostatic anomalies are both positive, then there is less compensation. The small compensation of Maat Mons is clearly evident in Figure 32 as it is by far the largest isostatic anomaly. Figure 33 shows the isostatic anomaly for the Mead Crater. Since Mead is a negative gravity and isostatic anomaly, it is also much less compensated than the global average.

The stability of the solution for gravity and other parameters is given in part by the correlations in the covariance of estimated parameters. Appendix H lists the correlations between the nongravity parameters and the first degree and order five gravity coefficients. The solution in this case solves for the Venus pole and rotation. The correlations are generally fairly small with the largest correlation being 0.6. Figure 34 is a contour plot of the rms of the correlations between the coefficients of a given degree for the covariance of MGNP90LSAAP. With the a priori constraint, the correlations remain fairly small with the majority of rms values below 0.1. The correlations are a maximum for the lower harmonic

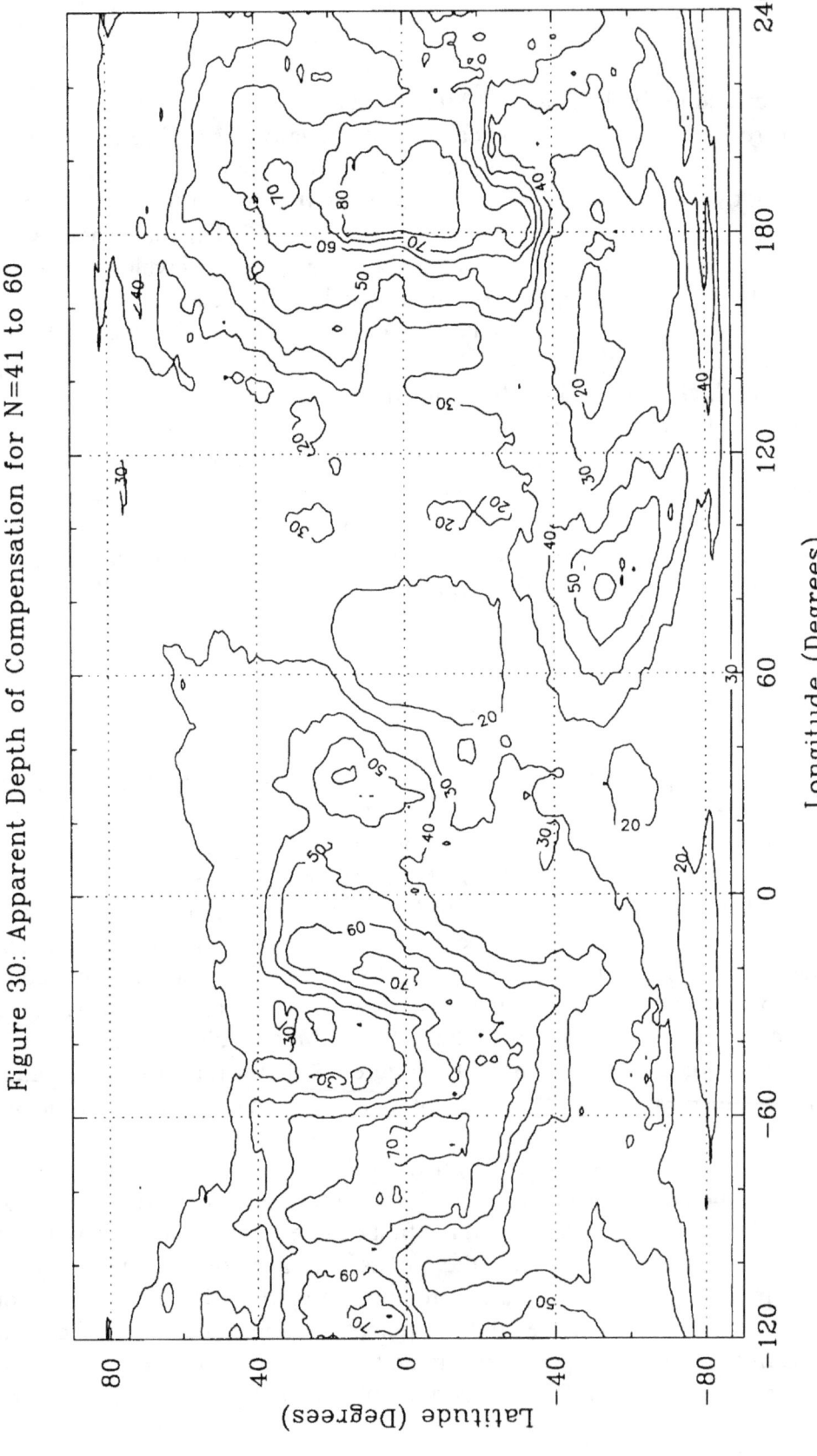

Figure 30: Apparent Depth of Compensation for N=41 to 60

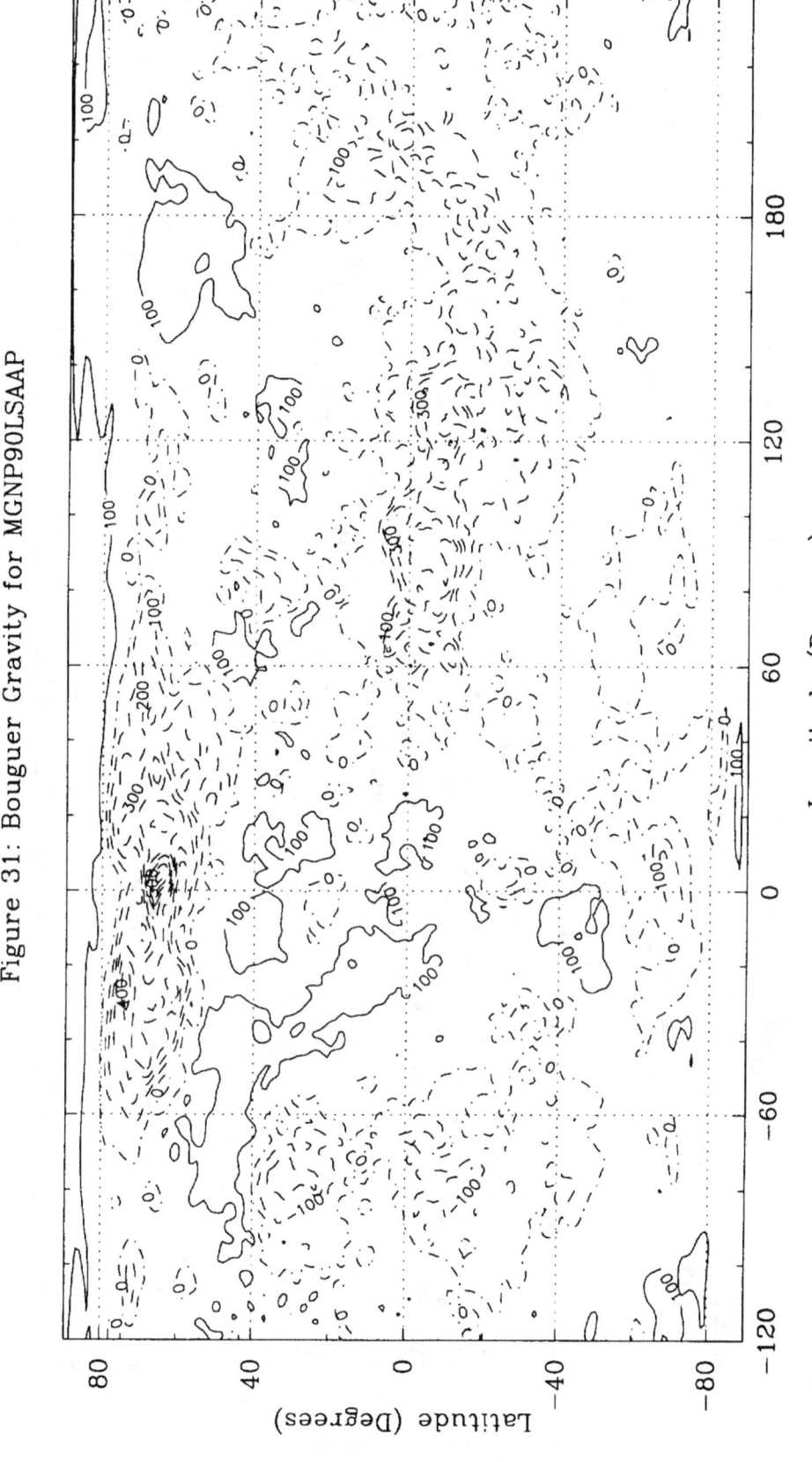

Figure 31: Bouguer Gravity for MGNP90LSAAP

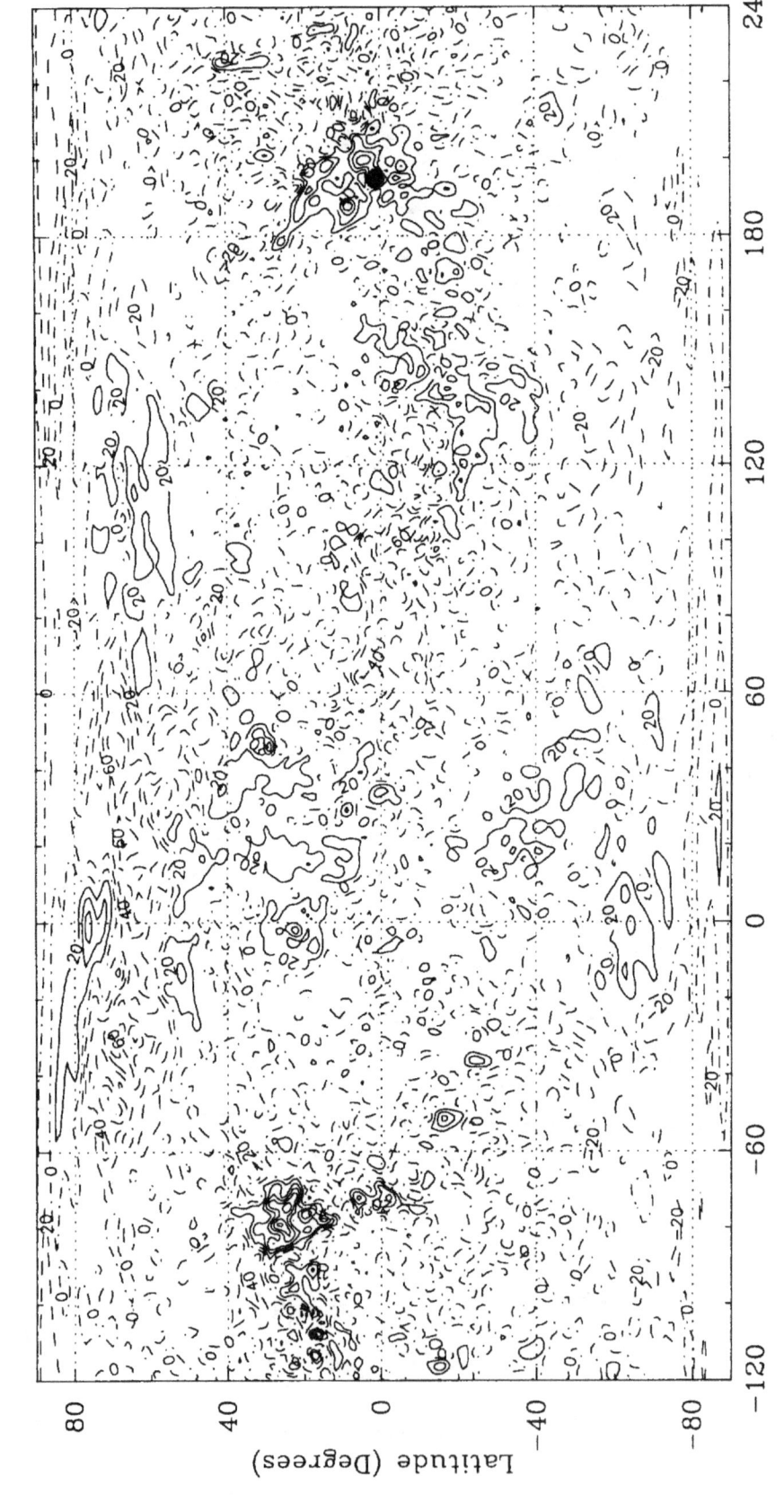

Figure 32: – Isostatic Perturbations for MGNP90LSAAP

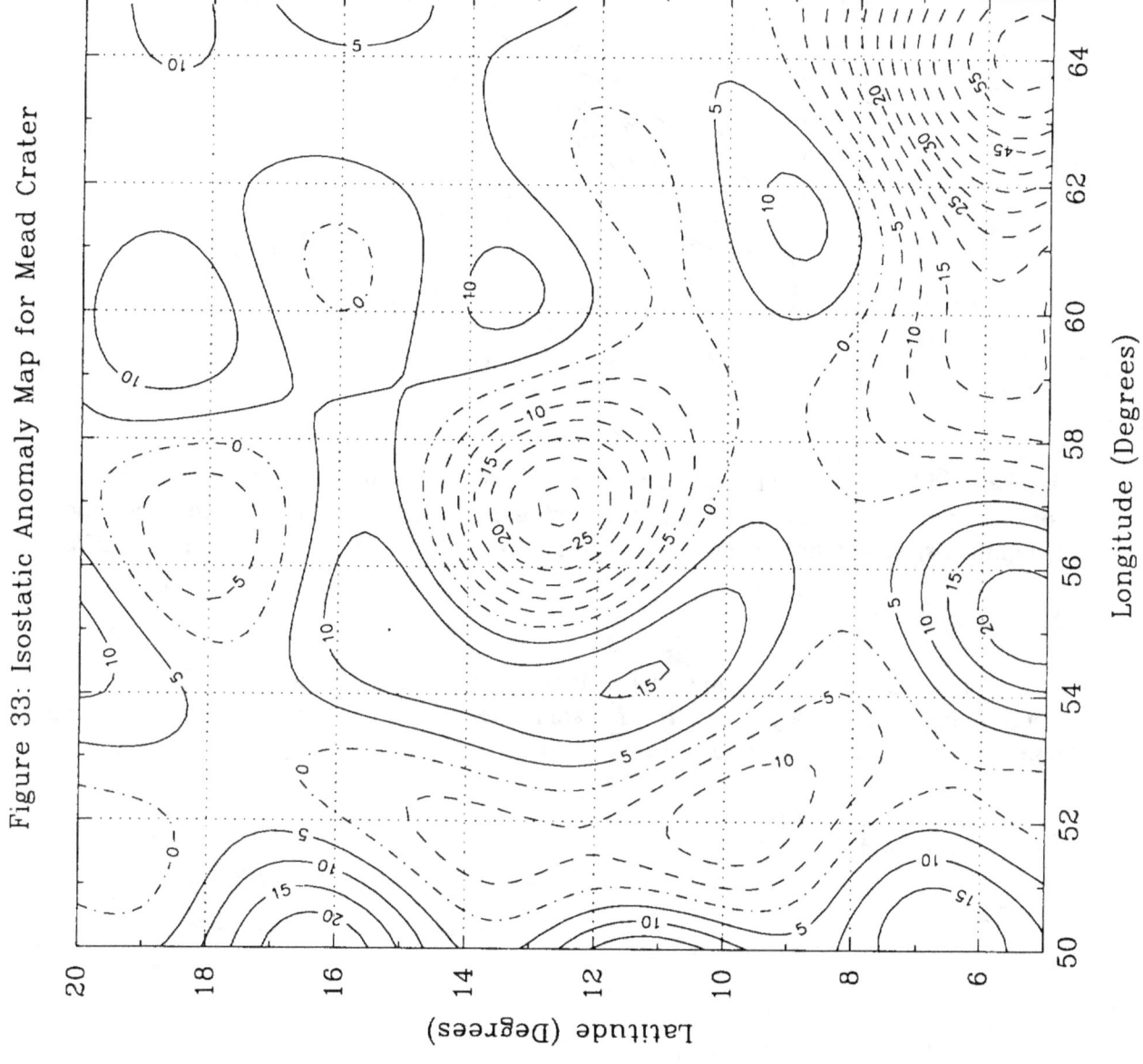

Figure 33: Isostatic Anomaly Map for Mead Crater

Figure 34: Gravity Coefficient Correlation Matrix

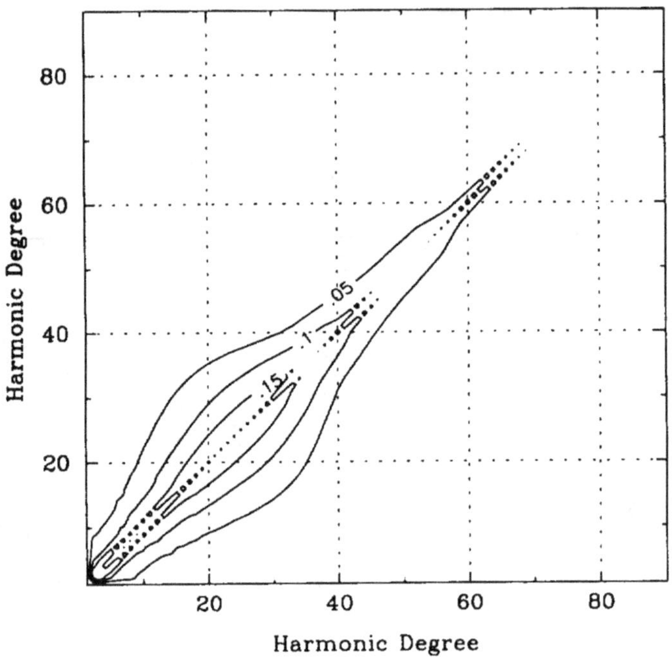

degrees and for degrees with the same approximate wavelength. We expect the correlations for the unconstrained covariance to be much greater for the higher degrees since the rms magnitude spectrum for the unconstrained spectrum is unrealistically large (Figure 22).

5b. Principal Axes

The principal axes of inertia are given by the second degree gravity harmonics. The unnormalized second degree coefficients are related to the moments and products of inertia about the body-fixed axes (1,2,3) by (Lambeck, 1988)

$$C_{20} = \frac{-1}{M_v a_e^2} \left[I_{33} - \frac{1}{2}(I_{11} + I_{22}) \right]$$

$$C_{21} = \frac{I_{13}}{M_v a_e^2} \qquad S_{21} = \frac{I_{23}}{M_v a_e^2} \qquad (14)$$

$$C_{22} = \frac{1}{4 M_v a_e^2} (I_{22} - I_{11}) \qquad S_{22} = \frac{I_{12}}{2 M_v a_e^2}$$

where M_v is the mass of Venus and a_e the reference radius. From equations (14), the inertia tensor with the moments of inertia along the diagonal and the negative products of inertia on the off diagonals is given by

Table 7. Principal Axes for Venus in Degrees

Axis	MGNP90LSAAP Lat	MGNP90LSAAP Lon	MGNP75ISAAP Lat	MGNP75ISAAP Lon	MGNP60FSAAP Lat	MGNP60FSAAP Lon
1	0.38	-3.33	0.37	-3.20	0.19	-2.89
2	0.36	86.67	0.36	86.81	0.42	87.11
3	89.48	-139.87	89.48	-139.17	89.54	-117.27

$$\mathbf{I} = M_v a_e^2 \begin{bmatrix} (I_{33}/M_v a_e^2)+C_{20}-2C_{22} & -2S_{22} & -C_{21} \\ -2S_{22} & (I_{33}/M_v a_e^2)+C_{20}+2C_{22} & -S_{21} \\ -C_{21} & -S_{21} & I_{33}/M_v a_e^2 \end{bmatrix}$$

(15)

From the inertia matrix, the principal moments of inertia cannot be determined because only two relationships exist between the three moments of inertia. The third relationship comes from observing the rotational behavior of the body (i.e., the precession rate). However for Venus, there has been no observation of the precession. The principal axes, however, can be determined because the eigenvectors are independent of any constant added to the diagonal elements of the inertia tensor. By setting $I_{33}=0$ in equation (15) and diagonalizing the matrix, the principal axes are determined and listed in Table 7.

Assuming a small offset ϕ of the inertial axis from the spin or z-axis, the offset in terms of the unnormalized coefficients is given by

$$\phi^2 = \frac{C_{21}^2}{(C_{20} - 2C_{22})^2} + \frac{S_{21}^2}{(C_{20} + 2C_{22})^2}$$

(16)

The error in ϕ is given by

$$\sigma_\phi^2 = \frac{\partial \phi}{\partial \mathbf{G}_2}^T \mathbf{P}_{\mathbf{G}_2} \frac{\partial \phi}{\partial \mathbf{G}_2}$$

where $\mathbf{G}_2$ are the second degree coefficients and $\mathbf{P}_{\mathbf{G}_2}$ is the covariance for the second degree coefficients. The partials are determined from equation (16). The formal sigma for the inertial offset from the spin axis is 0.01 degrees.

The pole offset to first order is given by the magnitude of the C_{21} and S_{21} gravity coefficients. Table 8 lists the normalized second degree coefficients and formal uncertainties for different data combinations. The variations are typically within three times the formal sigma except for the J_2 solution with Magellan cycle 4, which deviates from the nominal solution by eight times the formal uncertainty. Since the Magellan data are processed with only one-day arcs, the stability of the lower degree harmonics may improve

Table 8. Normalized Second Degree Coefficients with Formal Uncertainties ($\times 10^{10}$).

Data	Constraints	$\bar{J}_2$	$\bar{C}_{21}$	$\bar{S}_{21}$	$\bar{C}_{22}$	$\bar{S}_{22}$
P	40, K, lp	19776±48	280±48	-5±55	8569±79	-1114±76
4	40, K, lp	19932±24	320±19	173±12	8515±44	-1142±38
5	40, K, lp	19715±30	357±25	134±13	8610±24	-1016±21
P+4	40, K, lp	19799±16	289±14	160±9	8550±26	-1124±26
ALL	40, K, lp	19719±6	290±4	143±4	8543±9	-998±8
P	90, K, lp	19778±48	273±48	-15±56	8558±81	-1104±78
4	90, K, lp	19930±26	362±22	177±12	8487±46	-1091±41
5	90, K, lp	19706±35	316±30	117±14	8633±25	-1046±26
P+4	90, K, lp	19797±17	296±15	159±10	8547±26	-1113±27
ALL	90, K, lp	19718±7	291±5	145±5	8530±10	-1001±9
ALL	90, S, fp	19716±7	290±5	143±5	8547±9	-999±9

Data: P=PVO, 4=Magellan cycle 4, 5=Magellan cycle 5
Constraints: 90, 40 = degree and order of solution, K = Kaula rule, S = Spatial (SAAP)
fp = fixed IAU pole, lp = loose pole (estimated pole and rotation rate for Venus)

with increased arc lengths - the intent of future studies. Independent of the length of the data arc, the formal uncertainties for the low degree harmonics are too low as a result of the noise characteristics of the Doppler (as discussed above, i.e., it is not white noise). Using a factor of three times the formal error, the pole offset with a realistic error is 0.52± 0.03 degrees. Hence, these results give some confidence that there is a pole offset and that there is a wobble for Venus (Yoder and Ward, 1979).

5c. Love Number

The Love number is a time-varying part of the C_{22} and S_{22} coefficients as a function of the solar longitude. Since Venus rotates retrograde once every 243 days and orbits the Sun once every 221 days, the solar day on Venus is 117 Earth days. The highly eccentric PVO data include five coverages of all solar longitudes with respect to the body-fixed coordinates for the low-altitude periapse orbit and three coverages of all solar longitudes for the high-altitude PVO orbit. In local-solar-time (i.e., the longitude of the spacecraft with respect to the tidal bulge), the coverage is about one-half of the solar longitude coverage. The eccentric (cycle 4) Magellan data have two full coverages of solar longitude and the post-aerobraking data have about five full coverages of solar longitude. So multiple periods of the tidal effect are sampled by all data sets.

Table 9 gives the Love number solution (k_2) with the formal uncertainty for different data combinations as given by Konopliv and Yoder (1995c). From the 40th degree and order solution, the Love number estimate with a realistic error is k_2 = 0.295±0.066 (2 x formal uncertainty). This value indicates a liquid core for Venus (Yoder, 1995), although a solid core cannot be absolutely ruled out. Yoder (1995) gives a range for a liquid core of 0.23-0.29 and a value near 0.17 if the iron core has solidified.

Table 9. Love Number Solutions.

Data	Constraints	k_2
P	40, K, lp	0.230±0.244
4	40, K, lp	0.245±0.134
5	40, K, lp	0.301±0.062
P+4	40, K, lp	0.279±0.093
ALL	40, K, lp	0.295±0.033
P	90, K, lp	0.217±0.239
4	90, K, lp	0.309±0.138
4	90, K, tp	0.225±0.135
5	90, K, lp	0.337±0.070
5	90, K, tp	0.319±0.069
ALL	90, K, lp	0.320±0.035
ALL	90, S, fp	0.306±0.036

Data: P=PVO, 4=Magellan cycle 4, 5=Magellan cycle 5 and 6
Constraints: 90, 40 = degree and order of solution, K = Kaula rule, S = Spatial (SAAP)
fp = fixed IAU pole, lp = loose pole (estimated pole and rotation rate for Venus)

Future work will try to improve the Love number estimate by increasing the degree and order of the gravity field solution, increasing the length of the arcs, and improving the models of forces acting on the spacecraft. The ephemeris solution (discussed below) has shown sensitivity to the degree and order of the gravity solution and we expect the same sensitivity for the Love number. Appendix H shows that the Love number has fairly low correlations with the other global parameters with the maximum correlation being 0.31 with S_{44}. The Love number must be separated from the drag and albedo forces. The albedo force vanishes on the nightside of Venus and the drag force drops by an order of magnitude, which helps the determination of the tidal force.

6. Venus Constants

6a. Venus Ephemeris

The corrections to the Venus and Earth planetary ephemerides are estimated in the gravity solution using the Set III elements of Brouwer and Clemence (1961). The six elements for each planet are heliocentric and consist of changes in eccentricity (Δe) and semi-major axis ($\Delta a/a$) and corrections to the initial orientation of the orbit. The angle elements are $\Delta l_o + \Delta r$, Δp, Δq, and $e\Delta r$ where Δl_o is the change in mean anomaly at epoch, Δr is the rotation about the z-axis or ecliptic north, and Δp and Δq are the small rotations about the x and y axes (giving the inclination of the orbit). Table 10 lists the solutions for the corrections to the JPL ephemeris DE403 for different combinations of data. The a priori

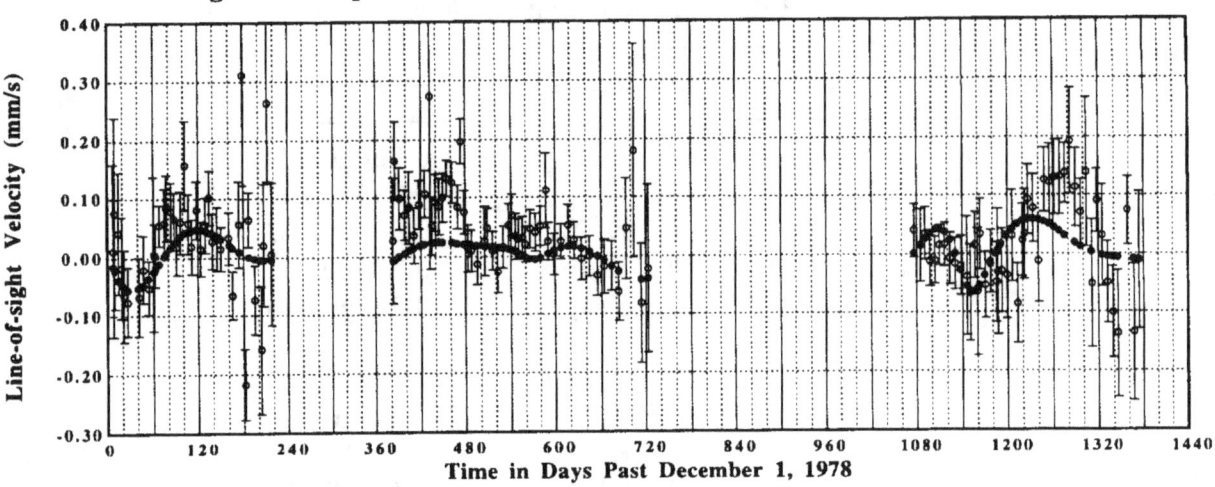

Figure 35: Ephemeris Solution for PVO wrt DE403

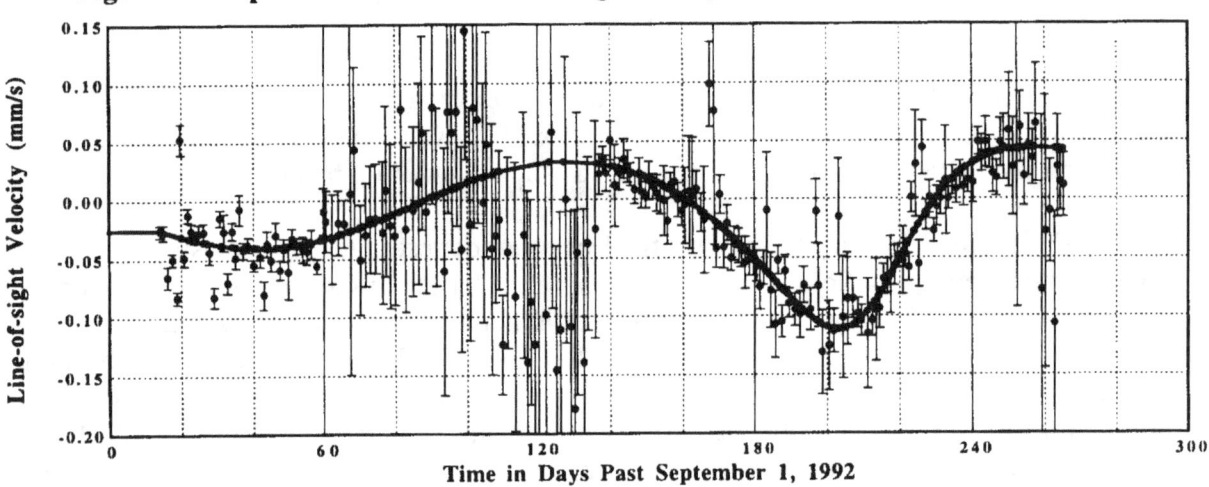

Figure 36: Ephemeris Solution for Magellan Cycle 4 wrt DE403

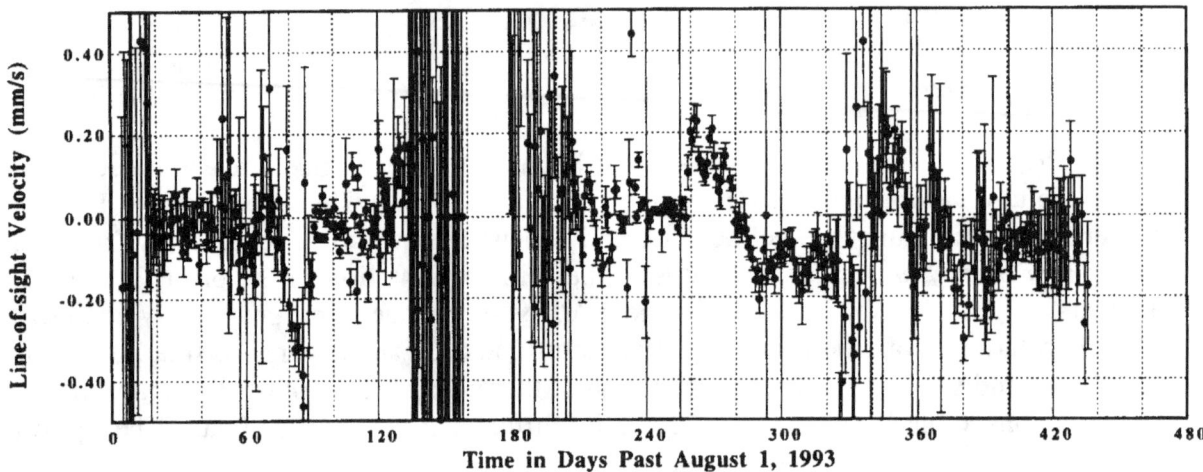

Figure 37: Ephemeris Solution for Magellan Cycle 5&6 wrt DE403

Table 10. Venus and Earth Ephemeris Solutions ($\times 10^9$).

Element	PVO	MGN Cycle 4	MGN Cycle 5	MGNP90LSAAP
Venus:				
$\Delta lo + \Delta r$	86.5±55.3	-13.5±97.4	-141.6±84.6	-80.9±13.9
$e\Delta r$	1.5±0.4	-1.9±1.2	1.3±1.4	0.9±0.2
$\Delta a/a$	0.6±0.3	-0.2±0.3	-0.5±0.2	-0.5±0.05
Δe	-0.2±0.3	1.1±1.3	-1.0±1.3	-0.8±0.2
Δp	27.4±44.1	-77.3±21.3	-17.3±28.5	-34.9±16.3
Δq	12.8±24.6	-78.2±40.2	-33.7±43.9	-14.6±16.4
Earth:				
$\Delta lo + \Delta r$	81.2±53.7	52.4±90.3	-7.8±97.0	-85.8±13.8
$e\Delta r$	1.0±0.5	0.9±0.9	2.8±1.2	0.9±0.2
$\Delta a/a$	1.0±0.5	0.0±0.4	-0.2±0.4	-0.8±0.1
Δe	0.8±0.4	2.7±1.2	-0.5±0.9	0.0±0.2
Δp	23.7±39.4	-18.6±25.2	-0.2±34.6	-20.2±15.4
Δq	24.8±31.2	-93.1±39.4	11.1±37.8	-30.8±17.2

constraint on the ephemeris is zero with an uncertainty of 10^{-6} for all the elements, and this is a minimal constraint.

The correlations of the Set III parameters are given in Appendix H and indicate that the absolute longitude and semi-major axes of the Earth and Venus are highly correlated (i.e., the Venus relative to Earth position is much better determined than absolute positions). The frame tie between the planetary reference frame and the International Earth Rotation Service (IERS) is known to about 40 nrad (Folkner et al, 1994a). The element Δr is a measure of this frame tie and the offset (80nrad) is greater than a realistic offset. This large value is due in part to the Magellan cycle 5 data (the value drops to 40 nrad without cycle 5 data) and the corrupting influence of higher order terms in the gravity field.

The Doppler data contain ephemeris information and this becomes more obvious when the relative velocity between the Earth and Venus (i.e., a Doppler bias) is estimated for each data arc (the MGNP90LSAAP gravity field is used and no global parameters are estimated). Figures 35, 36, and 37 show the solution for the Venus-to-Earth relative velocity with respect to DE403 for PVO, Magellan cycle 4, and Magellan cycle 5 and the formal uncertainties are given as error bars. The biases for the PVO data (Figure 35) are very sensitive to the ionospheric calibrations. The ionospheric calibrations for the high-altitude PVO data were not available for the generation of MGNP90LSAAP, but were for the solution of the Doppler biases. Plotted with the Doppler biases is the ephemeris solution. It was generated by Standish (private communication, JPL, 1995) using all the existing data for DE403 along with the Doppler biases for PVO and Magellan cycle 4 as observations (Magellan cycle 5 was not included due to systematic trends in the bias observations that are obviously not due to the ephemeris). The corrections to the initial state of Venus and Earth absorb the systematic trend in the Magellan cycle 4 data and some of the trends in the PVO data. The trend at 0.10 mm/s is determined to about 0.02 mm/s as given by the scatter in the bias solutions. The cycle 5 bias solutions contain trends that are

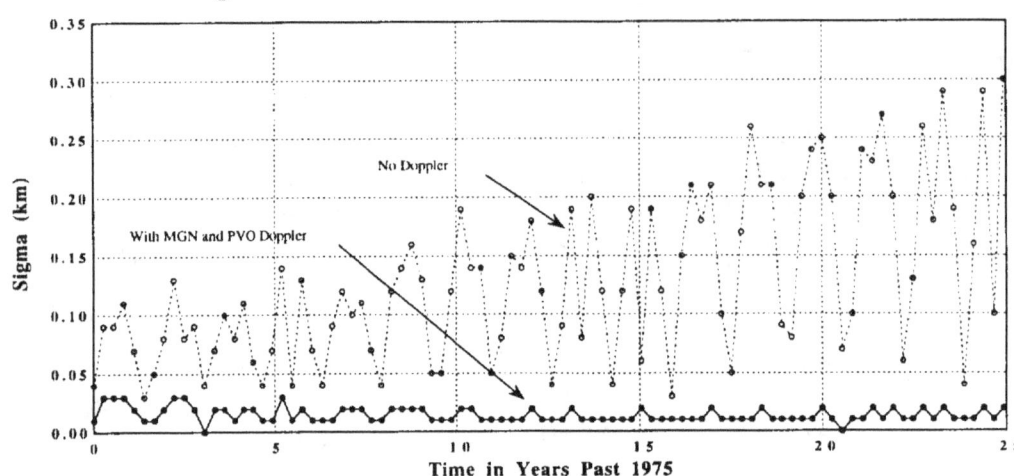

Figure 38: Venus wrt Earth Radial Position Error

most likely due to the higher than 90 degree field since the bias increases correspond to periapse passage over the Atla Regio and apoapse passage over Beta Regio (after apoapse was lowered). Figure 38 displays the errors in the Venus relative to Earth ephemeris in the solution by Standish. The errors have been reduced from hundreds of meters to tens of meters. Increased arc lengths (3 days plus) should help in determination of the Doppler bias since this will help average out the signature of the Earth's rotation in the Doppler.

6b. Venus Pole and Rotation

The Venus spin pole and rotation rate are fixed to the IAU 1991 values for the formal delivery of the MGNP90LSAAP gravity solution. However, solutions of the pole and rotation rate are determined along with the gravity field and other global parameters. Table 11 lists the solutions for different data combinations. The pole location is given by the right ascension and declination of the pole in the Earth mean equator at epoch J2000 coordinate system. The IAU 1991 values are used as the a priori values with an uncertainty of 0.05 degree for the pole and 0.005 day for the rotation rate. For all but the final solution, the pole rate along with the position was estimated. The a priori on the pole rate is zero with an uncertainty of 0.05 deg/century (about the expected amplitude for the Venus precession rate, Yoder 1995). The pole rate is not well enough determined to give a precession rate but estimating it gives a marked increase in the formal statistic due to the almost 1.0 correlation between the pole position and rate in the covariance matrix.

Increasing the formal statistic by a factor of three, the pole solution is 272.749±0.003 degrees for the right ascension, 67.160±0.003 degrees for the declination, and 243.0194±0.0006 days for the rotation rate. Our rotation period solution is slightly longer than that determined by Davies et al (1992b) of 243.0185 ± 0.0001. Both, however are consistently below the 1988 IAU value (243.025) based upon Earth radar measurements of Venus and more recent radar determinations by Slade et al (1990) of

Table 11. Venus Pole and Rotation Rate Solutions

Data	Constraint	Right Ascension	Declination	Rotation Rate
P	90,K,rate	272.735±0.047	67.194±0.040	243.0192±0.0014
4	90,K,rate	272.779±0.027	67.115±0.012	243.0204±0.0019
5	90,K,rate	272.759±0.004	67.160±0.003	243.0189±0.0010
P+4	90,K,rate	272.789±0.015	67.162±0.008	243.0201±0.0004
ALL	90,K,rate	272.751±0.003	67.152±0.003	243.0192±0.00024
ALL	90,S,no rate	272.749±0.002	67.160±0.001	243.0194±0.00018

243.022 ± 0.003. The pole solution from Davies et al (1992b) is 272.76±0.02 for the right ascension and 67.16±0.01 for the declination and these solutions agree well within the error bars.

7. Solutions for Auxiliary Forces

The auxiliary forces consist of solutions for the atmospheric drag, solar pressure, Venus albedo, Magellan momentum wheel desaturations, and Magellan hide information. The two most important forces (the drag and solar pressure) are addressed below. The solution trends for the other forces (mostly momentum wheel desaturations) have not been investigated in detail but will be checked for future models.

7a. Atmospheric Drag

PVO with its high velocity through periapse (9km/s) and 24 hour period provides an excellent measure of the atmospheric density. Using the VIRA scale height profiles, the solutions for the PVO densities are mapped to a constant altitude of 140 km in Figure 39 with the error bars also given (the error bars are generally so small that they are not visible). The densities on the nightside of Venus are generally an order of magnitude smaller than the dayside. The PVO spacecraft passes through periapse from north to south with the antenna leading. The axis of symmetry for the cylinder has an angle-of-attack of 22 degrees at periapse. A drag coefficient of 2.2 is used and at the given angle of attack the cross-sectional area of the cylinder is 5.858 m^2. The mass history was used for PVO and began at 362 kg and was 343 kg at the end of the low-altitude PVO coverage. These spacecraft values provide excellent agreement with the VIRA atmosphere model and this is not surprising since the VIRA model is, in part, based upon the PVO drag measurements (Keating et al, 1985). The differences with the VIRA model are displayed in Figure 40.

The solution for the lift-to-drag coefficient $C_{l/d}$ (one per arc) versus LST is given in Figure 41. At this point, not much can be said about the magnitude of the lift except that the correct direction is being determined. For PVO, the expected values are 0.015 to 0.044 for an accommodation coefficient of 1.0 and 0.182 to 0.195 for 0.8 (D. Rault, private communication, Langley Research Center, Aerothermodynamics Branch, 1995). For now, the lift coefficient is constrained to zero with an uncertainty of 0.05. If the uncertainties are

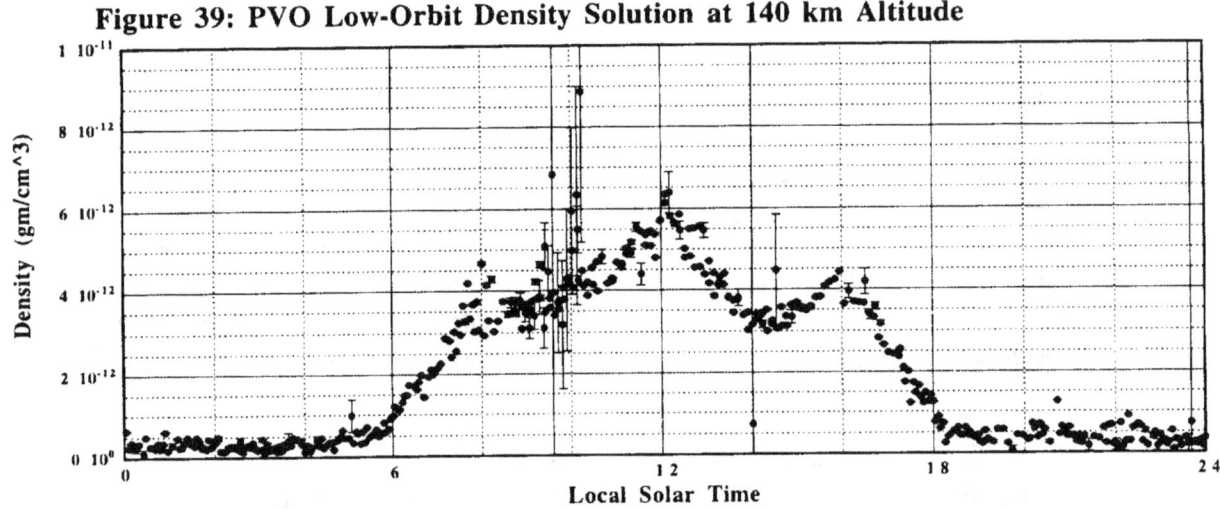

Figure 39: PVO Low-Orbit Density Solution at 140 km Altitude

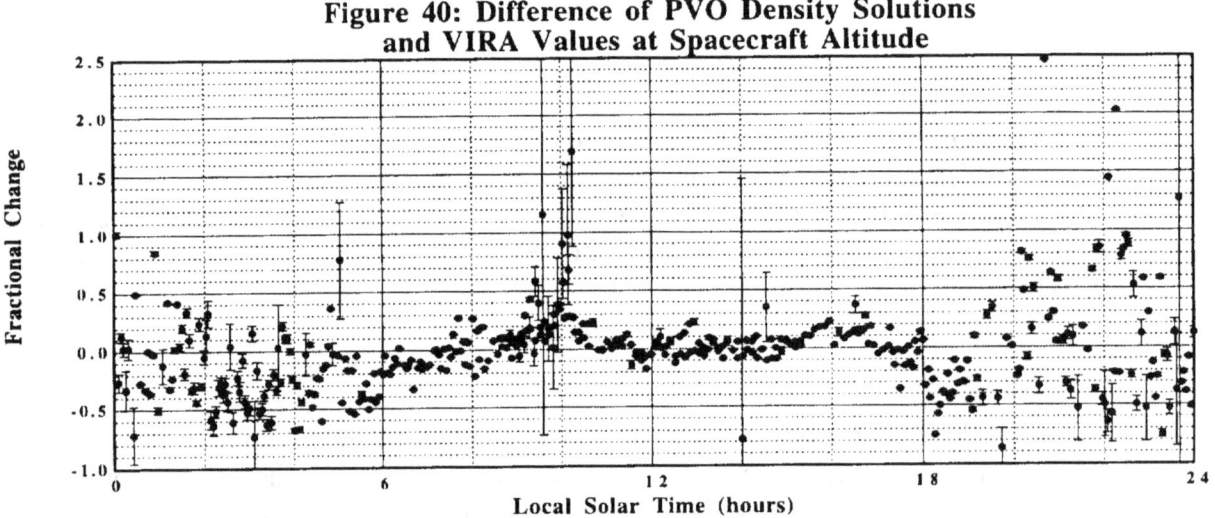

Figure 40: Difference of PVO Density Solutions and VIRA Values at Spacecraft Altitude

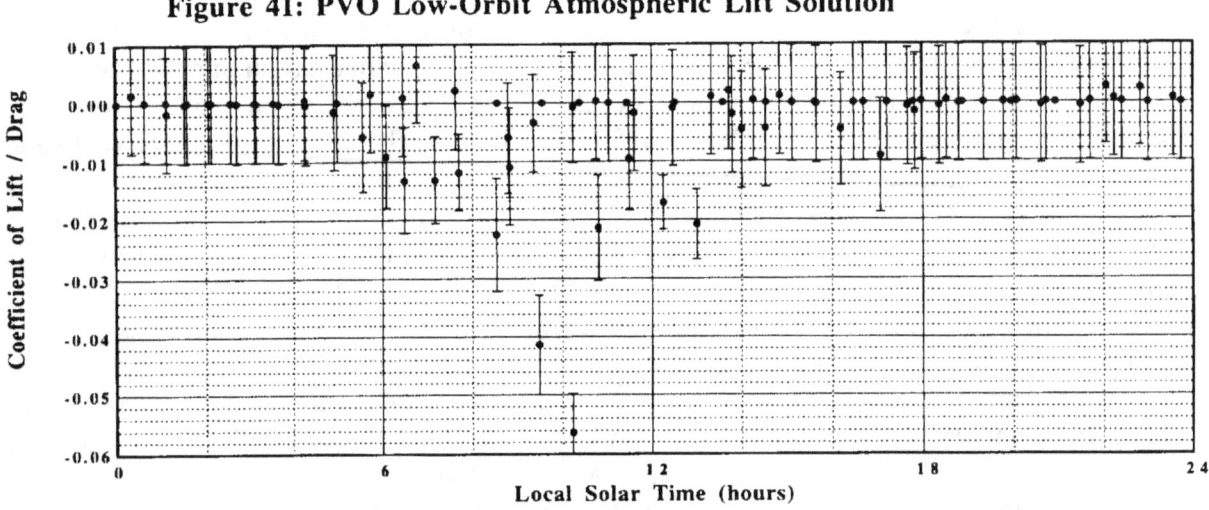

Figure 41: PVO Low-Orbit Atmospheric Lift Solution

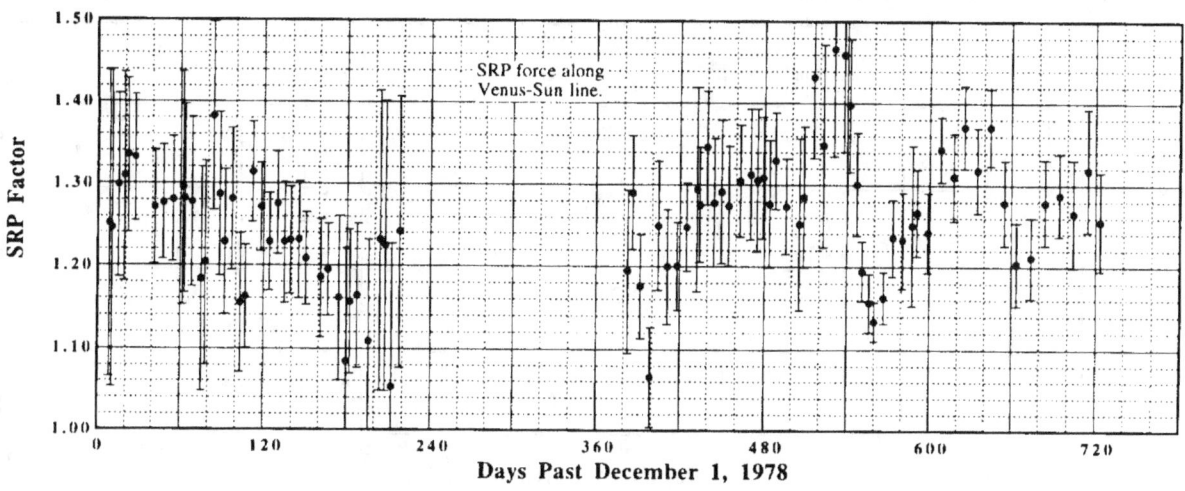

Figure 42: PVO Low-Orbit Solar Radiation Pressure (GR) Solution

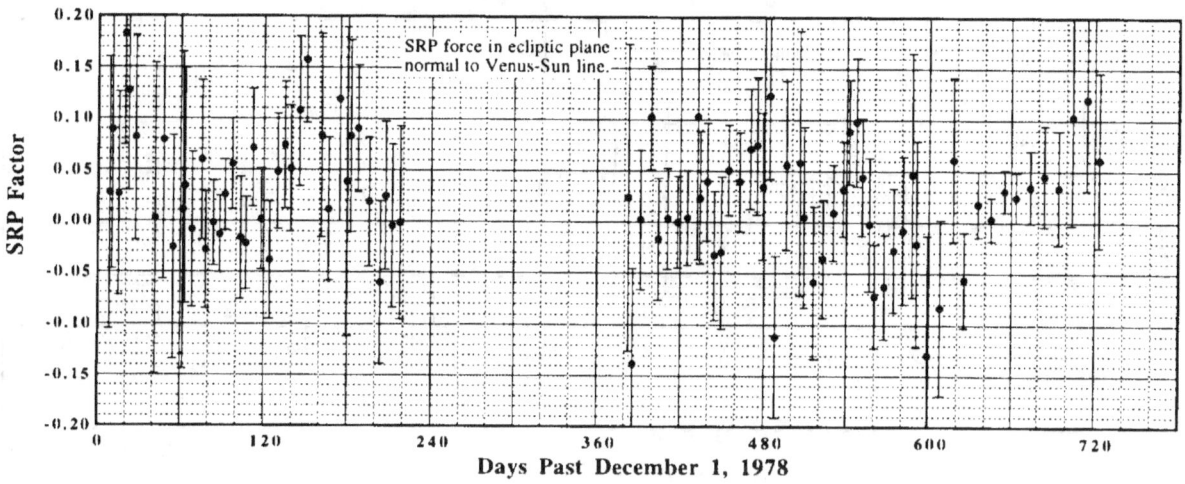

Figure 43: PVO Low-Orbit Solar Radiation Pressure (GX) Solution

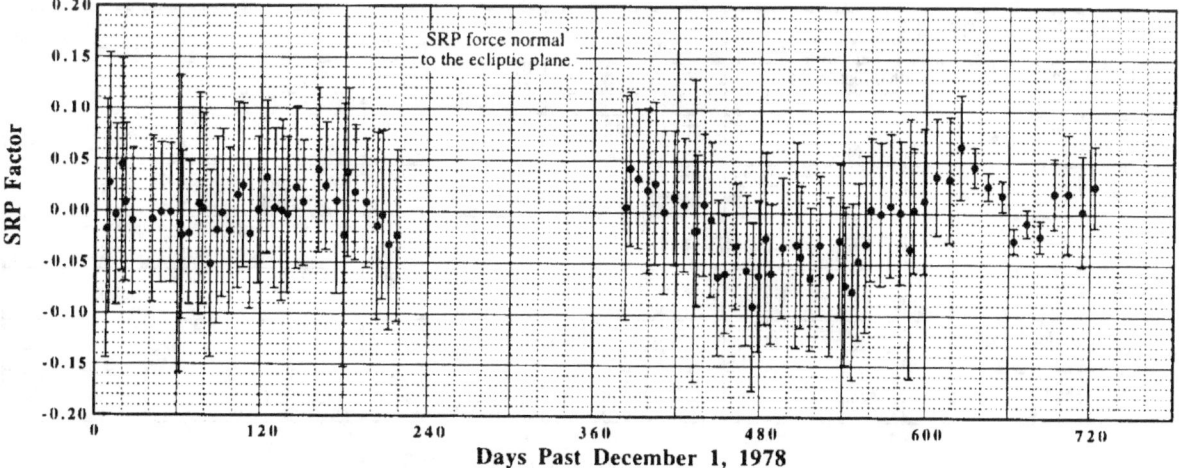

Figure 44: PVO Low-Orbit Solar Radiation Pressure (GY) Solution

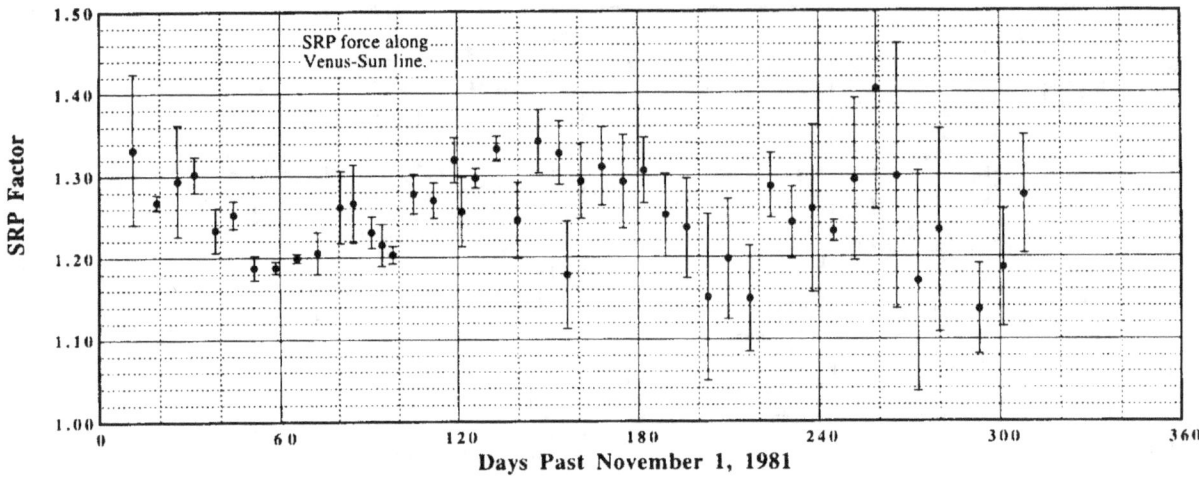

Figure 45: PVO High-Orbit Solar Radiation Pressure (GR) Solution

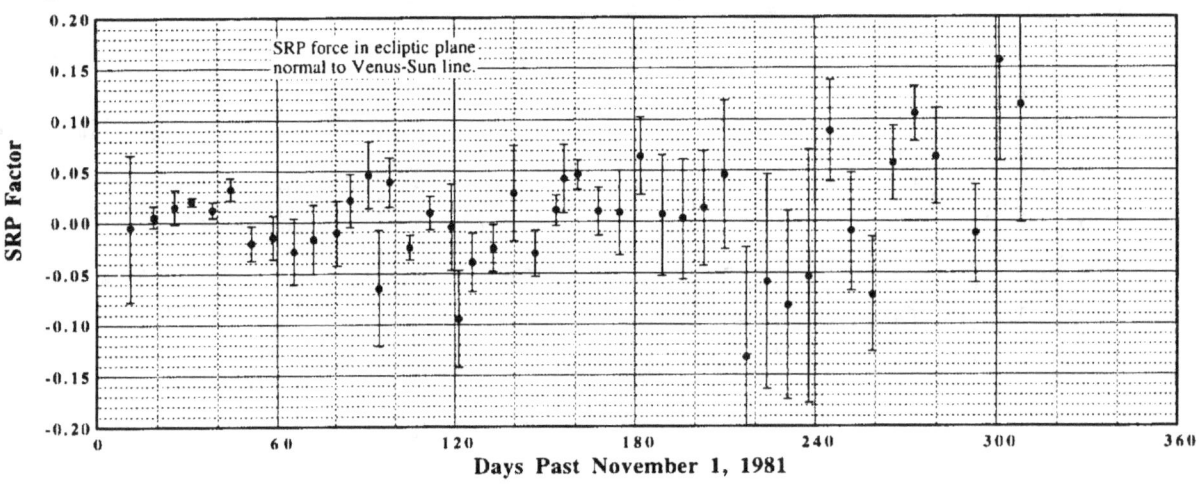

Figure 46: PVO High-Orbit Solar Radiation Pressure (GX) Solution

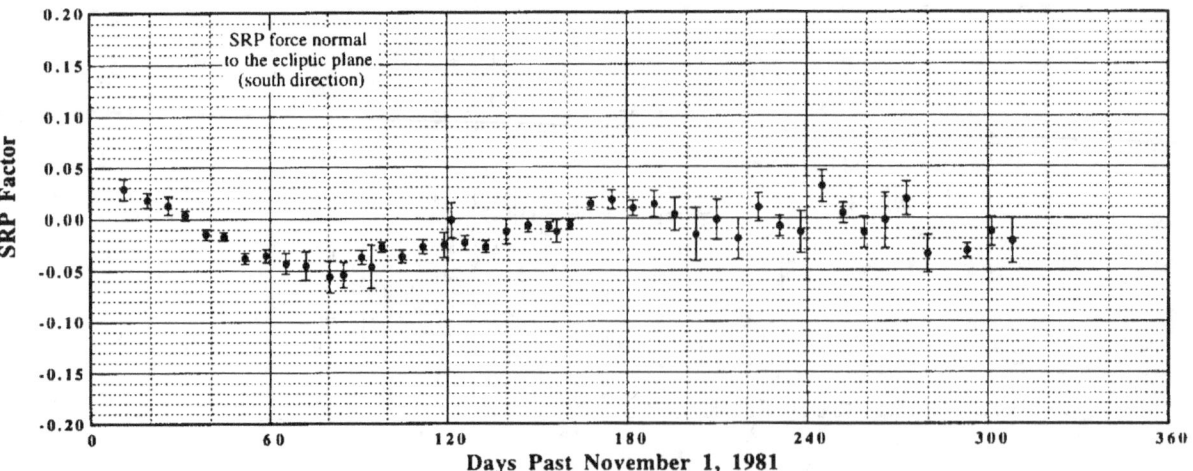

Figure 47: PVO High-Orbit Solar Radiation Pressure (GY) Solution

substantially relaxed then the values remain negative but fluctuate between -0.2 and -0.01. Also, they do not constrain the accommodation coefficient better than ±0.2. The $C_{l/d}$ is only determined on the dayside where the density at the spacecraft is large. Future studies may investigate a sideslip (out of orbit plane) force that may be due to a moving atmosphere or a spacecraft orientation in the atmosphere.

The Magellan densities are not as well determined as those for PVO due to the higher altitude and shorter orbit period. The densities are constrained to 1.0×10^{-12} gm/cm^3 with 100% uncertainty. The resulting density solutions have uncertainties of about 1×10^{-12} gm/cm^3 with higher density values on the dayside and lower density values on the nightside. There is not nearly the density information for Magellan as there is for PVO. Future gravity models will constrain the density solutions closer to the VIRA values. The Magellan spacecraft passes through the atmosphere with the solar array axis along the velocity vector and so the drag is mainly from the spacecraft bus and high gain antenna. The antenna is at a right angle to the direction of flow. With the solar panels titled by 10 degrees into the direction of flow (80 degree angle of attack) the cross sectional area of the spacecraft is about 10 m^2.

7b. Solar Pressure

The solar pressure solutions for the low-altitude and high-altitude PVO data are shown in Figures 42, 43, 44, 45, 46, and 47 for the three orthogonal solar pressure coefficients. The low-altitude uncertainties for the solar pressure are greater than the high-altitude due to the loss of information from the drag. The high-altitude solar pressure shows some systematic trends which are probably related to the modeling of solar pressure on the antenna as the antenna remains pointed to the Earth or to the albedo force from Venus. The albedo solution for the high-altitude PVO orbits (not shown) shows a systematic trend like the solar pressure and seems to be a function of the LST. It is probably related to varying reflectance properties of the Venusian clouds versus the incidence angles since the variations of 10% are greater than observed variations in the Venus albedo (which may be 1% at most). The albedo model assumes isotropic and diffuse reflectance from the Venus clouds. Taylor and Stowe (1984) have shown that for the Earth the reflectance is more specular for high incidence angles and the flux can vary by 10% to 20% depending on the zenith angle of the Sun and spacecraft and the relative azimuth. These systematic trends will be investigated in more detail in future analysis. The solar coefficients have an a priori uncertainty of 0.2 and are fairly well determined. The solar pressure solutions for the Magellan arcs are very poorly determined (the a priori uncertainty is not reduced) because of the increased number of atmospheric drag passages and the shorter data arcs.

Future efforts will attempt to improve the solar pressure model (maybe orientation, reflectivity coefficients) and increase the arc lengths - especially the high-altitude PVO orbits. With a better model and increased arcs, this should improve the information in the low degree harmonics and other global parameters such as the Love number.

8. Summary

With the gravity software now running on the massively parallel JPL Cray T3D Supercomputer, we have been able to increase the resolution of the Venus gravity field with dramatic improvements in the amount of time required to generate the solution (on the order of weeks for the supercomputer versus years with workstations). The Venus gravity model MGNP90LSAAP which is complete to degree and order 90 represents the best gravity solution to date. It shows increased correlation with topography over the last solution MGNP75ISAAP with minor improvements in the lower to medium degree harmonics and more improvement in the higher degree. Interpretation for smaller features (e.g., Mead Crater and Maat Mons) shows substantial improvement. The topography and this gravity model, which has been provided to the science community, provide the basis of geophysical interpretation.

Other parameters were estimated along with the gravity field. For the first time, we are sensing a Love number for another planet, perhaps indicating a liquid core. The Venus spin pole solution is most likely the best solution and the rotation rate of Venus is determined to a comparable level of the Magellan SAR data. The GM estimate for Venus is the best estimate to date and slightly better than previous determinations.

The models affecting the gravity solution have been reviewed. Many parameters and geometries that are useful in evaluating the Doppler data have been provided in the Appendices, hopefully making this a useful handbook.

References

Akim, E.L. Z.P. Vlasova, and I.V. Chuiko, "Determination of the Dynamical Flattening of Venus from Measurements of the Trajectories of Its First Artificial Satellites, Venera 9 and 10," *Sov. Phys. Dokl.*, 23, 313-315, 1978.

Ananda, M.P., W.L. Sjogren, R.J. Phillips, R.N. Wimberly, and B.G. Bills, "A Low-Order Global Gravity Field of Venus and Dynamical Implications," *J. Geophys. Res.* 85, 8303-8318, 1980.

Anderson, J.D., and L. Efron, "The Mass and Dynamical Oblateness of Venus," *Bull. Amer. Astron. Soc.* 1, 231, 1969.

Banerdt, W.B., A.S. Konopliv, N.J. Rappaport, W.L. Sjogren, R.E. Grimm, and P.G. Ford, "The Isostatic State of Mead Crater," *Icarus* 112, 117-129, 1994.

Barriot, J.P. and G. Balmino, "An Analysis of LOS Gravity Data Set from Cycle 4 of the Magellan Probe Around Venus", *Icarus* 112, 34-41, 1994.

Battin, R.H., *An Introduction to the Mathematics and Methods of Astrodynamics*, AIAA Education Series, New York, 1987.

Bent, R.B., S.K. Llewellyn, G. Nesterczuk, and P.E. Schmid, "The Development of a Highly Successful Worldwide Empirical Ionospheric Model," in J. Goodman (Ed.), Effect of the Ionosphere on Space Systems and Communications, Springfield, VA, National Technical Information Service, 13-28, 1976.

Bierman, G. J., *Factorization Methods for Discrete Sequential Estimation*, Academic Press, New York, 1977.

Bills, B.G., W.S. Kiefer, and R.L. Jones, "Venus Gravity: A Harmonic Analysis," *J. Geophys. Res.*, 92, 10335-10351, 1987.

Boucher, C., Z. Altamimi, L. Duhem, "Results and Analysis of the ITRF93," IERS Technical Note 18, Central Bureau of IERS, Paris, France, October, 1994.

Brouwer, D. and G. Clemence, *Methods of Celestial Mechanics*, Academic Press, New York, 241, 1961.

Davies, M.E., V.K. Abalakin, A. Brahic, M. Bursa, B.H. Chovitz, J.H. Lieske, P.K. Seidelmann, A.T. Sinclair, and Y.S. Tjuflin, "Report of the IAU/IAG/COSPAR Working Group on Cartographic Coordinates and Rotational Elements of the Planets and Satellites: 1991," *Celes. Mech.* 53, 377-397, 1992a.

Davies, M.E., T.R. Colvin, P.G. Rogers, P.W. Chodas, W.L. Sjogren, E.L. Akim, V.A. Stepanyantz, Z.P. Vlasova, and A.I. Zakharov, "The Rotation Period, Direction of the North Pole, and Geodetic Control Network of Venus," *J. Geophys. Res.*, 97, 13141-13151, 1992b.

Ellis, J., "Large Scale State Estimation Algorithms for DSN Tracking Station Location Determination", *J. Astronaut. Sci.*, 28, 15-30, 1980.

Esposito, P.B., W.L. Sjogren, N.A. Mottinger, B.G. Bills, and E. Abbott, "Venus Gravity: Analysis of Beta Regio," *Icarus* 51, 448-459, 1982.

Estefan, J.A. and O.J. Sovers, "A Comparative Survey of Current and Proposed Tropospheric Refraction-Delay Models for DSN Radio Metric Data Calibration," JPL Publication 94-24, Jet Propulsion Laboratory, California Institute of Technology, Pasadena, CA, October, 1994.

Folkner, W.M., "Station Location Covariance for Mars Observer," JPL IOM 335.1-92-004 (internal document), Jet Propulsion Laboratory, California Institute of Technology, Pasadena, CA, January 27, 1992a.

Folkner, W.M., "DE234 Station Locations and Covariance for Mars Observer," JPL IOM 335.1-92-013 (internal document), Jet Propulsion Laboratory, California Institute of Technology, Pasadena, CA, May 26, 1992b.

Folkner, W.M., P. Charlot, M.H. Finger, J.G. Williams, O.J. Sovers, XX Newhall, and E.M. Standish, "Determination of the Extragalactic-Planetary Frame Tie from Joint Analysis of Radio Interferometric and Lunar Laser Ranging Measurements," *Astron. Astrophys.* 287, 279-289, 1994a.

Folkner, W.M., "Effect of Uncalibrated Charged Particles on Doppler Tracking," JPL IOM 335.1-94-005 (internal document), Jet Propulsion Laboratory, California Institute of Technology, Pasadena, CA, March 1, 1994b.

Goltz, G.L., "DSN Tracking System Interfaces: Orbit Data File Interface Mark IVA," TRK-2-18 of DSN Document 820-13, Rev. A (internal document), Jet Propulsion Laboratory, Pasadena, CA, 1988a (currently under revision).

Goltz, G.L., "DSN Tracking System Interfaces: Archival Tracking Data File Interface," TRK-2-25 of DSN Document 820-13, Rev. A (internal document), Jet Propulsion Laboratory, Pasadena, CA, 1988b.

Gross, R.S., "A Combination of Earth Orientation Data: Space91," In *IERS Technical Note 11, Earth orientation reference frames and atmospheric excitation functions submitted for the 1991 IERS Annual Report* (P. Charlot, Ed.), Central Bureau of IERS, Paris, France, 1992.

Heiskanen, W.A. and H. Moritz, *Physical Geodesy*, W.H. Freeman, San Francisco, 1967.

Howard, H.T., G.L. Tyler, G. Fjeldbo, A.J. Kliore, G.S. Levy, D.L. Brunn, R. Dickinson, R.E. Edelson, W.L. Martin, R.B. Postal, B. Seidel, T.T. Sesplaukis, D.L. Shirley, C.T. Stelzried, D.N. Sweetnaum, A.I. Zygielbaum, P.B. Esposito, J.D. Anderson, I.I. Shapiro, and R.D. Reasenberg, "Venus: Mass, Gravity Field, Atmosphere and Ionosphere as Measured by Mariner 10 Dual Frequency Radio System," *Science* 183, 1297-1301, 1974.

International Earth Rotation Service (IERS), 1994 IERS Annual Report, Central Bureau of IERS, Paris, France, July, 1995.

Kaula, W.M., *Theory of Satellite Geodesy*, Blaisdell, Waltham, MA, 1966.

Kaula, W.M., "Regional Gravity Fields on Venus from Tracking of Magellan Cycles 5 and 6," *J. Geophys. Res. Planets*, in press, 1995.

Keating, G.M., J.L. Bertaux, S.W. Bougher, T.E. Cravens, R.E. Dickinson, A.E. Hedin, V.A. Krasnopolsky, A.F. Nagy, J.Y. Nicholson III, L.J. Paxton, U. von Zahn, "Models of Venus Neutral Upper Atmosphere: Structure and Composition," *Adv. Space Res.* **5**, 117-171, 1985.

Knocke, P. and J. Ries, "Earth Radiation Pressure Effects on Satellites," Center for Space Research Technical Memorandum, University of Texas at Austin, Sept. 1987.

Konopliv, A.S., B.G. Williams, E.J. Christensen, "A Venus Gravity Solution to Degree and Order 42 from PVO Data Only," presented at *AGU Spring Meeting*, Montreal, Canada, 1992.

Konopliv, A. S., N. J. Borderies, P. W. Chodas, E. J. Christensen, W. L. Sjogren, B. G. Williams, G. Balmino, and J. P. Barriot, "Venus Gravity and Topography: 60th Degree and Order Model", *Geophys. Res. Lett.* 20, No. 21, pp. 2403-2406, 1993a.

Konopliv, A. S., W. L. Sjogren, R. N. Wimberly, R. A. Cook, A. Vijayaraghavan, "A High Resolution Lunar Gravity Field and Predicted Orbit Behavior," AAS Paper 93-622 in *Proceedings of the AAS/AIAA Astrodynamics Specialist Conference held August 16-19, 1993, Victoria, British Columbia, Canada*, Univelt, San Diego, 1275-1294, 1993b.

Konopliv, A. S., and W. L. Sjogren, "Venus Spherical Harmonic Gravity Model to Degree and Order 60", *Icarus* 112, 42-54, 1994a.

Konopliv, A. S., W. L. Sjogren, E. Graat, J. Arkani-Hamed, "Venus Gravity Data Reduction", Presentation at Fall 1994 Meeting, American Geophysical Union, San Francisco, CA, December 5-9, 1994b.

Konopliv, A.S. and W.L. Sjogren, "The JPL Mars Gravity Field, Mars50c, Based Upon Viking and Mariner 9 Doppler Tracking Data," JPL Publication 95-5, Jet Propulsion Laboratory, California Institute of Technology, Pasadena, CA, February, 1995a.

Konopliv, A.S., "DDF Final Report," JPL IOM 312.D-95-103 (internal document), Jet Propulsion Laboratory, California Institute of Technology, Pasadena, CA, Oct. 5, 1995b.

Konopliv, A.S., and C.F. Yoder, "Venusian k_2 Tidal Love Number from Magellan and PVO Tracking Data," submitted to *Geophys. Res. Lett.*, 1995c.

Lambeck, K., *Geophysical Geodesy: The Slow Deformation of the Earth*, Clarendon Press, Oxford, 1988.

Lawson, C. L. and R. J. Hanson, *Solving Least Squares Problems*, SIAM Classics in Applied Mathematics, Vol. 15, Society for Industrial and Applied Mathematics, Philadelphia, 1995.

Lemoine, F.G., "Mars: The Dynamics of Orbiting Satellites and Gravity Model Development," Ph.D. thesis, Univ. of Colorado, Boulder, 1992.

Lemoine, F.G., D.E. Smith, M.T. Zuber, and G.A. Neumann, "High Degree and Order Spherical Harmonic Models for the Moon and Historic S-Band Doppler Data," IUGG, XXI General Assembly, Boulder, CO, 1995.

Lieske, J.H., T. Lederle, W. Fricke, B. Morando, "Expression for the Precession Quantities Based upon the IAU (1976) System of Astronomical Constants," *Astron. Astrophys.* 58, 1-16, 1977.

McCarthy, D.D., et al, IERS Technical Note 3, Central Bureau of IERS, Paris, France, November 1989.

McKenzie, D. and F. Nimmo, "Elastic Thickness Estimates for Venus from Line-of-Sight Accelerations," *Icarus*, in press, 1995.

McNamee, J.B., G.R. Kronschnabl, S.K. Wong, and J.E. Ekelund, "A Gravity Field to Support Magellan Navigation and Science at Venus," *J. Astron. Sci.*, 40, 107-134, 1992.

McNamee, J.B., N.J. Borderies and W.L. Sjogren, "Venus: Global Gravity and Topography," *J. Geophys. Res. Planets*, 98, E5, 9113-9128, 1993.

Mottinger, N.A., W.L. Sjogren and B.G. Bills, "Venus Gravity: A Harmonic Analysis and Geophysical Implications," *J. Geophys. Res.* 90, 739-756, 1985.

Moyer, T. D., Mathematical Formulation of the Double-Precision Orbit Determination Program (DPODP). JPL Technical Report 32-1527. Jet Propulsion Laboratory, California Institute of Technology, Pasadena, CA, 1971.

Moyer, T.D., "Changes for Voyager Jupiter Encounter / Pioneer Venus Orbiter Version of Regres," JPL IOM 314.7-122 (internal document), Jet Propulsion Laboratory, California Institute of Technology, Pasadena, CA, November 7, 1977.

Moyer, T.D., "Transformation from Proper Time on Earth to Coordinate Time in Solar System Barycentric Space-Time Frame of Reference, Parts 1 and 2," *Celes. Mech.* 23, 33-68, 1981.

Moyer, T.D., "Station Location Sets LS111B and LS118," JPL IOM 314.5-724 (internal document), Jet Propulsion Laboratory, California Institute of Technology, Pasadena, CA, October 26, 1983.

Moyer, T.D., "Changes to the ODP and the ODE for Processing X-Band Uplink Data," JPL EM 314-430 (internal document), Jet Propulsion Laboratory, California Institute of Technology, Pasadena, CA, October 15, 1987.

Moyer, T.D., "Station Location Sets Referred to the Radio Frame," JPL IOM 314.5-1334 (internal document), Jet Propulsion Laboratory, California Institute of Technology, Pasadena, CA, February 24, 1989.

Moyer, T.D., "Relativistic Equations of Motion for Earth Satellites in Geocentric and Solar System Barycentric Frames of Reference," JPL EM 314-476 (internal document), Jet Propulsion Laboratory, California Institute of Technology, Pasadena, CA, January 15, 1990.

Nerem, R.S., "An Improved Gravity Model for Venus Using Tracking Data from Pioneer Venus Orbiter," *Eos Trans. AGU*, 72(17), 174-175, 1991.

Nerem, R.S., B.G. Bills and J.B. McNamee, "A High Resolution Gravity Model for Venus: GVM-1," *Geophys. Res. Lett.*, 20, 7, 599-602, 1993.

Nerem, R.S., F.J. Lerch, J.A. Marshall, E.C. Pavlis, B.H. Putney, J.C. Chan, S.M. Klosko, S.B. Luthcke, G.B. Patel, N.K. Pavlis, R.G. Williamson, B.D. Tapley, R.J. Eanes, J.C. Ries, B.E. Schutz, C.K. Shum, M.M. Watkins, R.H. Rapp, R. Biancale, and F. Nouel, "Gravity Model Development for TOPEX/Poseidon: Joint Gravity Models 1 and 2," *J. Geophys. Res.*, 99, 24421-24447, 1994.

Nerem, R.S., C. Jekeli, W.M. Kaula, "Gravity Field Determination and Characteristics: Retrospective and Prospective," *J. Geophys. Res.*, 100, 15053-15074, 1995.

Phillips, R.J., W.L. Sjogren, E.A. Abbott, J.C. Smith, R.N. Wimberly, and C.A. Wagner, "Gravity field of Venus: A preliminary analysis," *Science*, 205, 93-96, 1979.

Rapp, R.H., Y.M. Wang, and N.K. Pavlis, "The Ohio State 1991 Geopotential and Sea Surface Topography Harmonic Coefficient Models," Rep. 410, Dep. of Geod. Sci. and Surv., Ohio State Univ., Columbus, Aug. 1991.

Rappaport, N. J., and J. J. Plaut, "A 360-Degree and -Order Model of Venus Topography", *Icarus* 112, 27-33, 1994.

Reasenberg, R.D., Z.M. Goldberg, P.E. MacNeil, and I.I. Shapiro, "Venus Gravity: A High Resolution Map," *J. Geophys. Res.*, 86, 7173-7179, 1981.

Reasenberg, R.D., Z.M. Goldberg, and I.I. Shapiro, "Venus: Comparison of Gravity and Topography in the Vicinity of Beta Regio," *Geophys. Res. Lett.*, 9, 637-640, 1982.

Reasenberg, R.D. and Z.M. Goldberg, "High-Resolution Gravity Model of Venus," *J. Geophys. Res.*, 97, E9, 14681-14690, 1992.

Seidelmann, P.K. "1980 IAU Nutation: The Final Report of the IAU Working Group on Nutation," *Celest. Mech.* 27, 79-106, 1982.

Simons, M., B.H. Hager, and S.C. Solomon, "Global Variations in the Geoid/Topography Admittance of Venus," *Science,* 264, 798-803, 1994.

Simpson, R.A, "Magellan Spherical Harmonic ASCII Data Record (SHADR)," Magellan Software Interface Specification MGN-SHADR, Ver. 1.0, Washington University PDS Geosciences Node, St. Louis, MO, 1993a.

Simpson, R.A, "Magellan Spherical Harmonic Binary Data Record (SHBDR)," Magellan Software Interface Specification MGN-SHBDR, Ver. 1.0, Washington University PDS Geosciences Node, St. Louis, MO, 1993b.

Simpson, R.A, "Line-of-Sight Acceleration Profile Data Record (LOSAPDR)," Magellan Software Interface Specification NAV-138, Ver. 1.13.1, Washington University PDS Geosciences Node, St. Louis, MO, 1995a.

Simpson, R.A, "Radio Science Digital Map (RSDMAP)," Magellan Software Interface Specification SU-MGN-RSDMAP, Ver. 1.0.2, Washington University PDS Geosciences Node, St. Louis, MO, 1995b.

Sjogren, W.L., R.J. Phillips, P.W. Birkeland, and R.N. Wimberly, "Gravity Anomalies on Venus," *J. Geophys. Res.*, 85, 8295-8302, 1980.

Sjogren, W.L., B.G. Bills, P.W. Birkeland, P.B. Esposito, A.S. Konopliv, N.A. Mottinger, R.J. Phillips, and S.J. Ritke, "Venus Gravity Anomalies and Their Correlation With Topography," *J. Geophys. Res.*, 88, 1119-1128, 1983.

Sjogren, W.L., B.G. Bills and N.A. Mottinger, "Venus: Ishtar Gravity Anomaly," *Geophys. Res. Lett.*, 11, No. 5, 489-491, 1984.

Sjogren, W.L., G.B. Trager, and G.R. Roldan, "Venus: A Total Mass Estimate," *Geophys. Res. Lett.*, 17, 1485-1488, 1990.

Slade, M.A., S. Zohar, and R.F. Jurgens, "Venus: Improved Spin Vector from Goldstone Radar Observations," *Astron. J.*, *100*, 1369-1374, 1990.

Smith, D.E., F.J. Lerch, R.S. Nerem, M.T. Zuber, G.B. Patel, S.K. Fricke, and F.G. Lemoine, "An Improved Gravity Model for Mars: Goddard Mars Model 1," *J. Geophys. Res.*, 98, 20871-20889, 1993.

Smrekar, S.E., "Evidence for Active Hotspots on Venus from Analysis of Magellan Gravity Data", *Icarus* 112, 1994.

Standish, E.M., XX Newhall, J.G. Williams, and W.M. Folkner, "JPL Planetary and Lunar Ephemerides, DE403/LE403," JPL IOM 314.10-127 (internal document), Jet Propulsion Laboratory, California Institute of Technology, Pasadena, CA, May 22, 1995.

Standish, E.M., "The Observational Basis for JPL's DE 200, the Planetary Ephemerides of the Astronomical Almanac," *Astron. Astrophys.* 233, 252-271, 1990.

Steppe, J.A., S.H. Oliveau, and O.J. Sovers, "Earth Rotation Parameters from DSN VLBI: 1991," Earth orientation and reference frame determinations, atmospheric excitation functions, up to 1990, IERS Technical Note 8, Central Bureau of IERS, Paris, France, 47-60, October, 1991.

Taylor, F.W., D.M. Hunten, L.V. Ksanfomaliti, "The Thermal Balance of the Middle and Upper Atmosphere of Venus," *Venus*, D.M. Hunten et al (eds.), U. of Arizona, Tucson, AZ, 650-679, 1983.

Taylor, V. R., and L.L. Stowe, "Reflectance Characteristics of Uniform Earth and Cloud Surfaces Derived from NIMBUS-7 ERB," *J. Geophys. Res.*, 89, 4987-4996, 1984.

Turcotte, D.L., and D.C. McAdoo, "Geoid Anomalies and the Thickness of the Lithosphere," *J. Geophys. Res.*, 84, 2381-2387, 1979.

Williams, B.G., N.A. Mottinger, and N.D. Panagiotacopulos, "Venus Gravity Field: Pioneer Venus Orbiter Navigation Results," *Icarus* 56, 578-589, 1983.

Yoder, C.F., and W.R. Ward, "Does Venus Wobble?" *Astrophys. J.* 233, 33-37, 1979.

Yoder, C.F., "Venus' Free Obliquity," Icarus, 117, 250-286, 1995.

Appendix A
PVO Low-Altitude Periapse Information

The following plots are included in this appendix:

1. Semi-major axis
2. Eccentricity
3. Inclination
4. Latitude at periapse
5. Longitude at periapse
6. Altitude at periapse
7. Plane-of-sky inclination
8. One-way light time from Venus to Earth
9. Sun-Earth-Venus angle
10. Earth-Venus-Probe at periapse angle
11. Local solar time at periapse
12. Altitude vs. latitude profile

PVO Low Orbit Semi-Major Axis

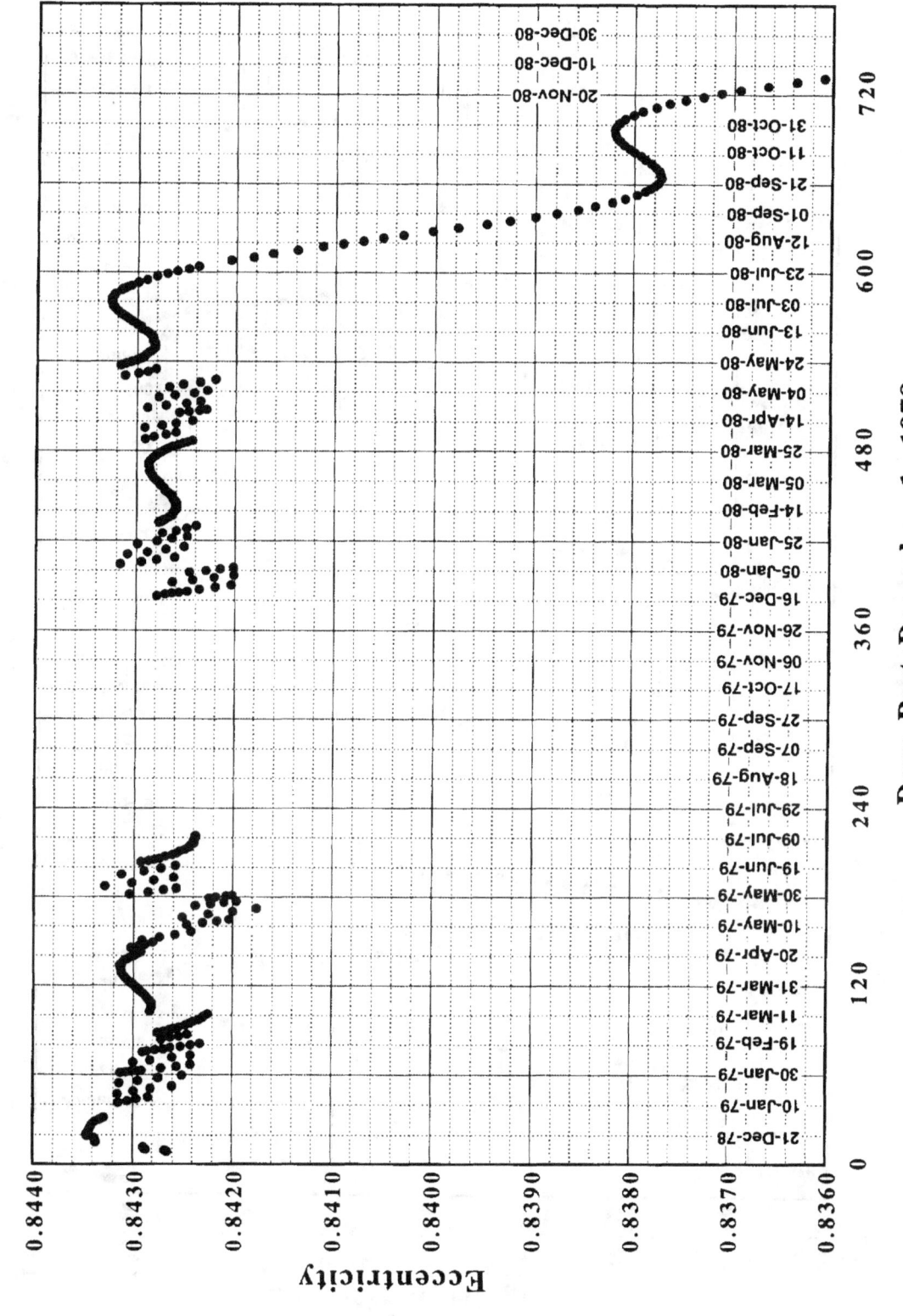

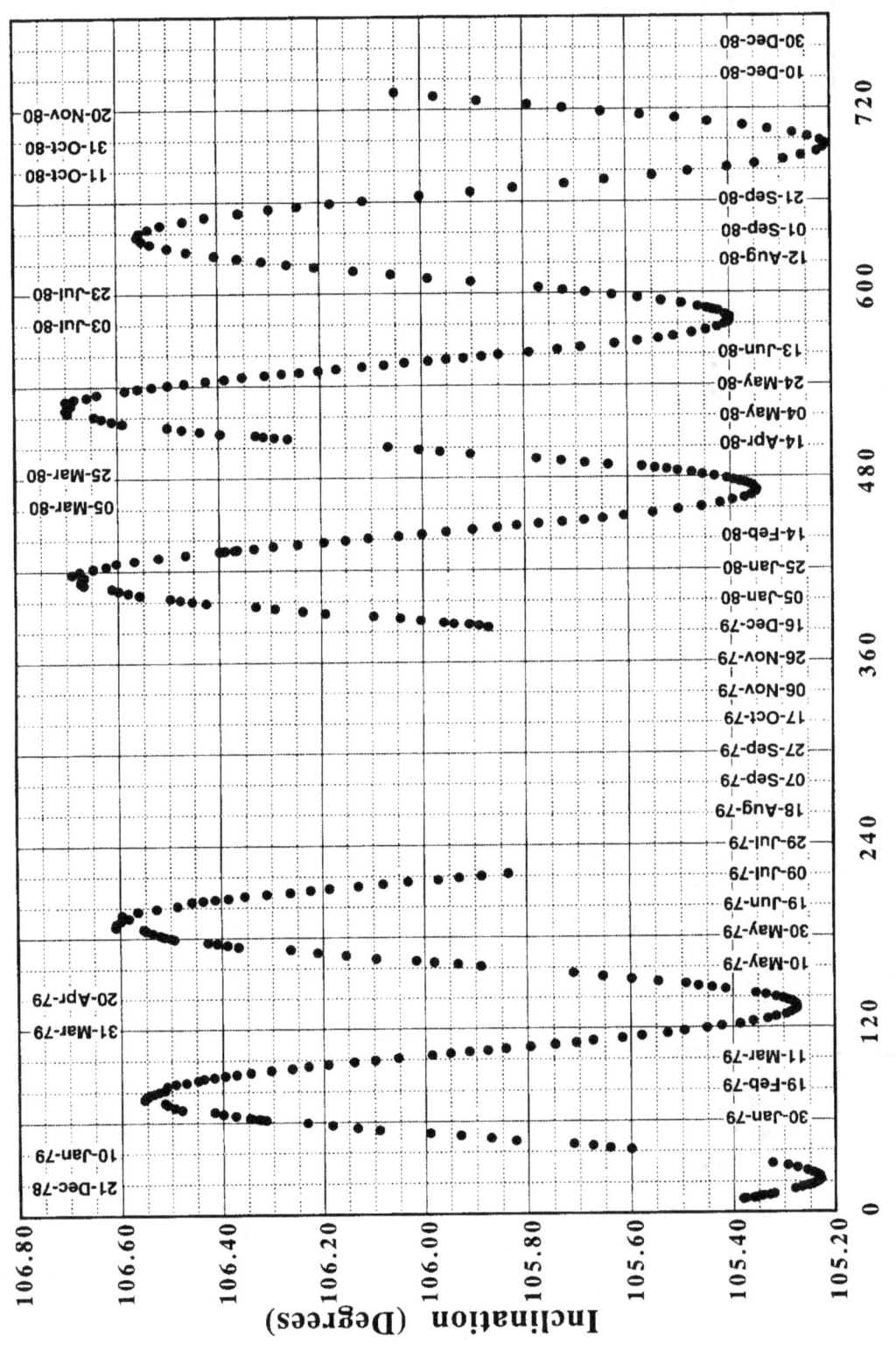

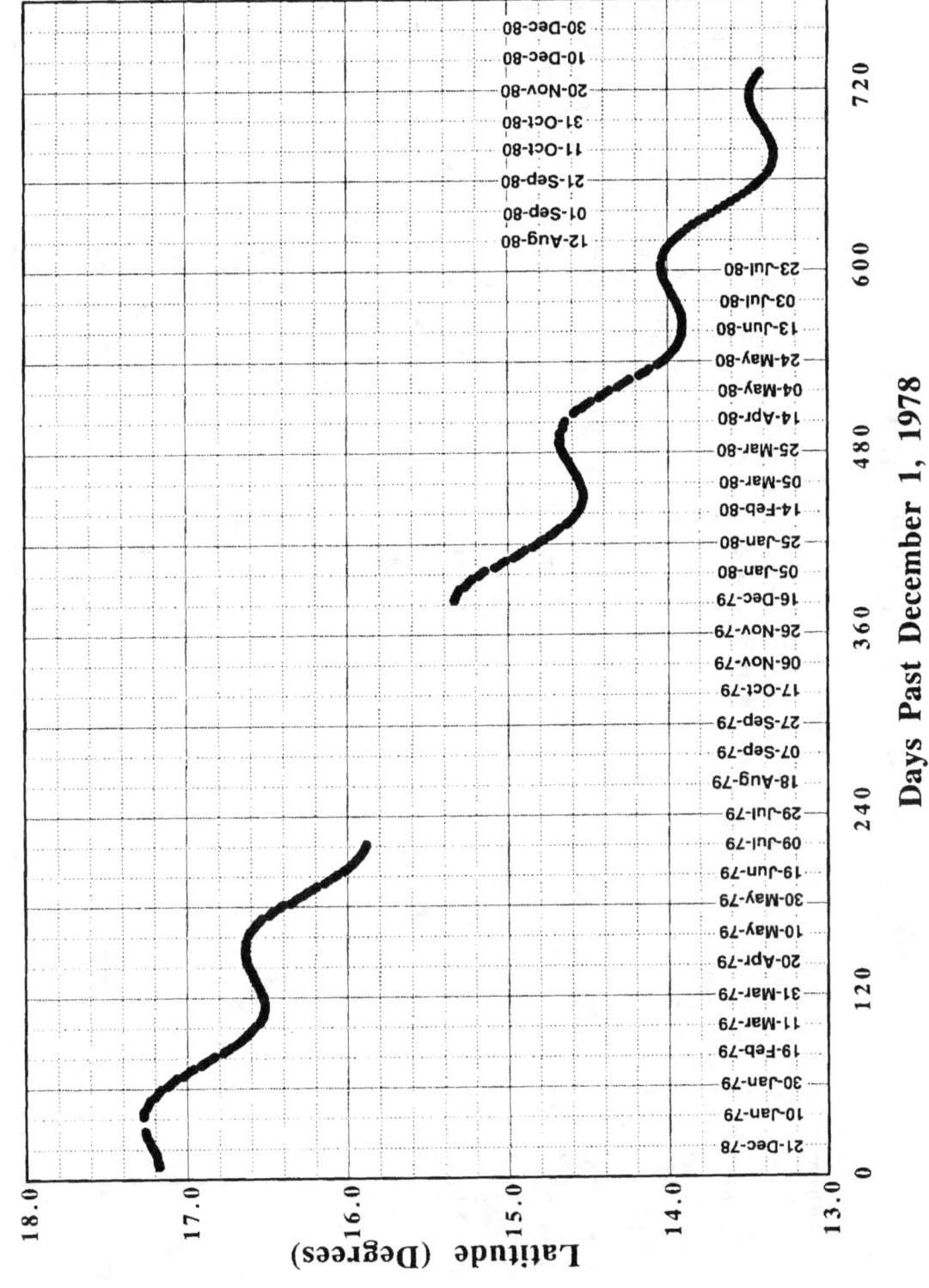

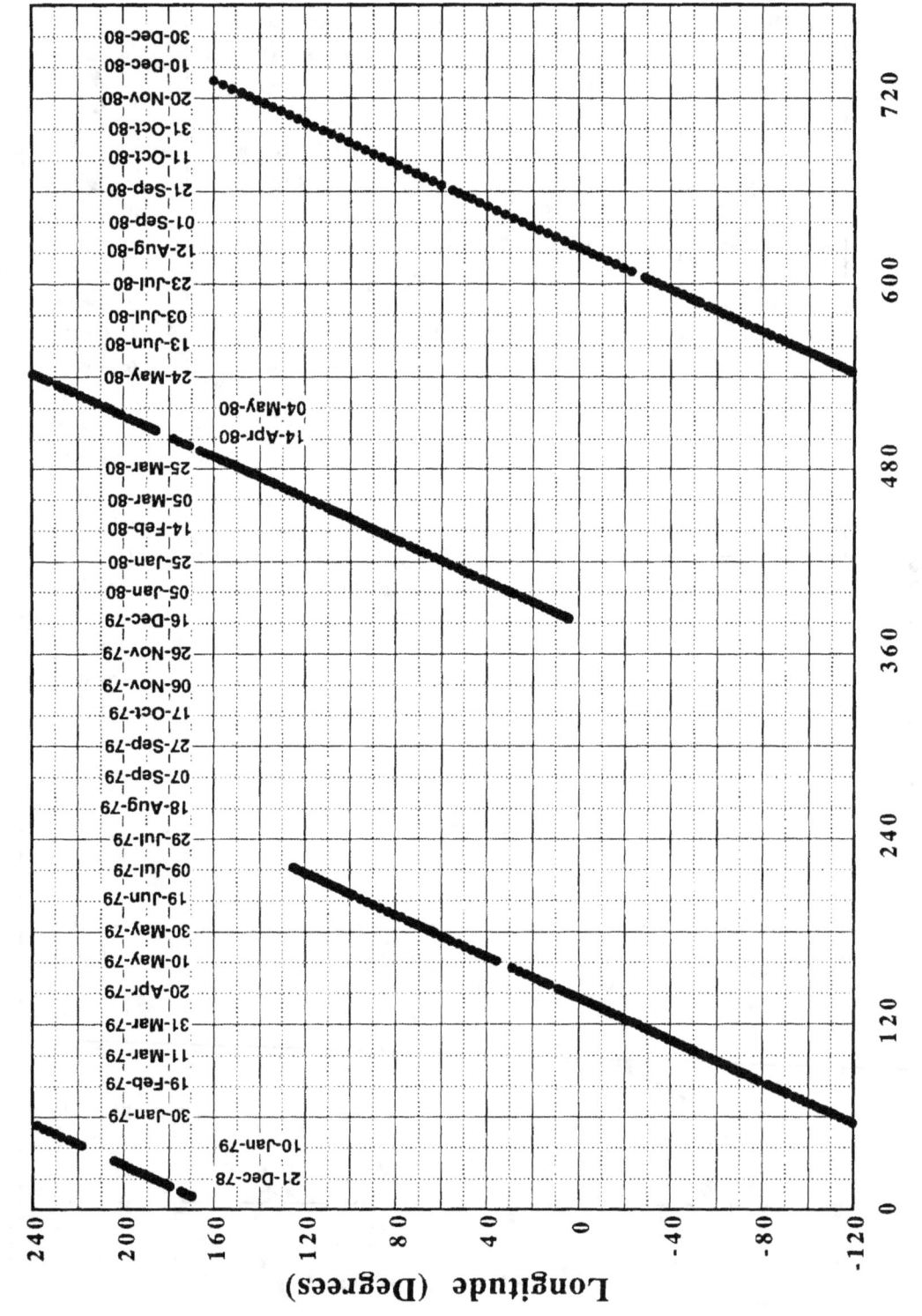

A-6

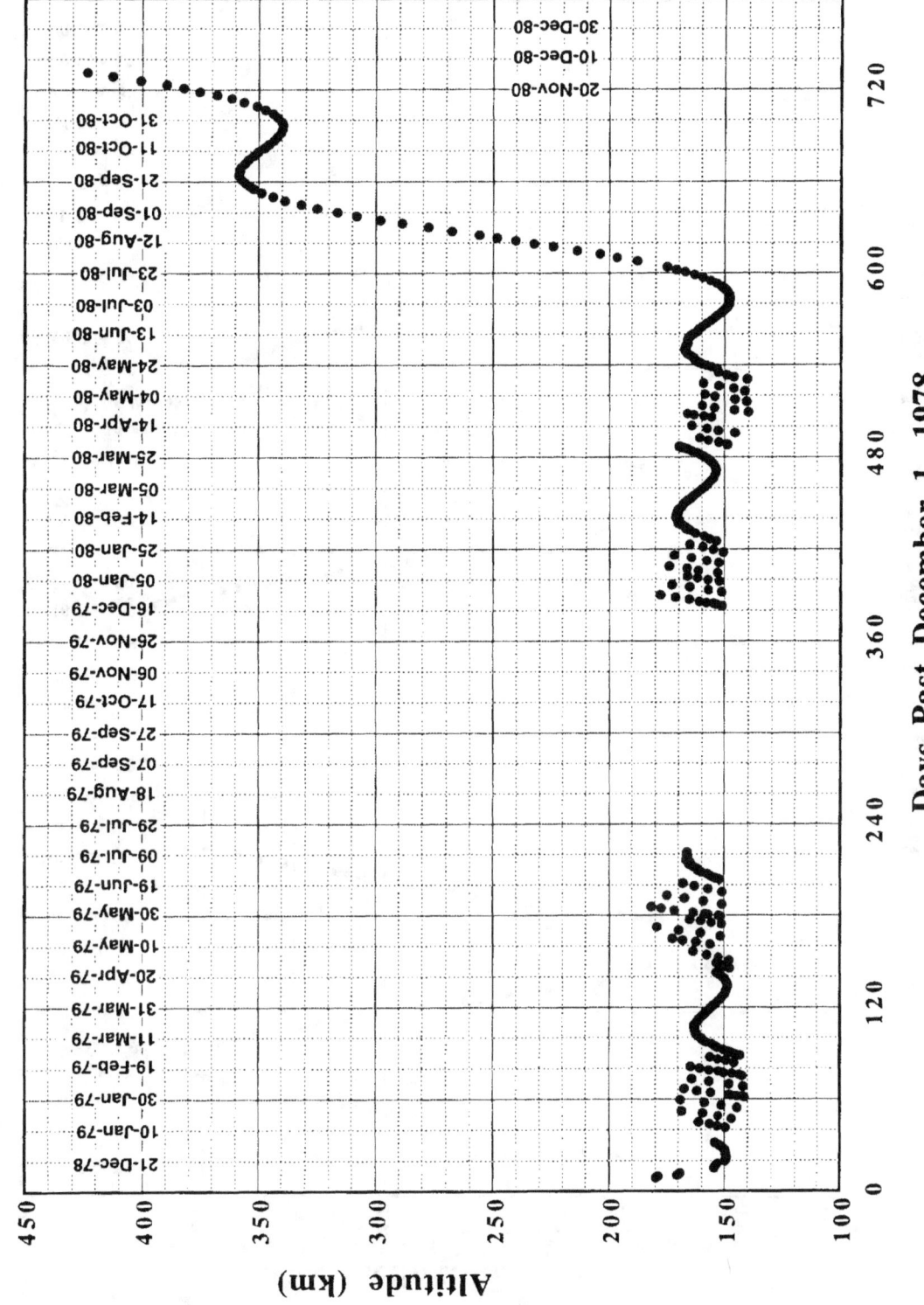

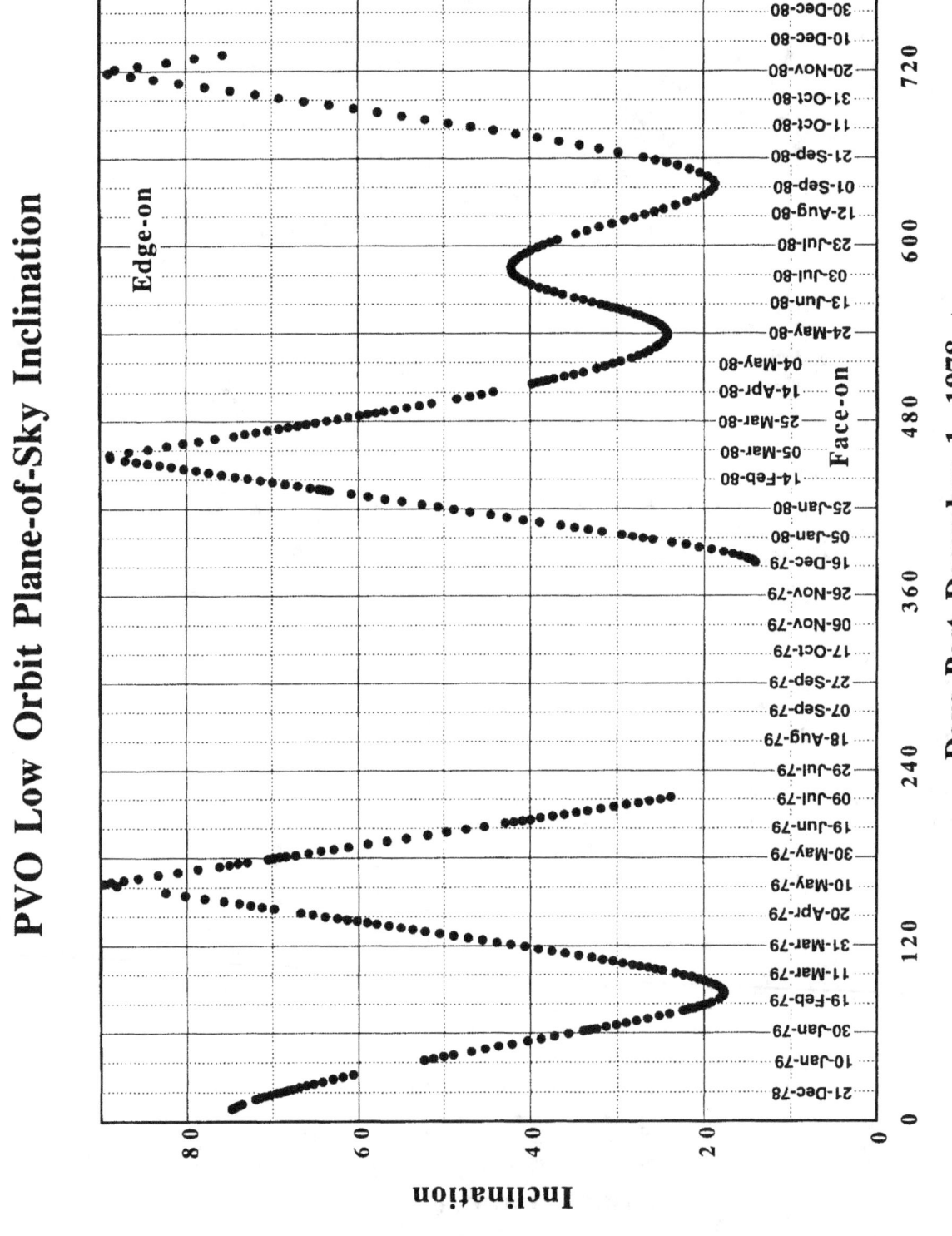

PVO Low Orbit One-Way-Light-Time

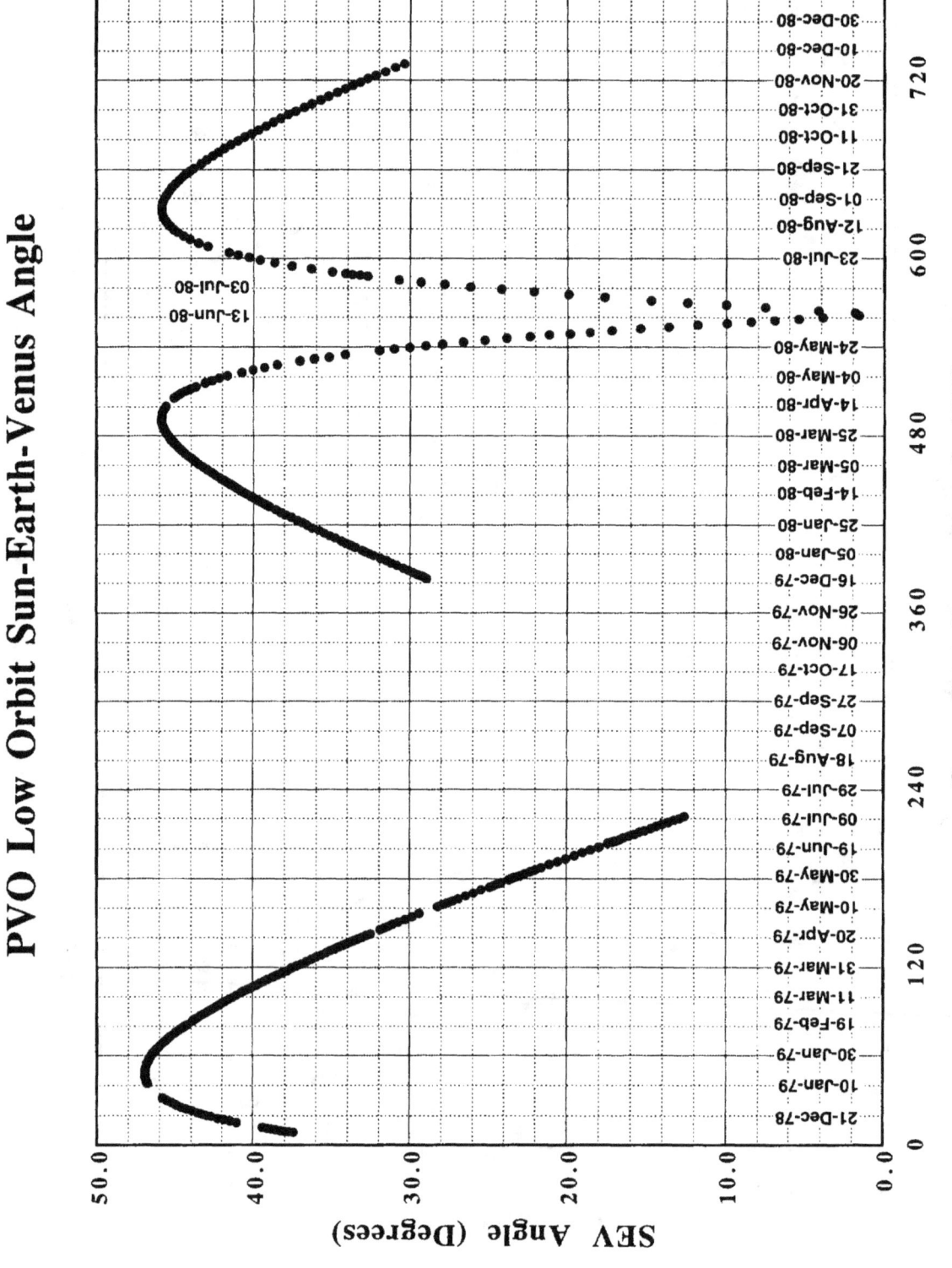

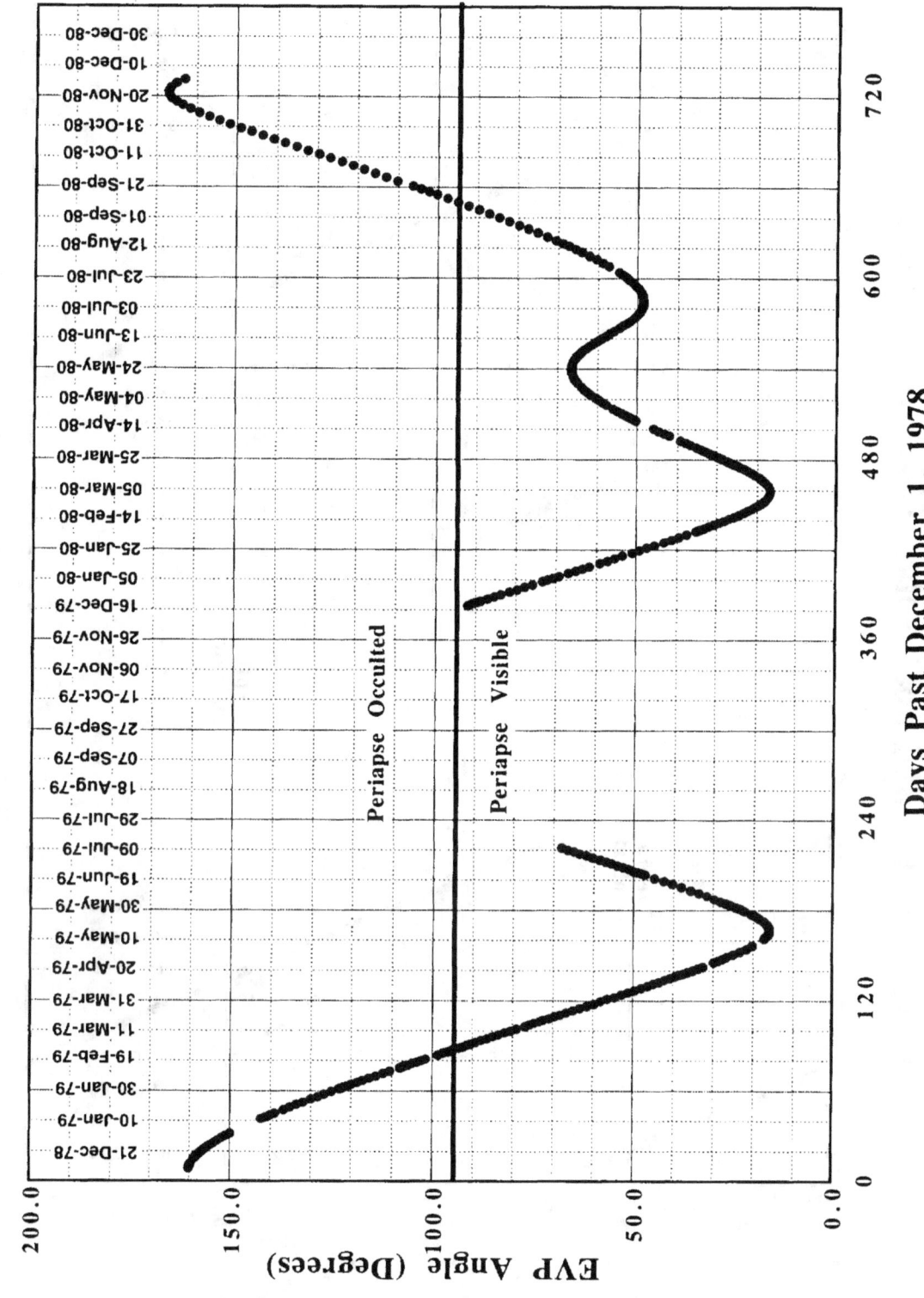

A-11

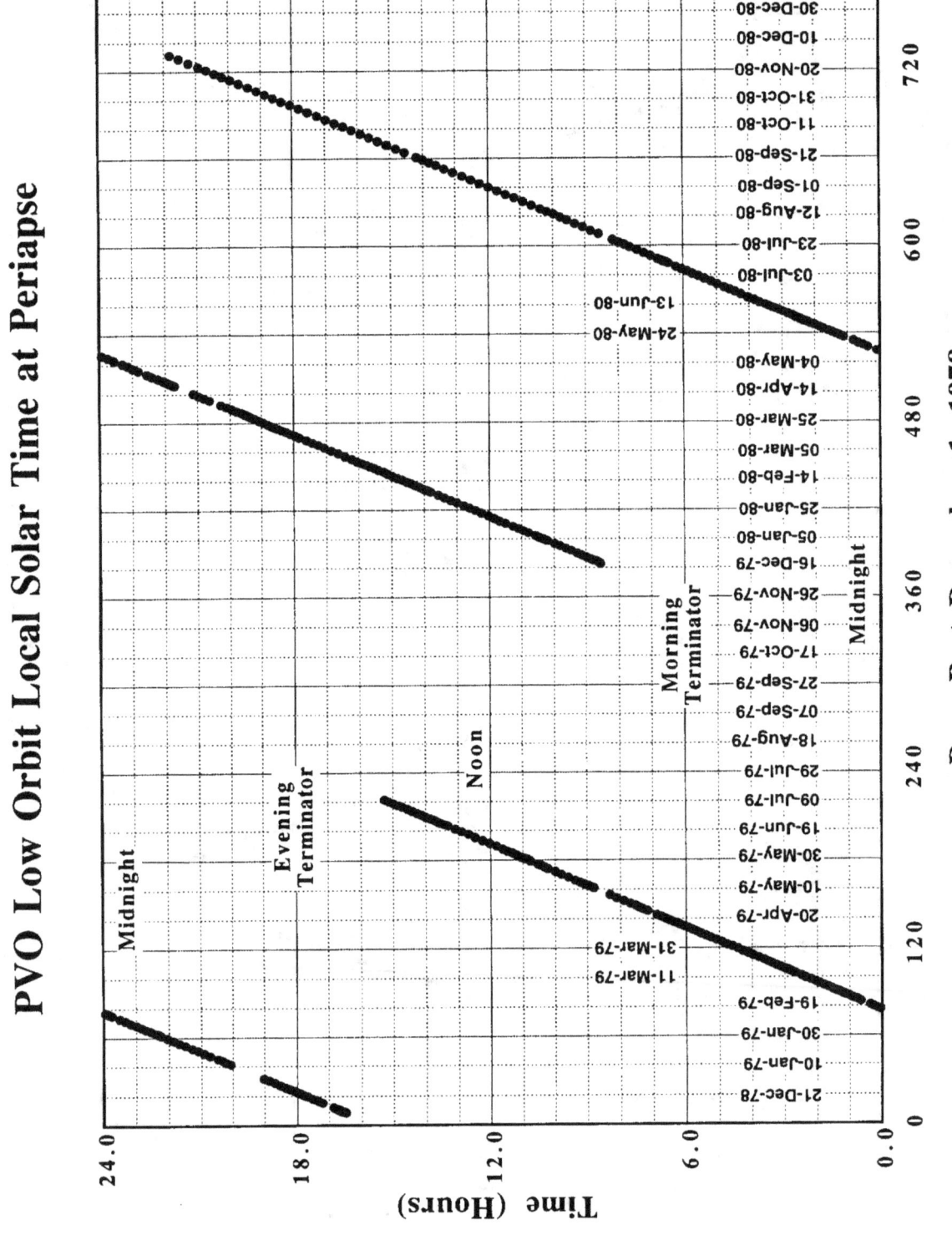

A-12

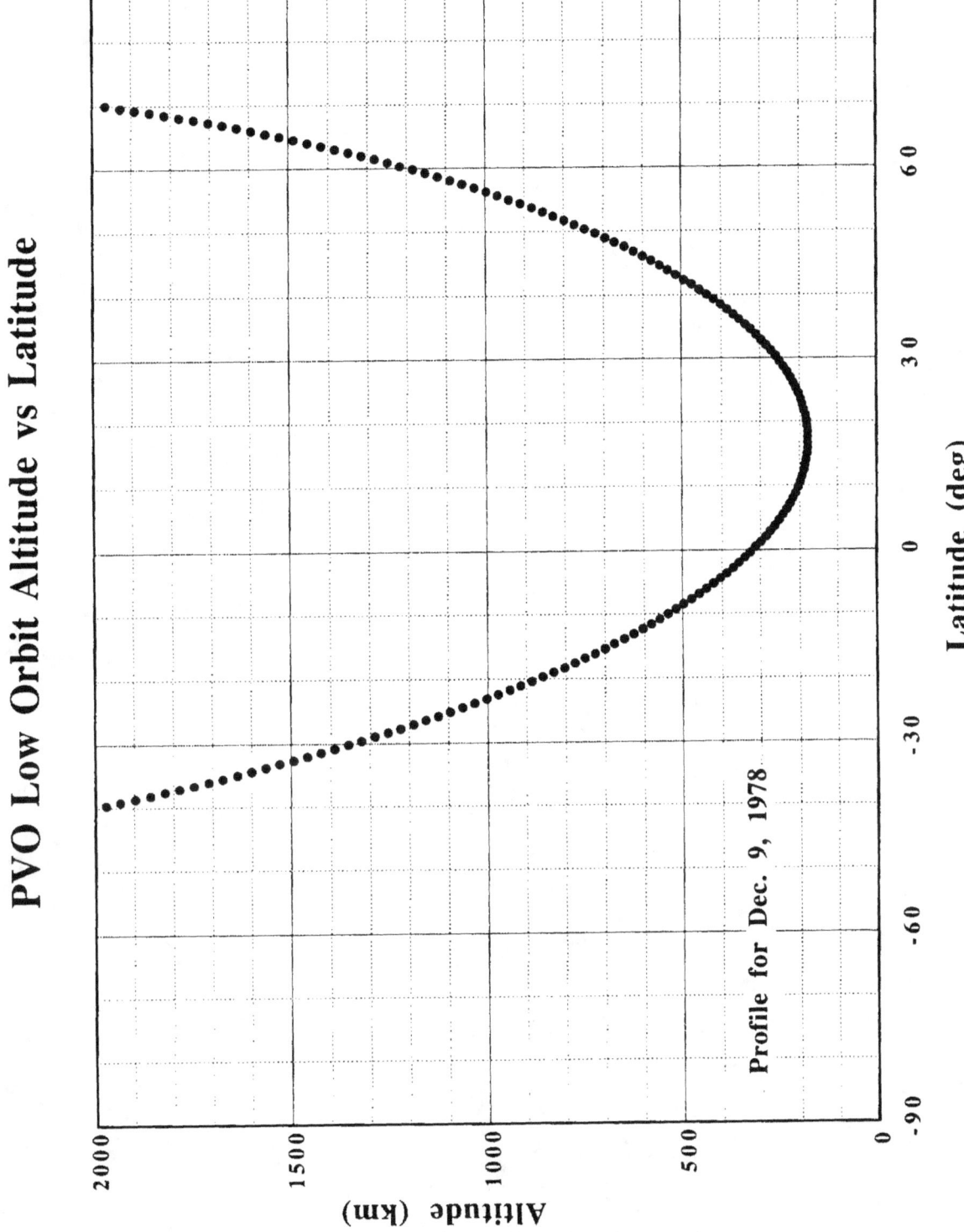

Appendix B
PVO High-Altitude Periapse Information

The following plots are included in this appendix:

1. Semi-major axis
2. Eccentricity
3. Inclination
4. Latitude at periapse
5. Longitude at periapse
6. Altitude at periapse
7. Plane-of-sky inclination
8. One-way light time from Venus to Earth
9. Sun-Earth-Venus angle
10. Earth-Venus-Sun angle
11. Earth-Venus-Probe at periapse angle
12. Local solar time at periapse

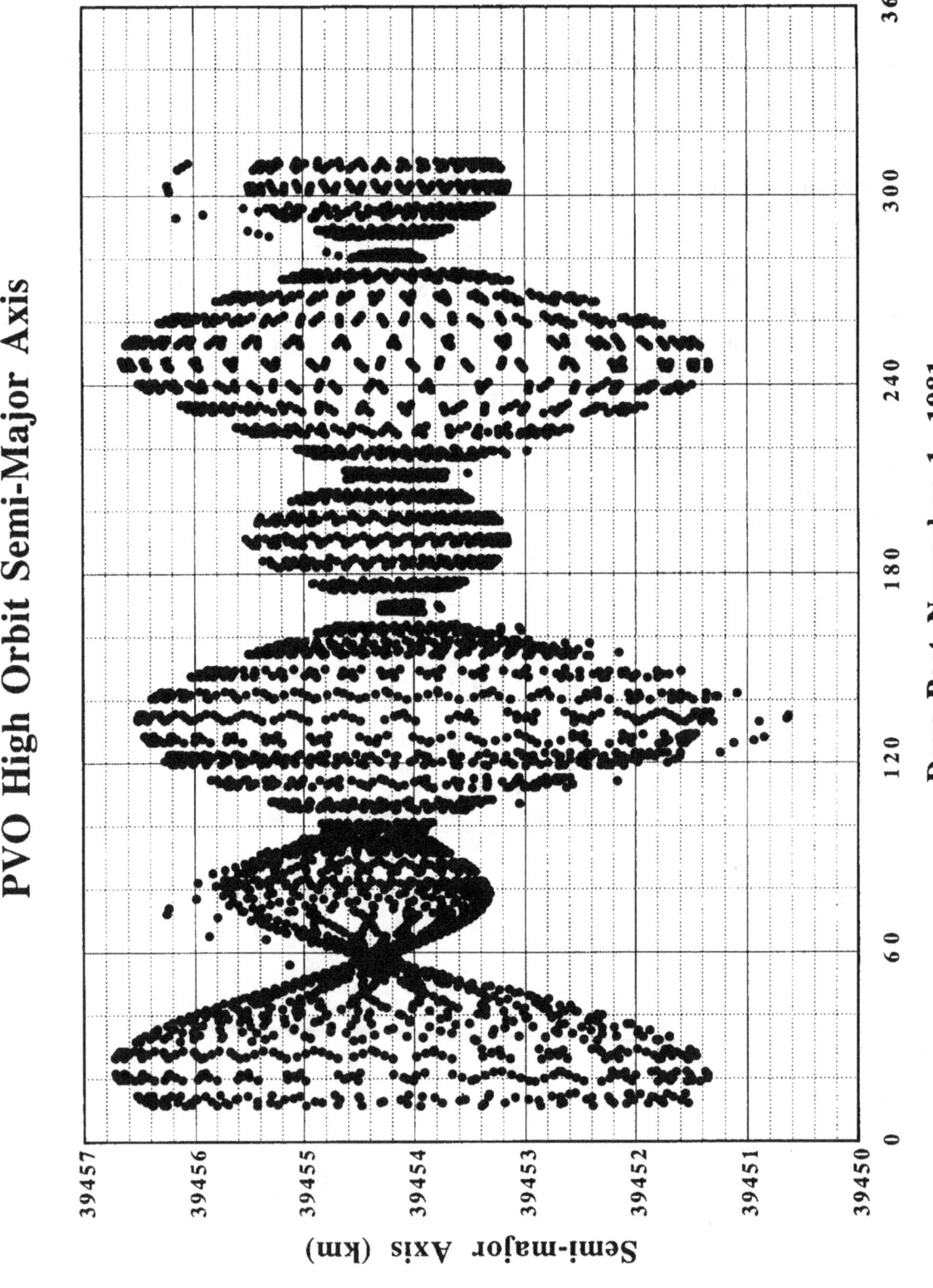

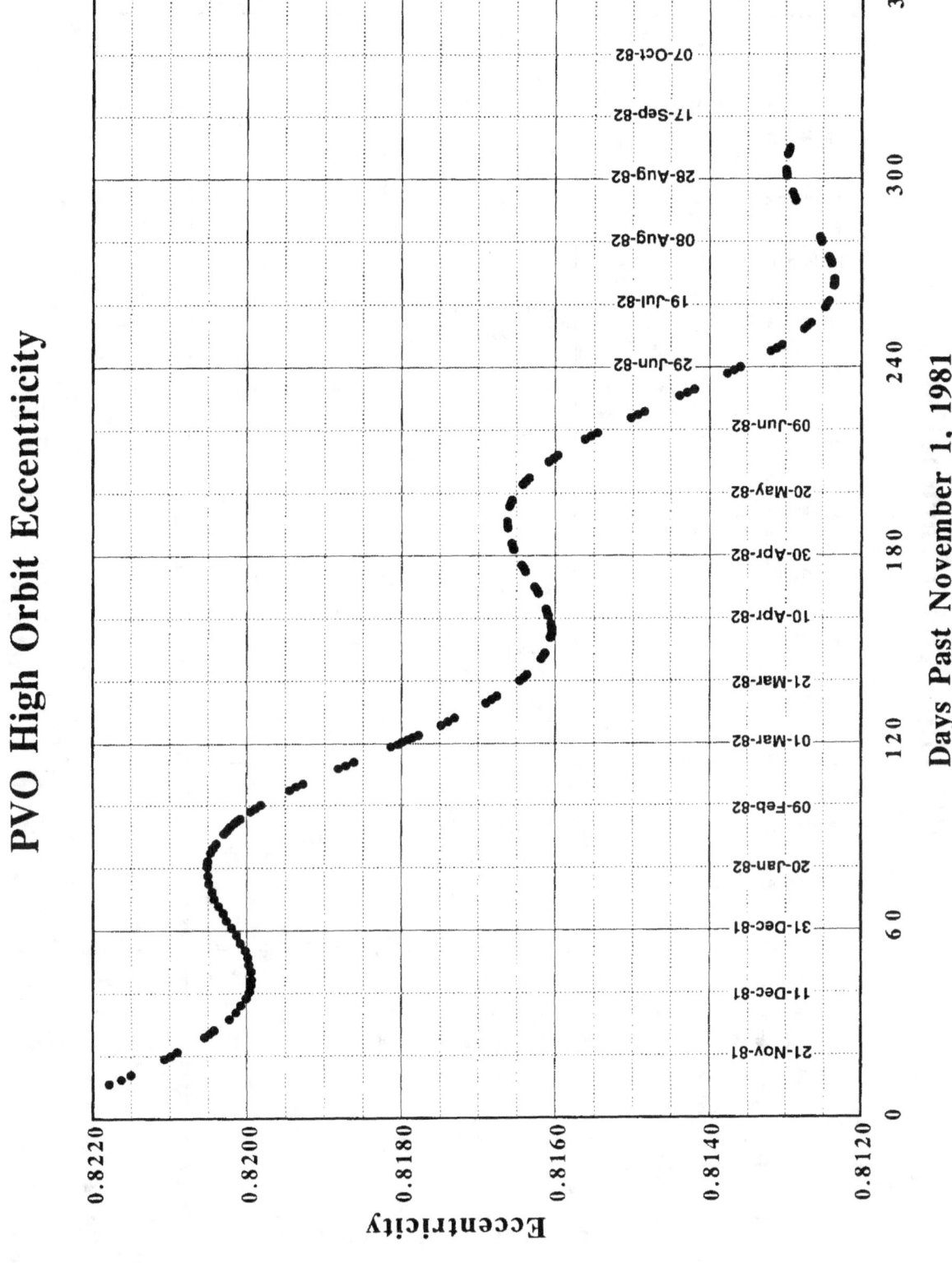

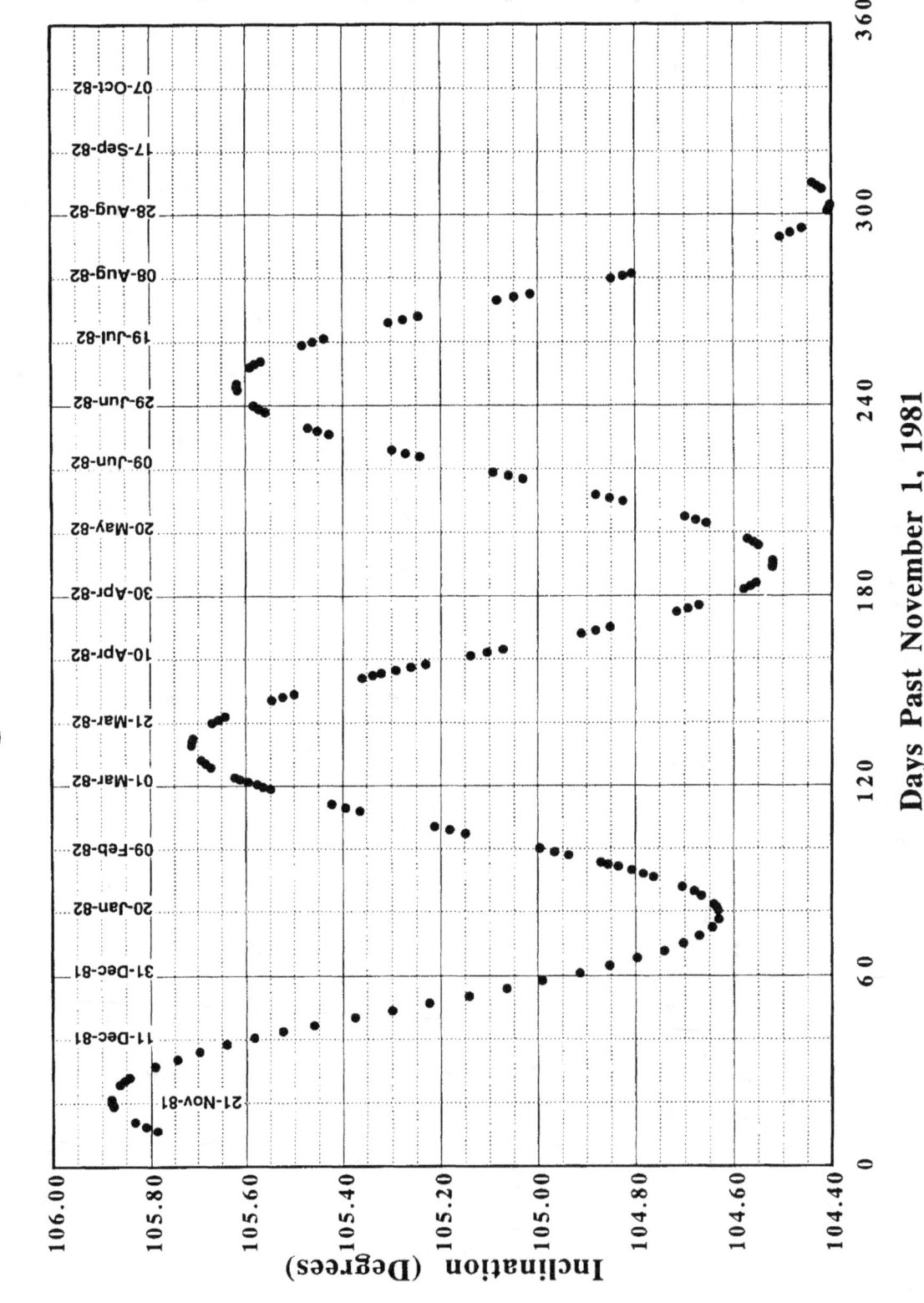

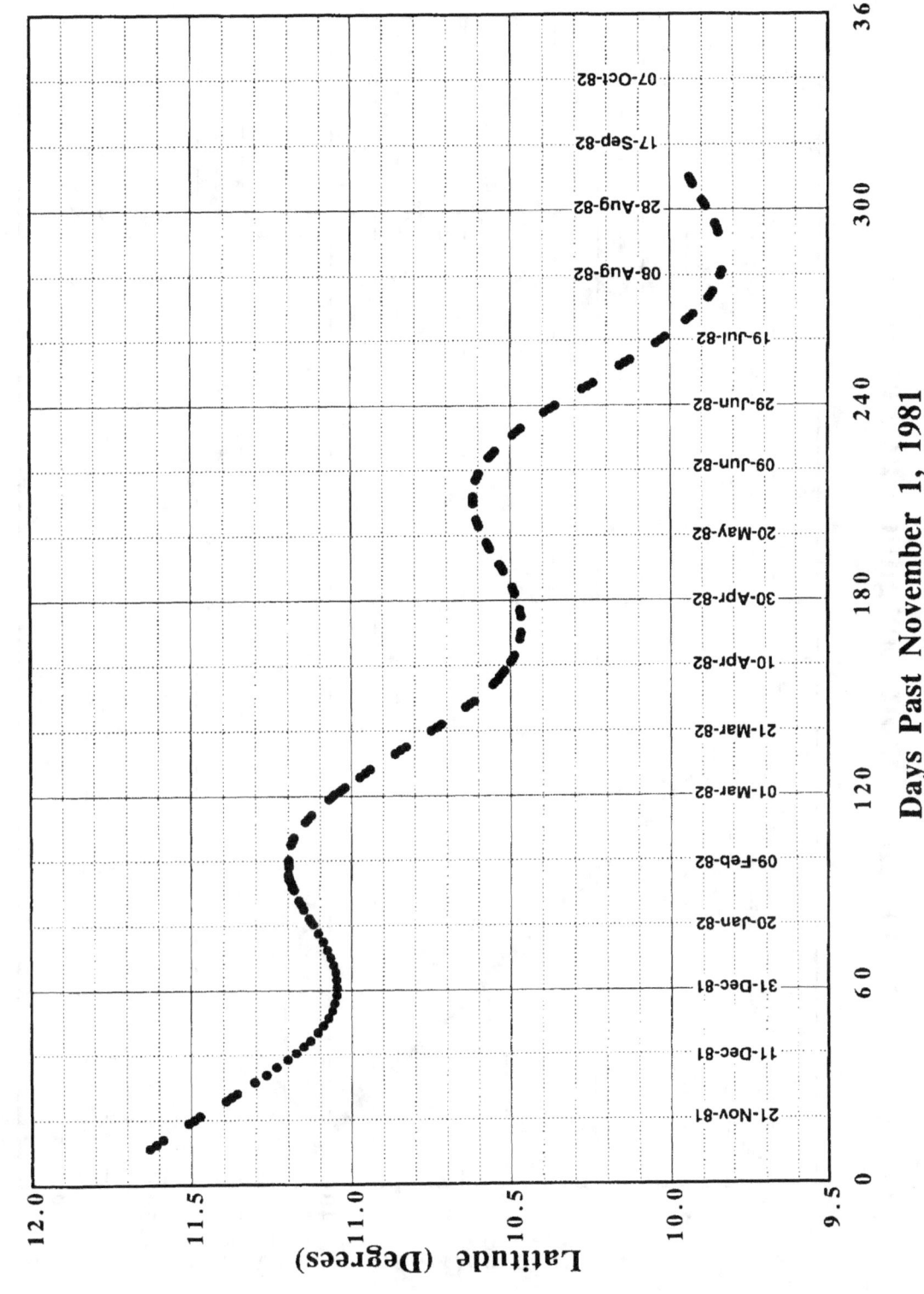

B-5

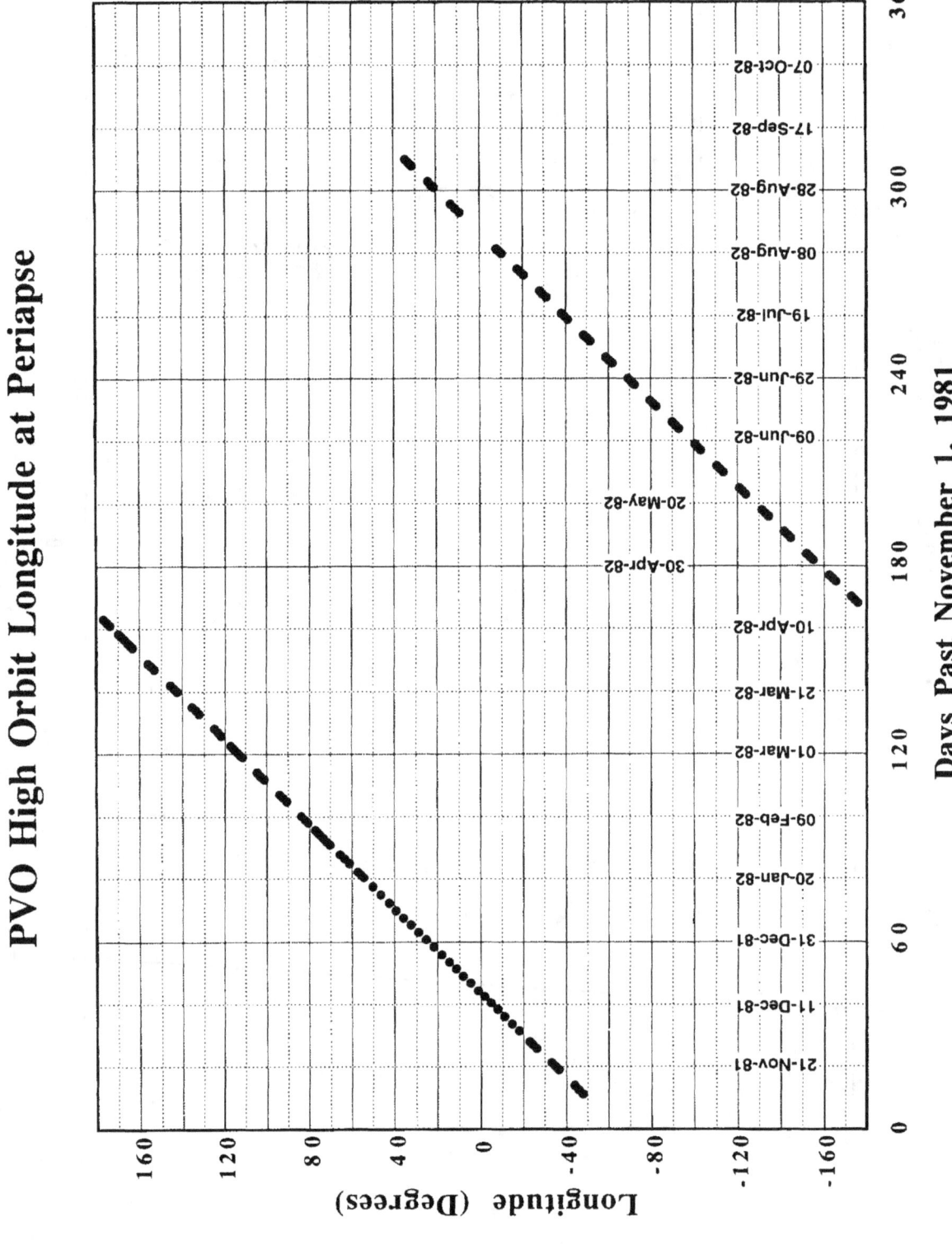

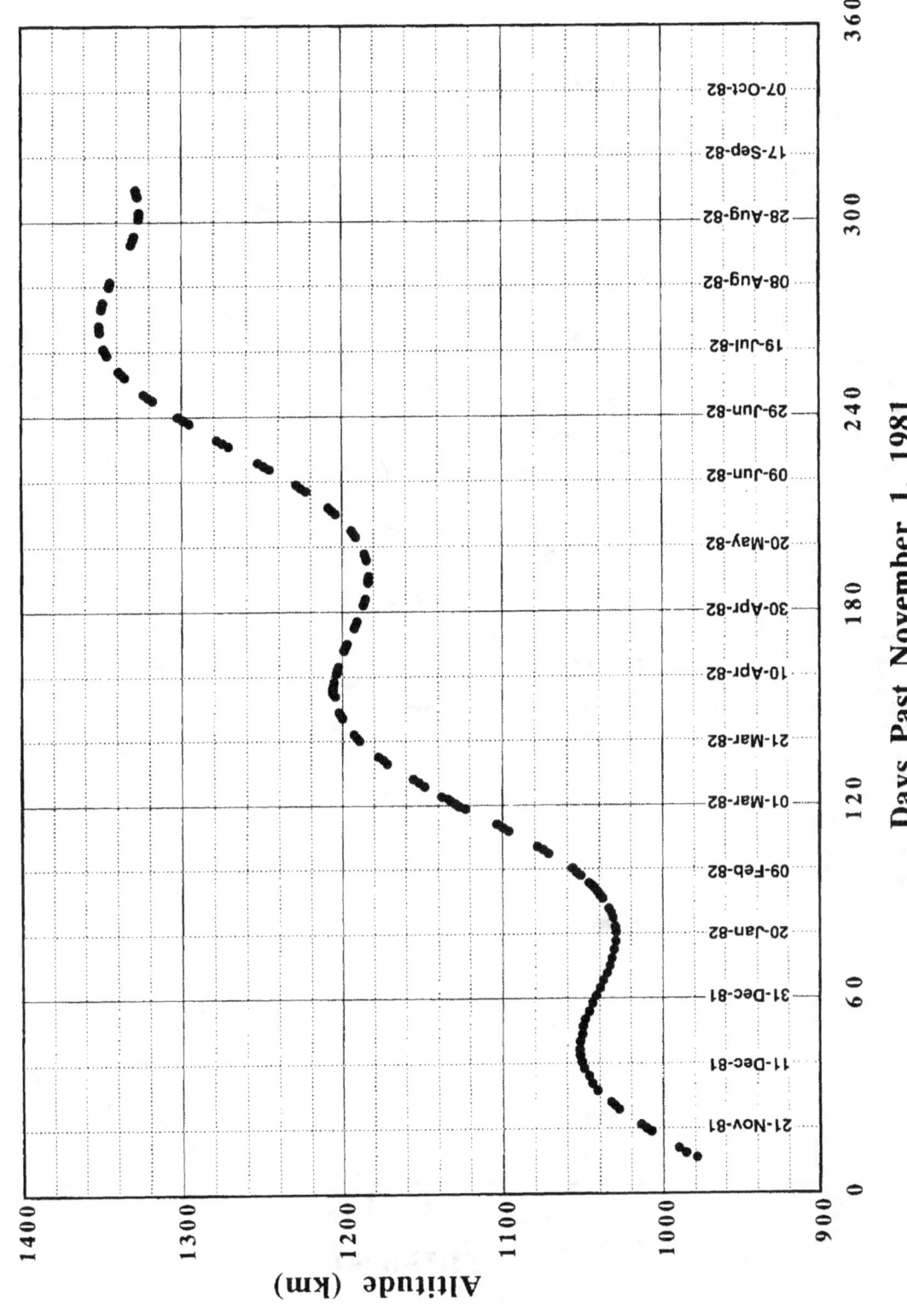

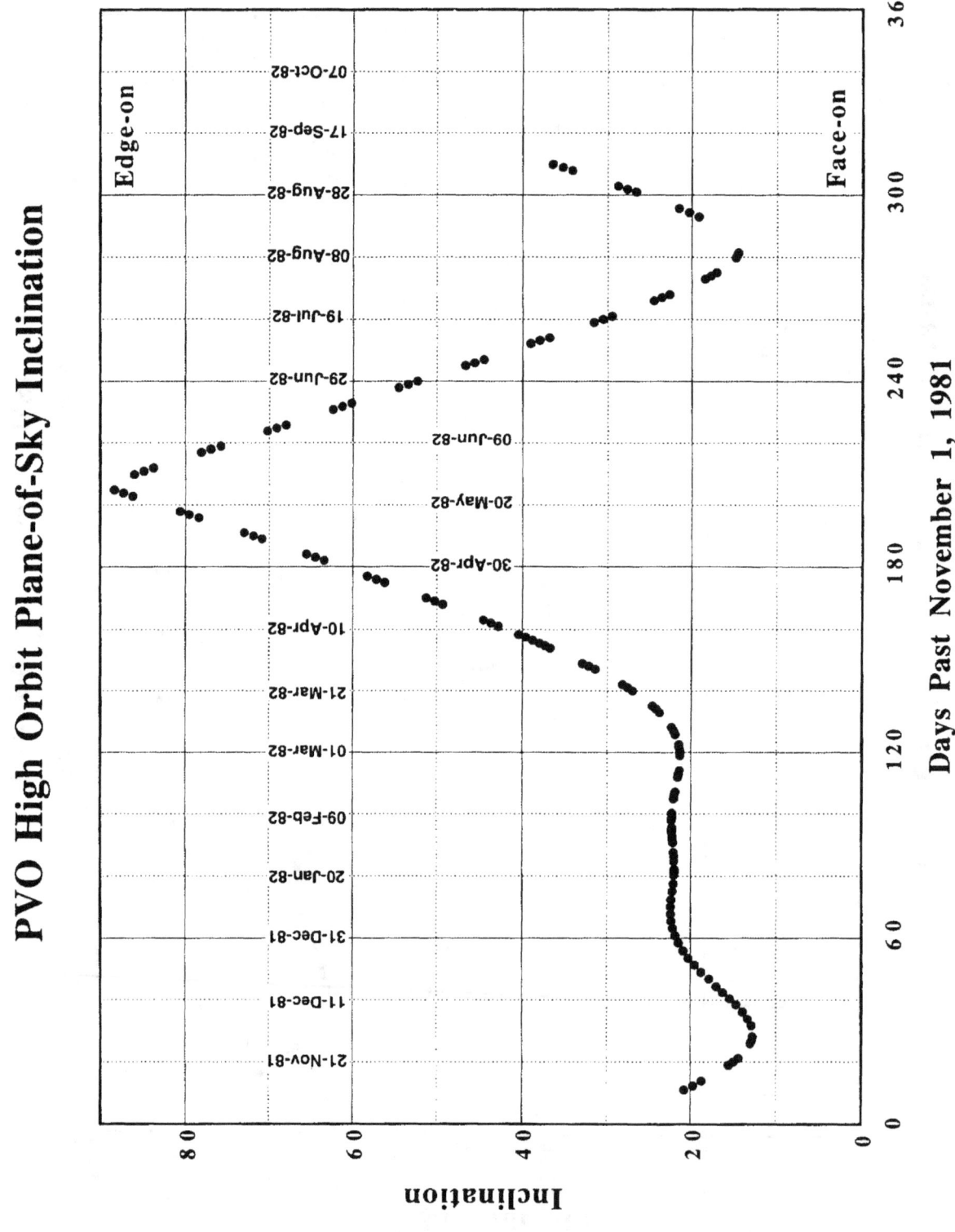

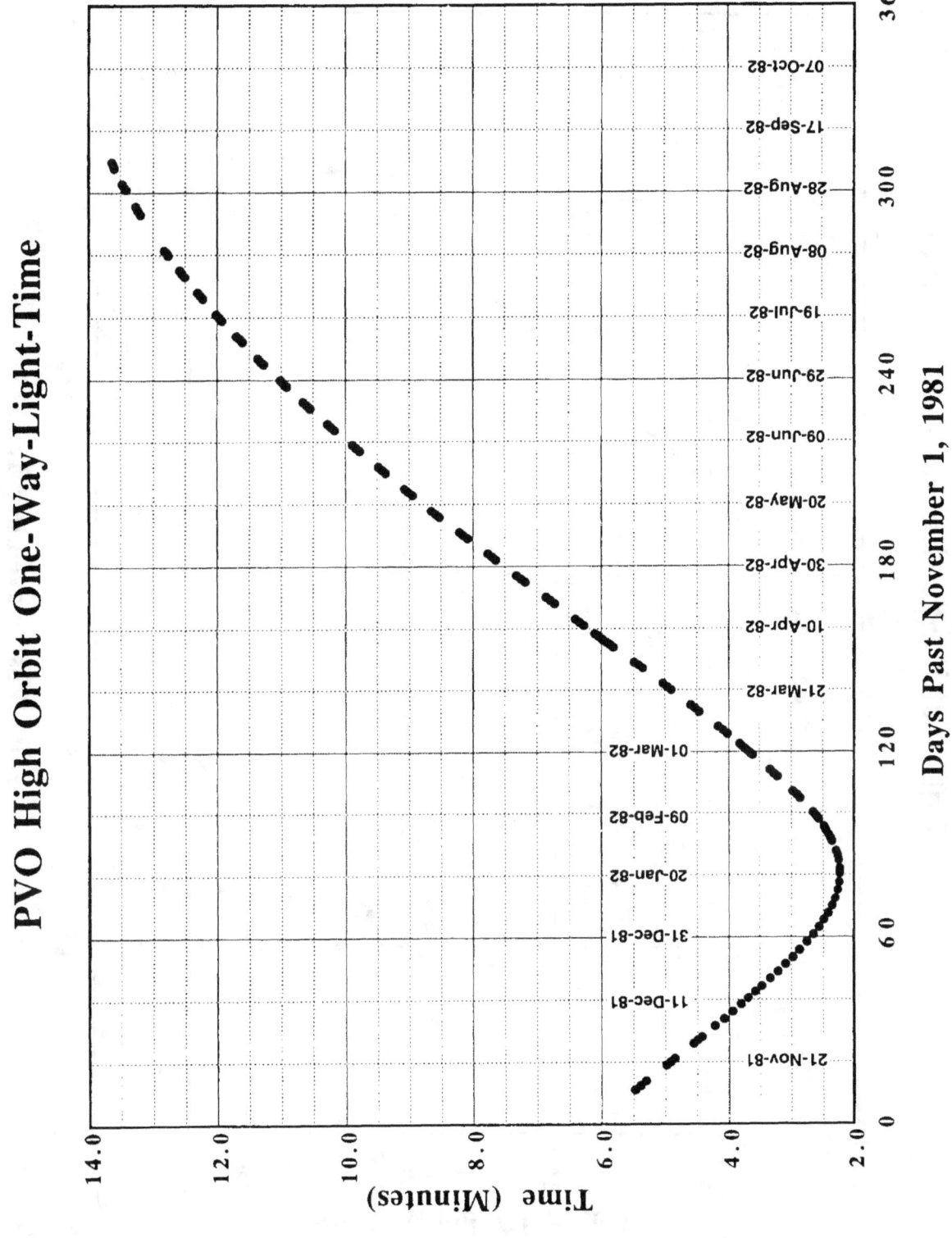

B-9

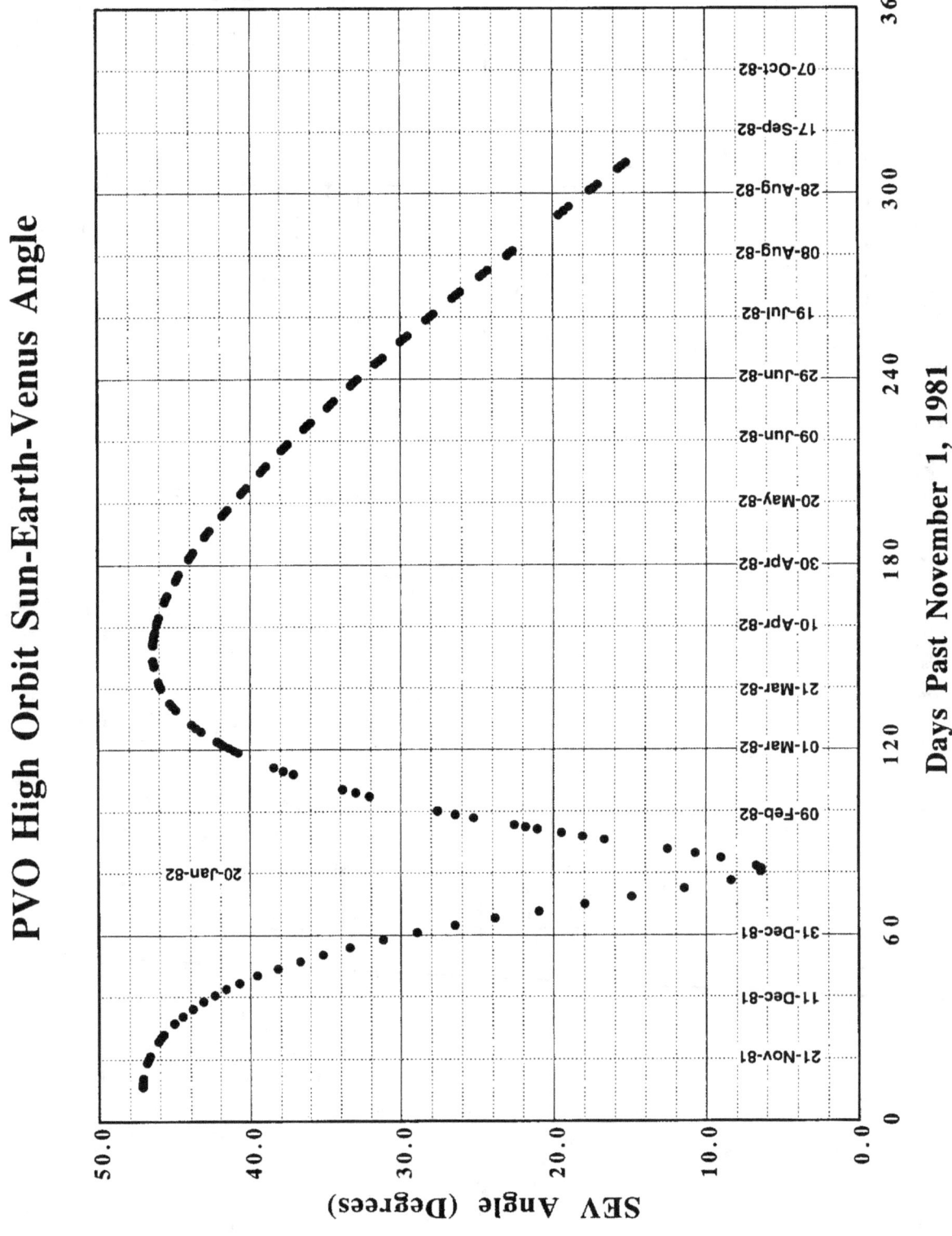

B-10

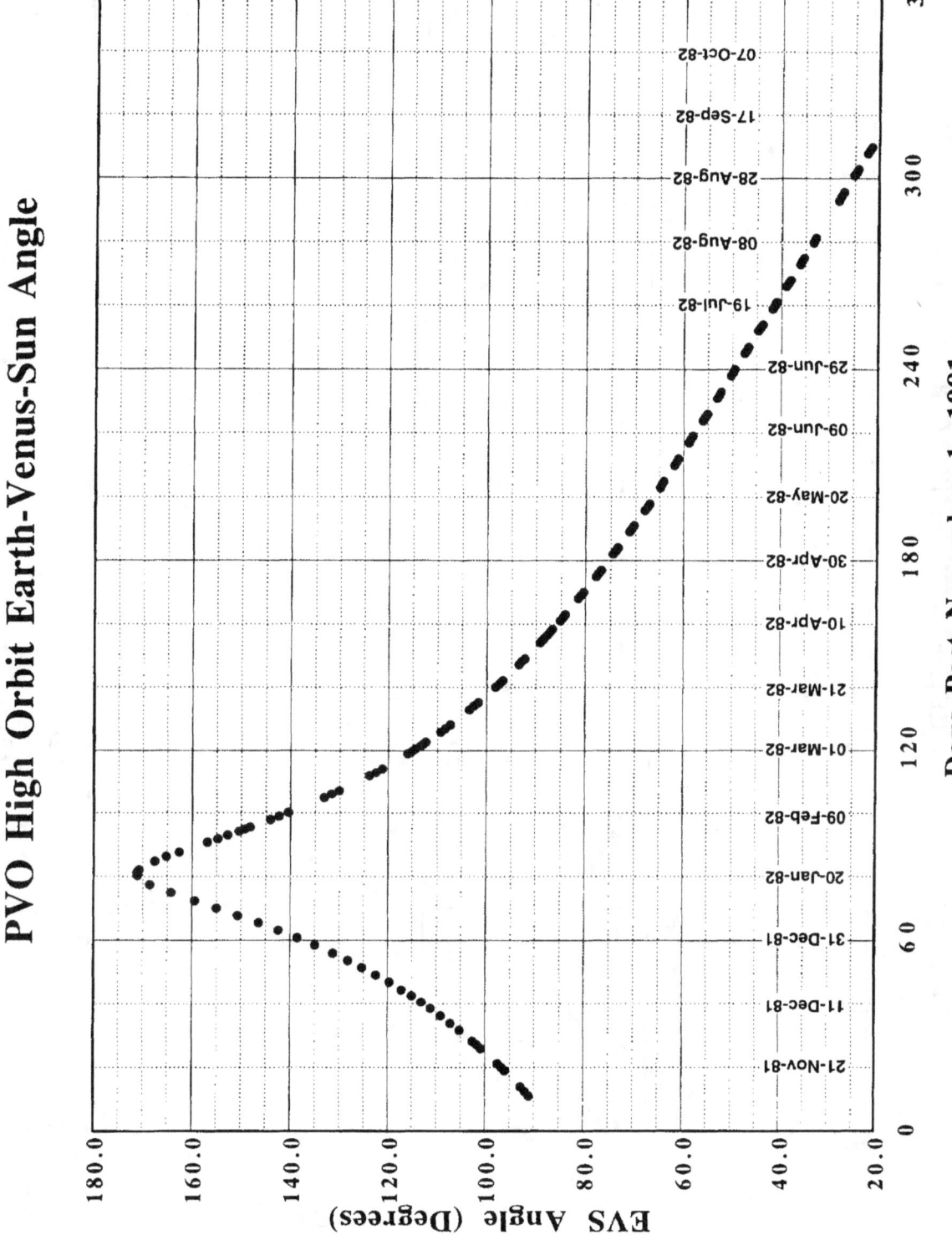

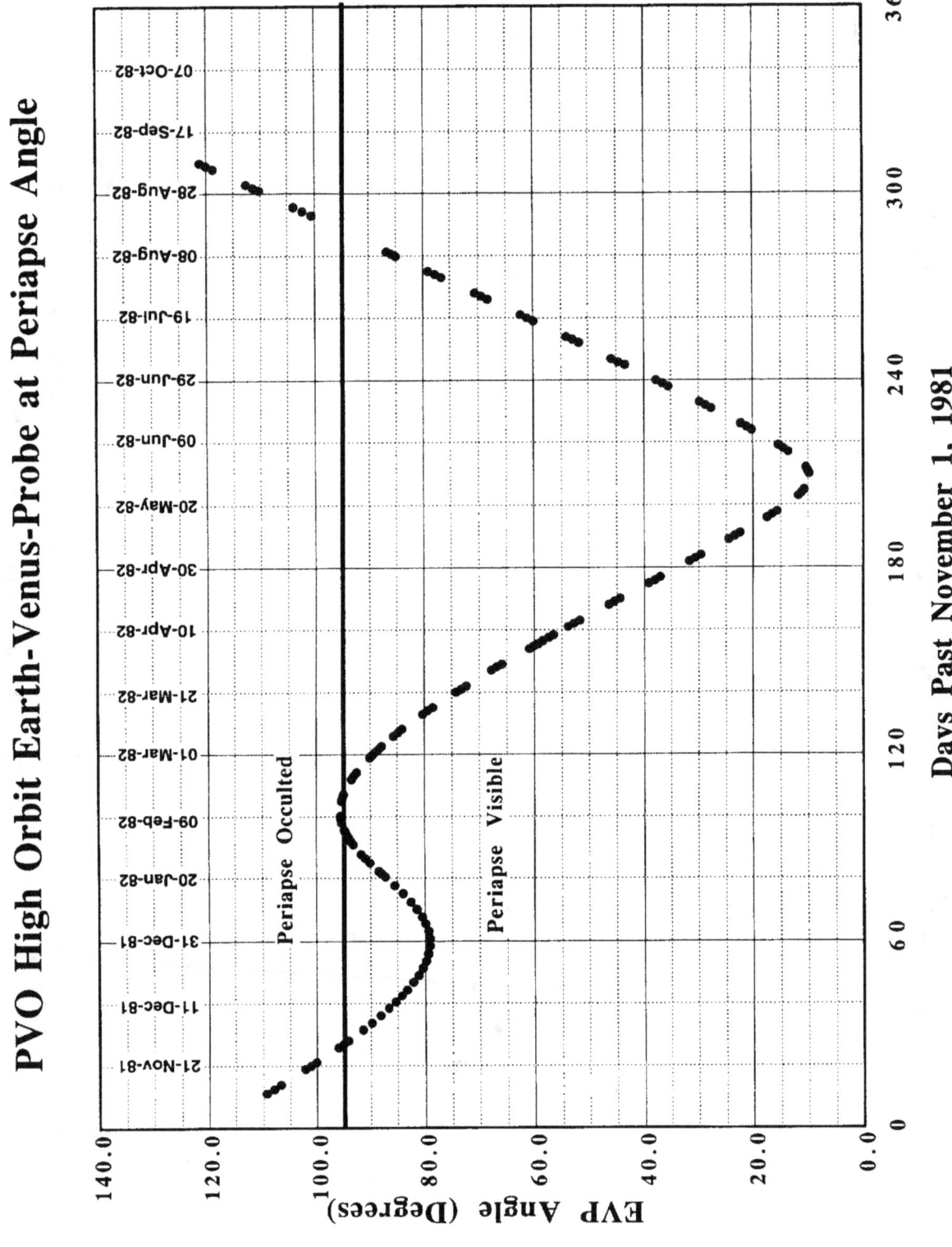

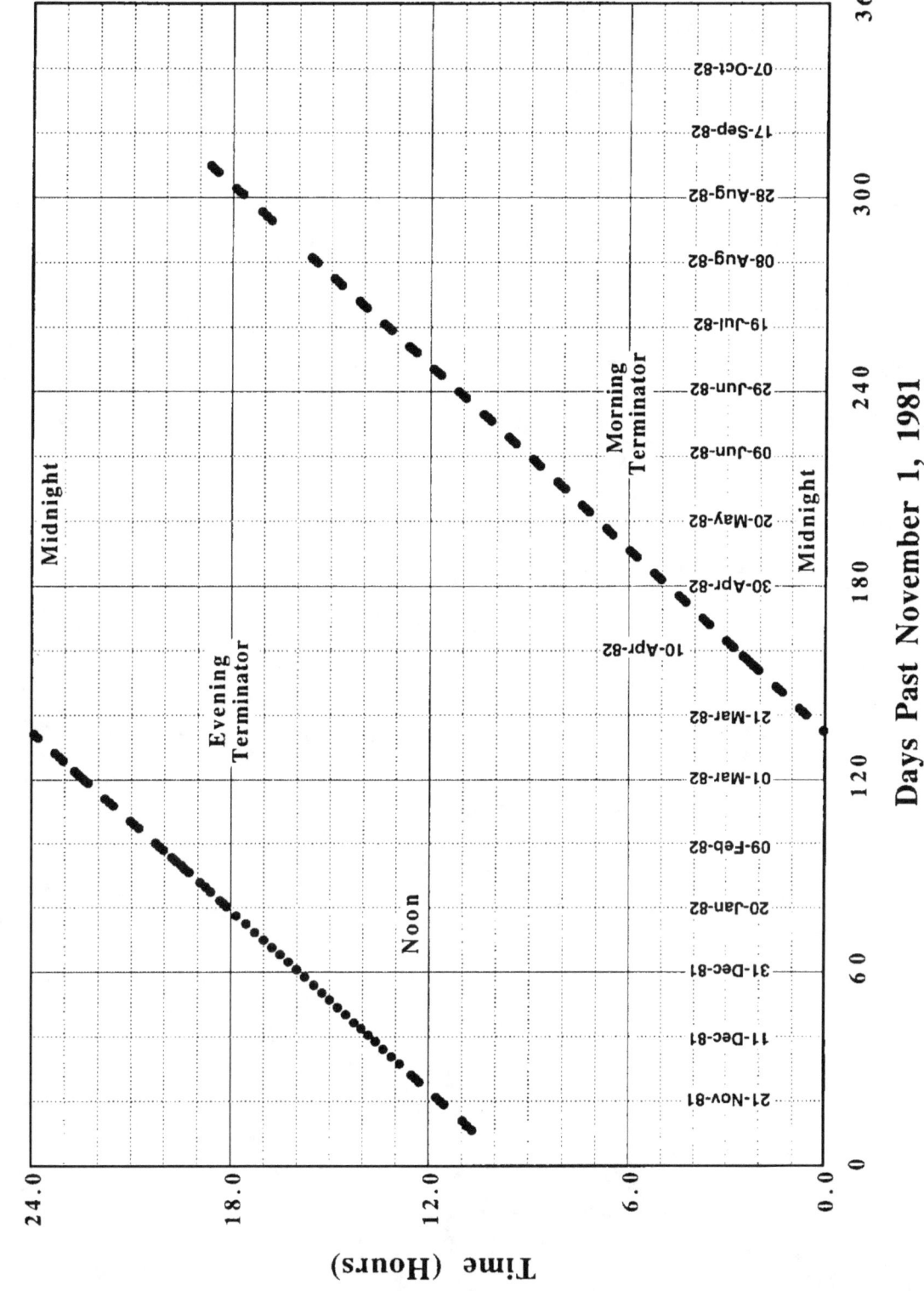

B-13

Appendix C
Magellan Cycle 4 (Elliptical) Information

The following plots are included in this appendix:

1. Semi-major axis
2. Eccentricity
3. Inclination
4. Latitude at periapse
5. Longitude at periapse
6. Altitude at periapse
7. Plane-of-sky inclination vs time
8. Plane-of-sky inclination vs longitude
9. One-way light time from Venus to Earth
10. Sun-Earth-Venus angle
11. Earth-Venus-Probe at periapse angle
12. Local solar time at periapse
13. Altitude vs. latitude profile

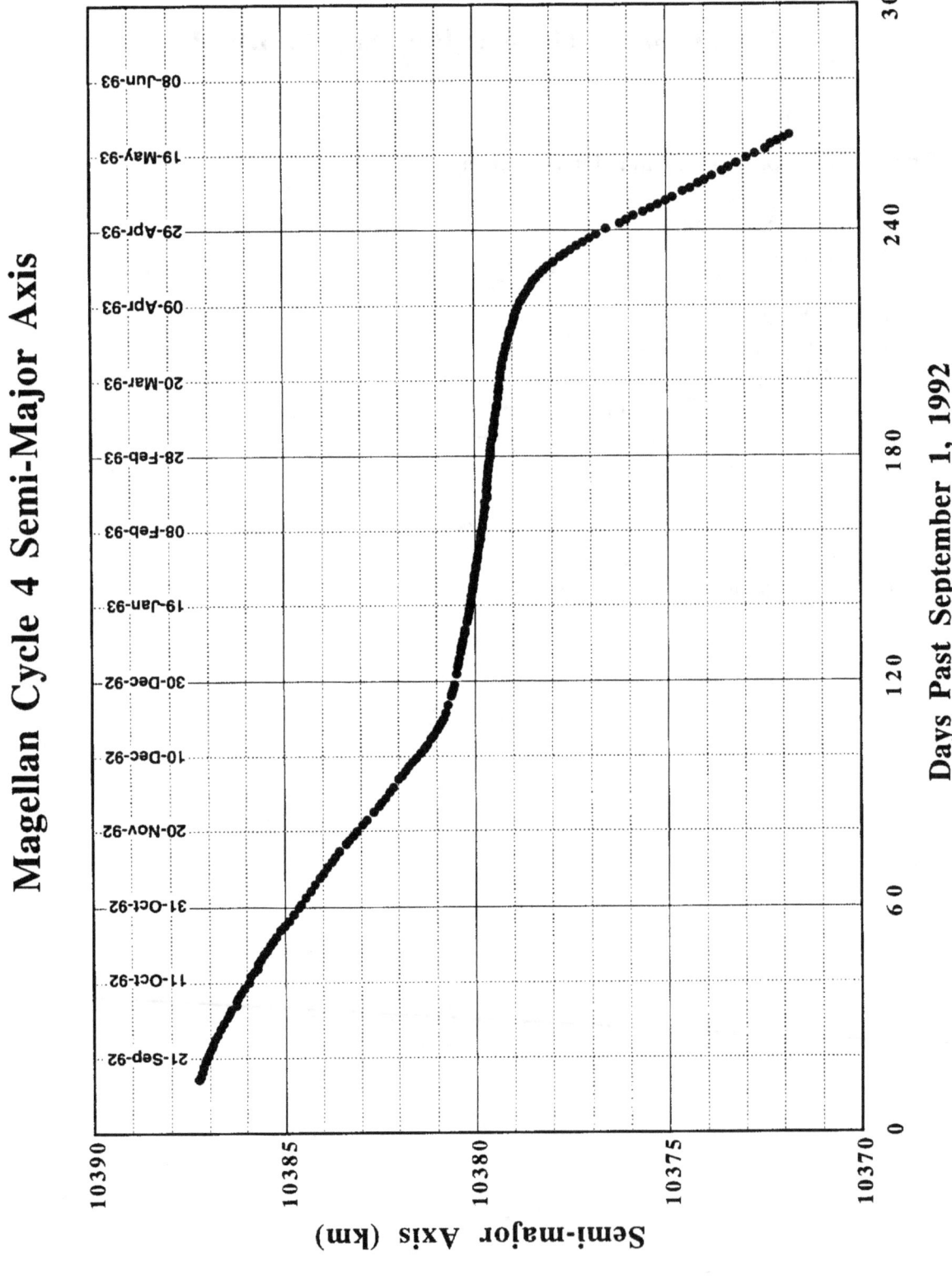

C-2

Magellan Cycle 4 Eccentricity

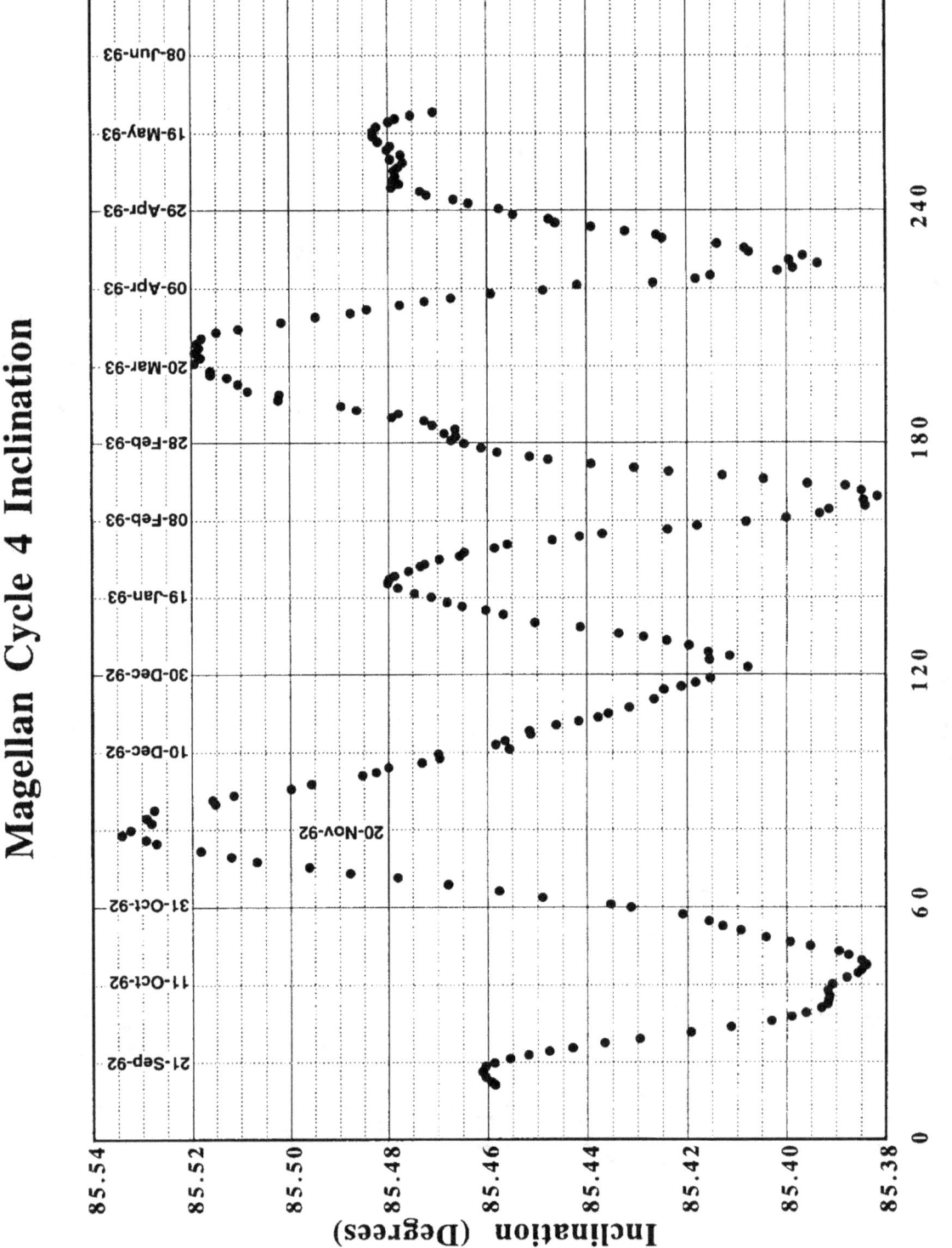

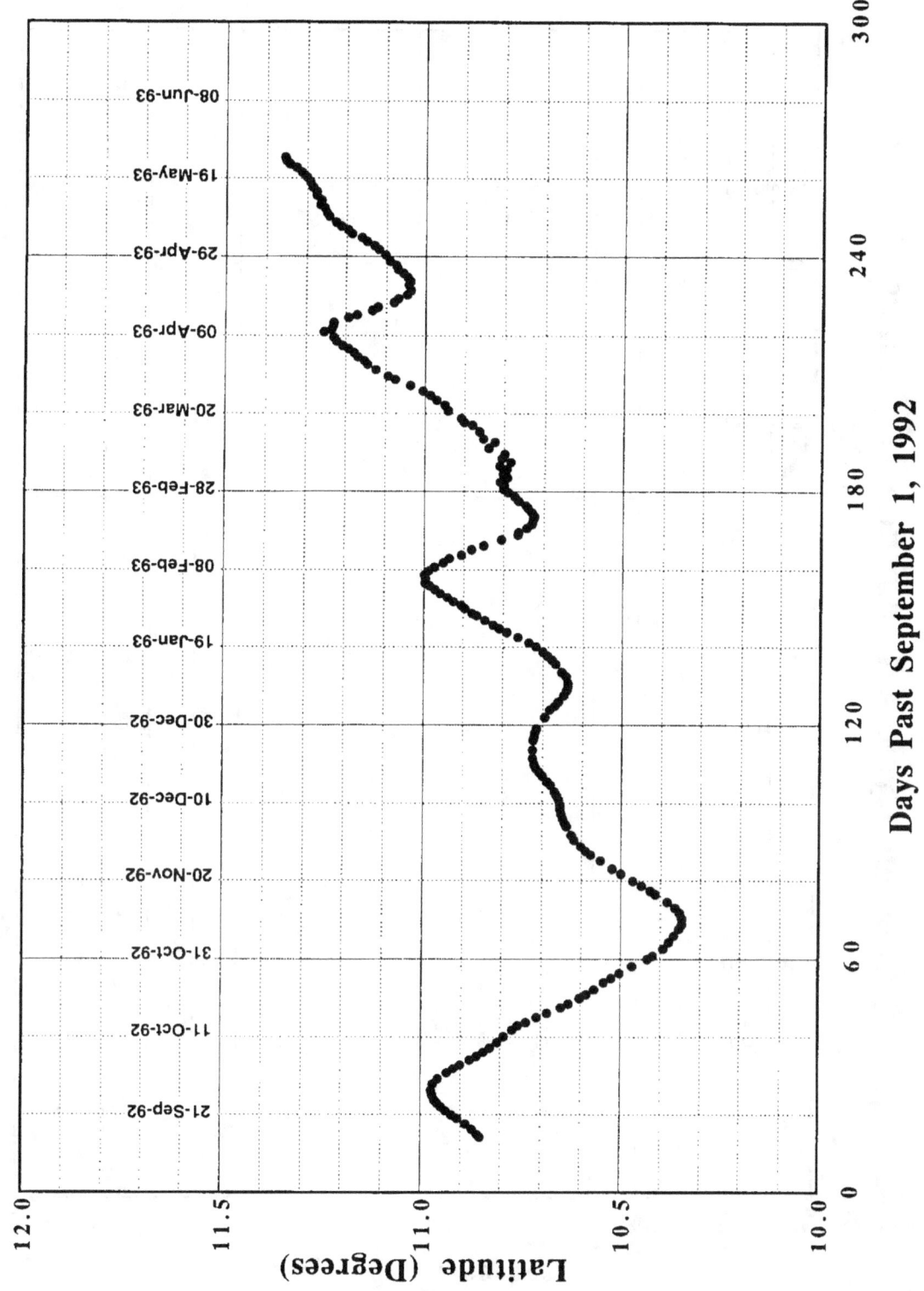

Magellan Cycle 4 Longitude at Periapse

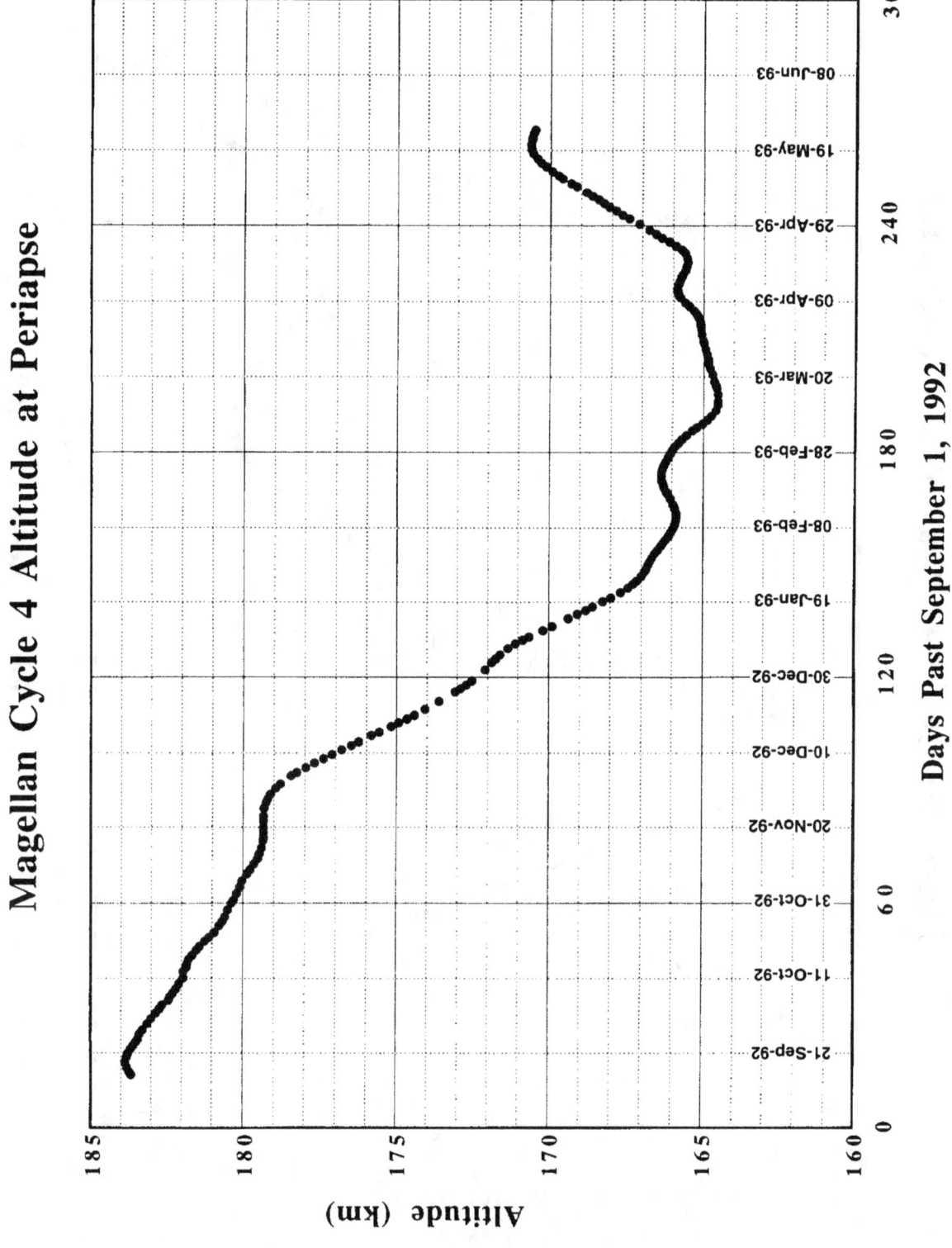

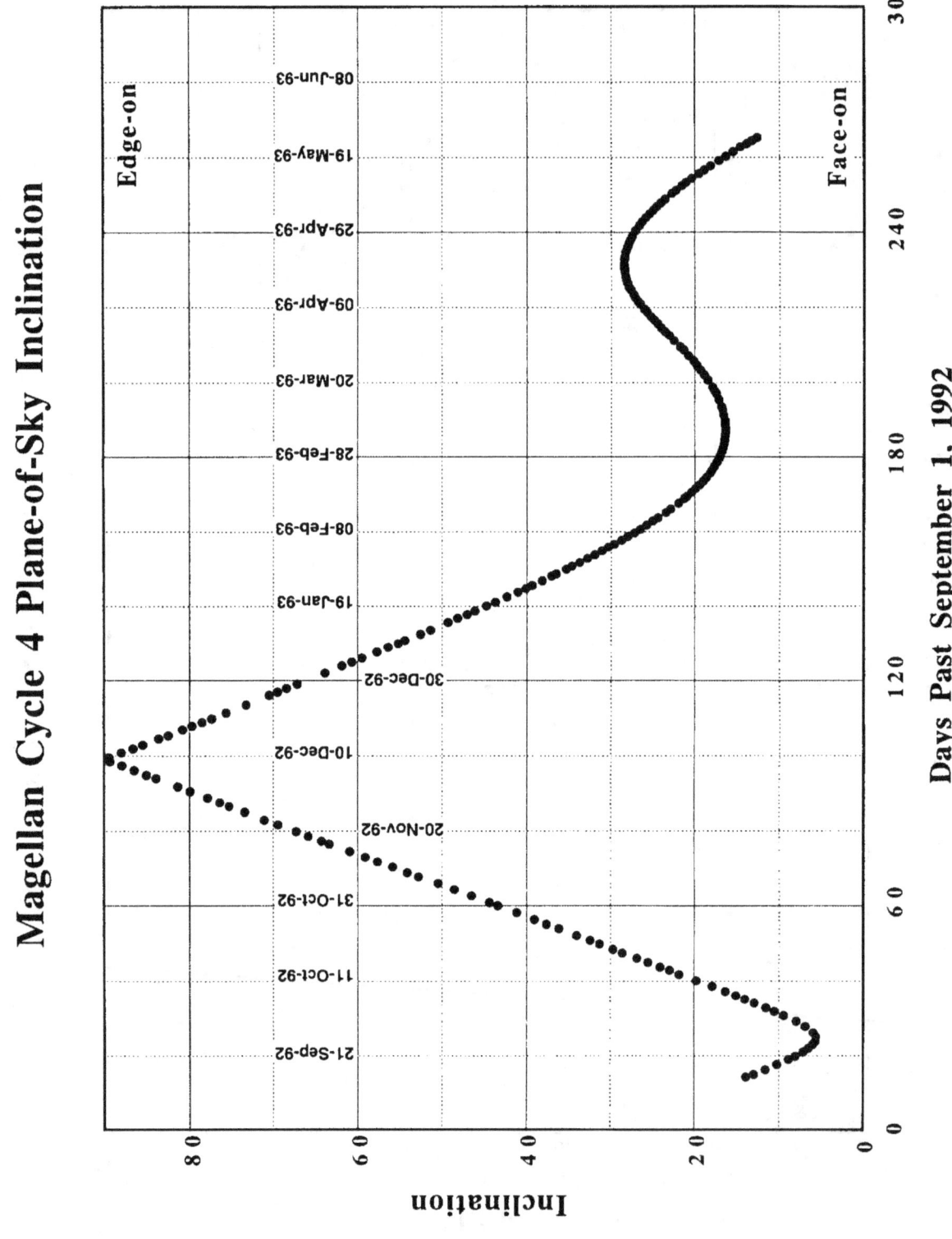

Magellan Cycle 4 Plane-of-Sky Inclination

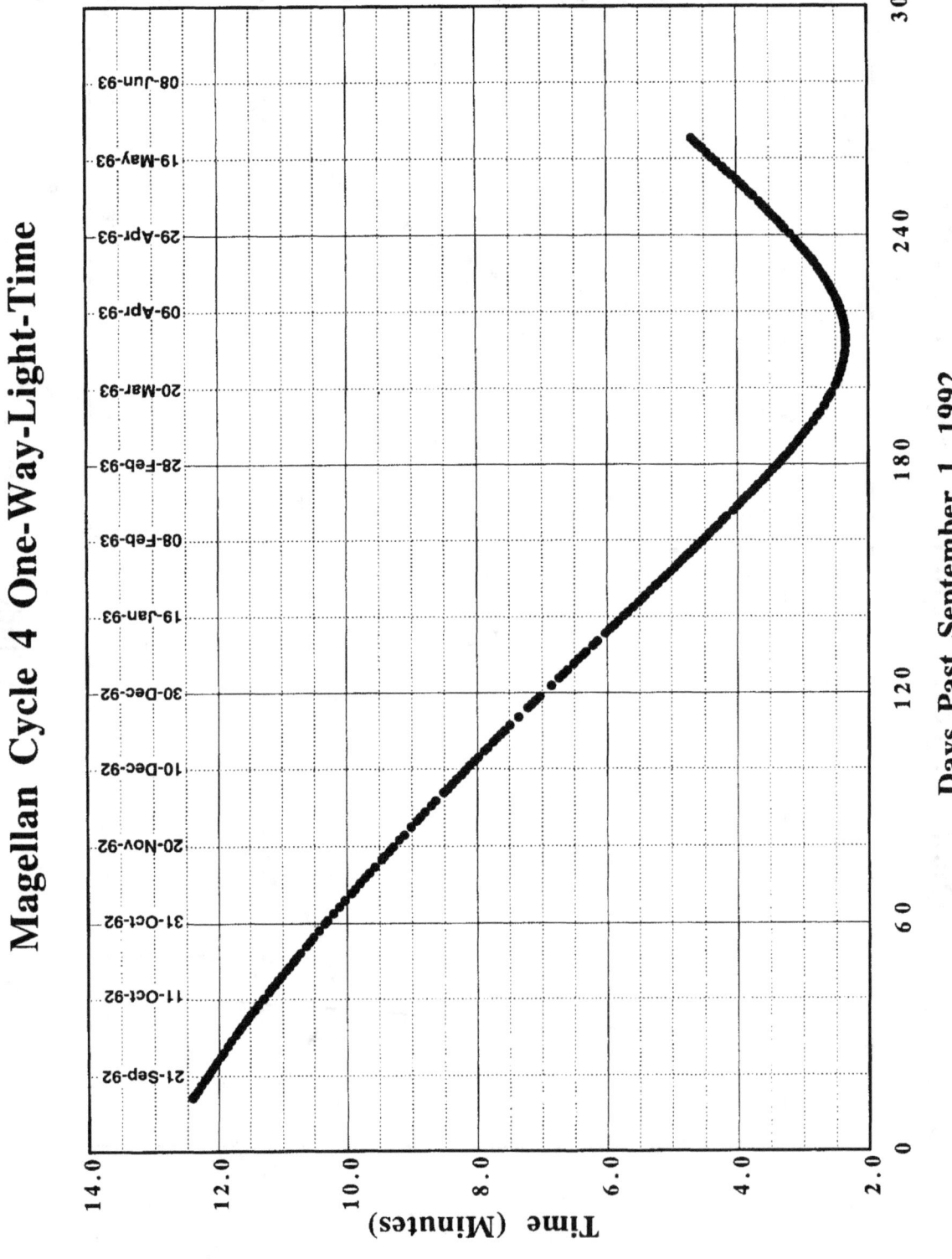

Magellan Cycle 4 Sun-Earth-Venus Angle

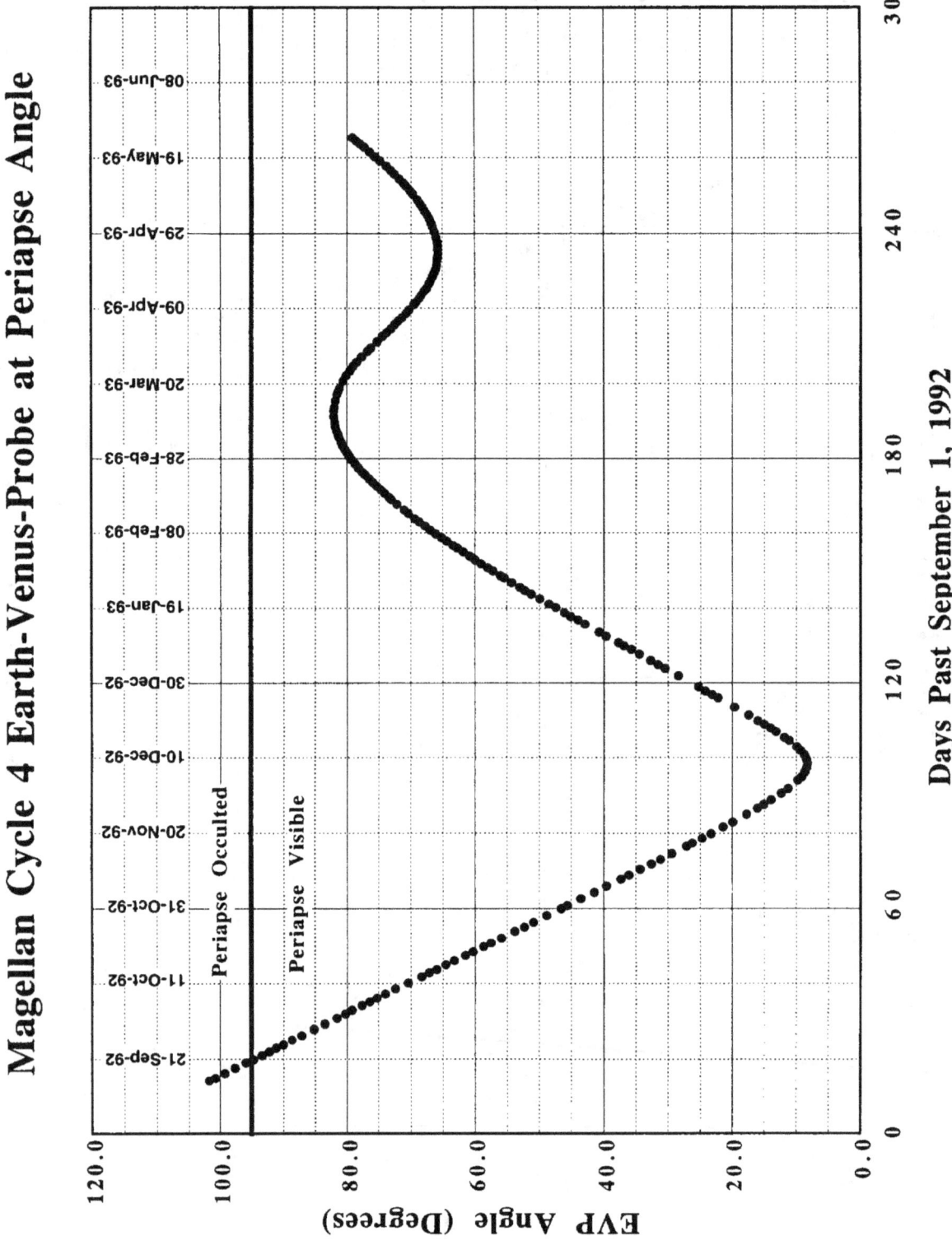

C-12

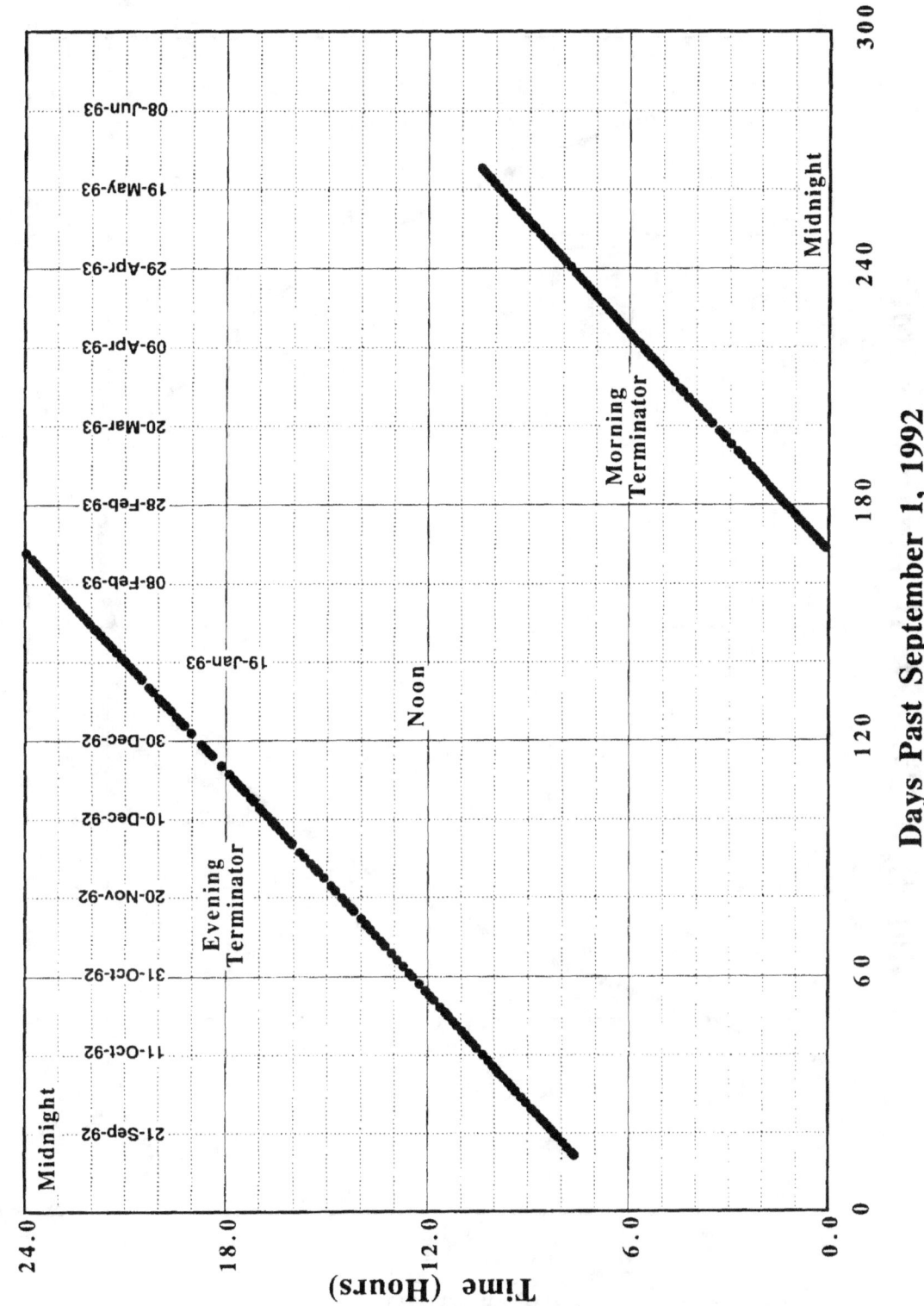

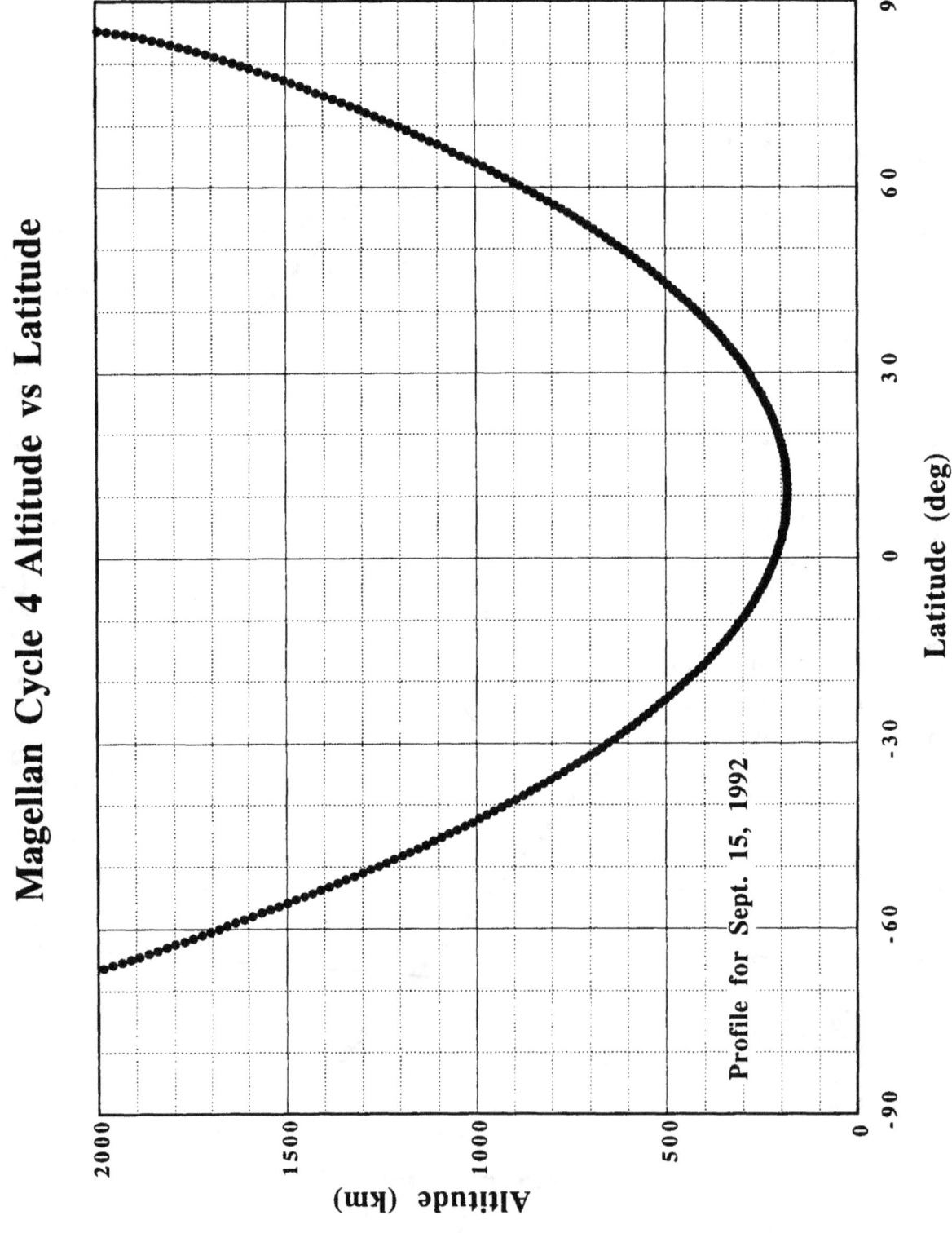

C-14

Appendix D

Magellan Cycle 5 and 6 (Circular) Information

The following plots are included in this appendix:

1. Semi-major axis
2. Eccentricity
3. Inclination
4. Latitude at periapse
5. Longitude at periapse
6. Altitude at periapse
7. Altitude at apoapse
8. Plane-of-sky inclination vs time
9. Plane-of-sky inclination vs longitude
10. One-way light time from Venus to Earth
11. Sun-Earth-Venus angle
12. Earth-Venus-Probe at periapse angle
13. Local solar time at periapse
14. Altitude vs. latitude profile

Magellan Cycle 5&6 Semi-Major Axis

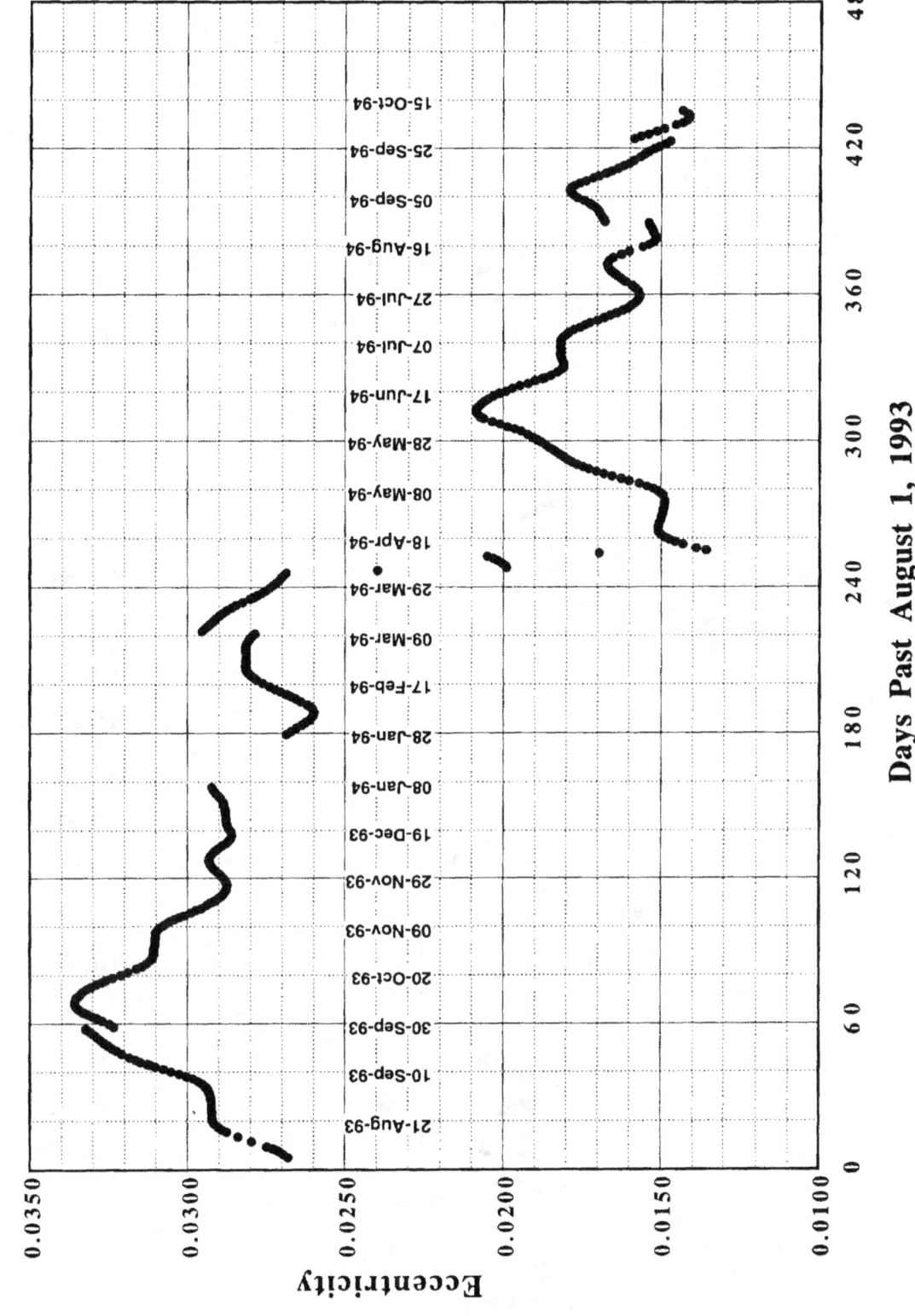

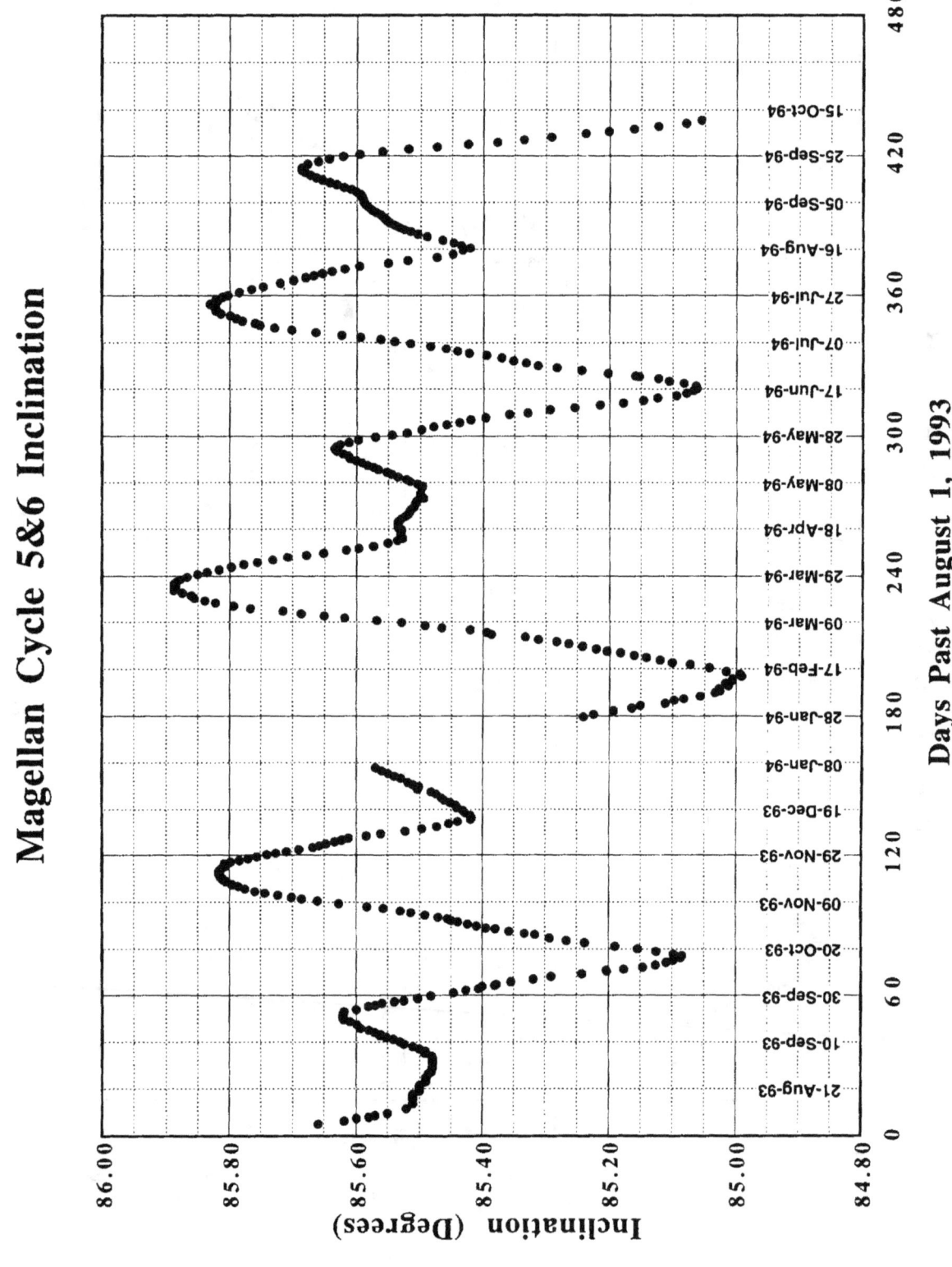

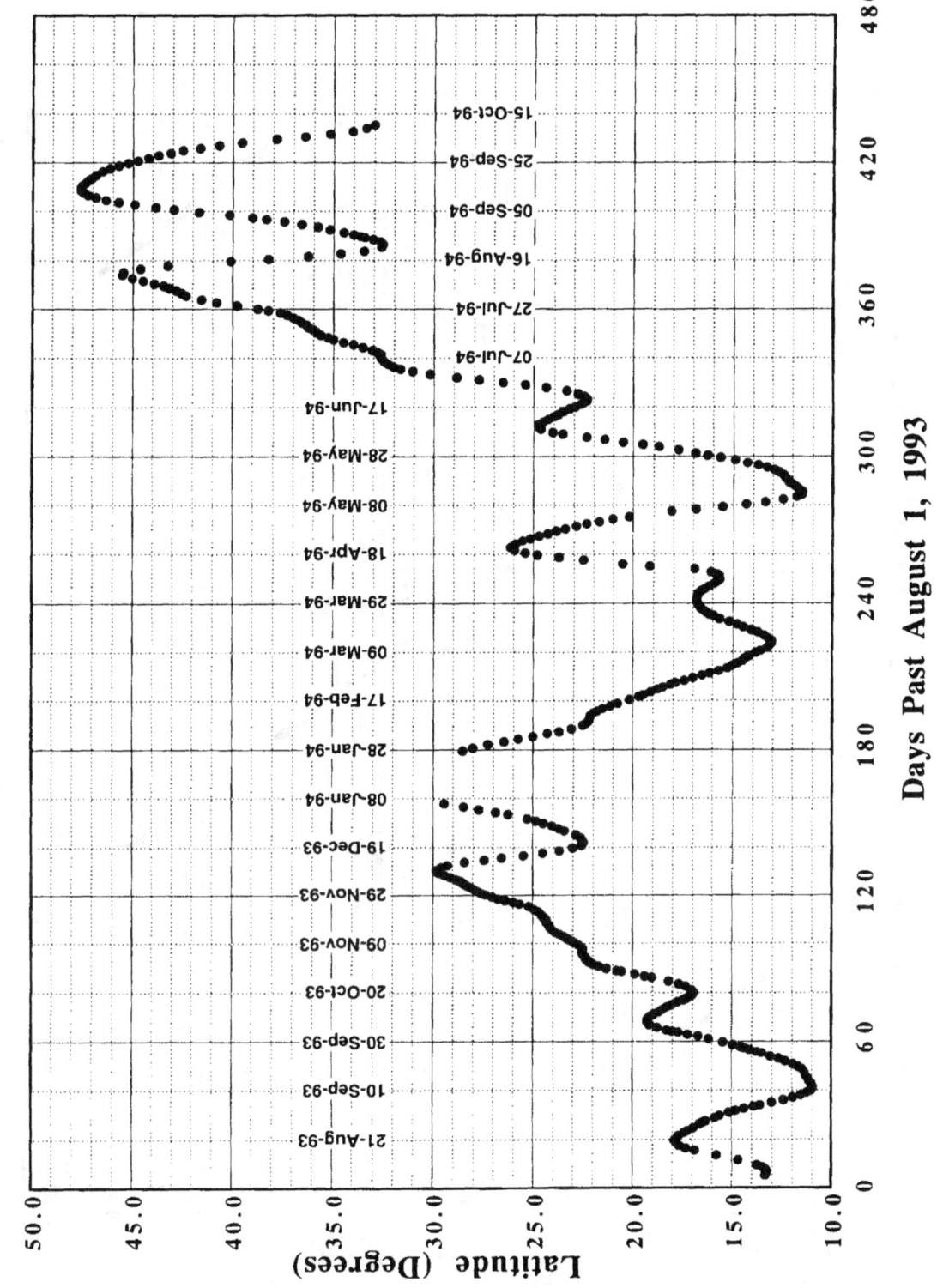

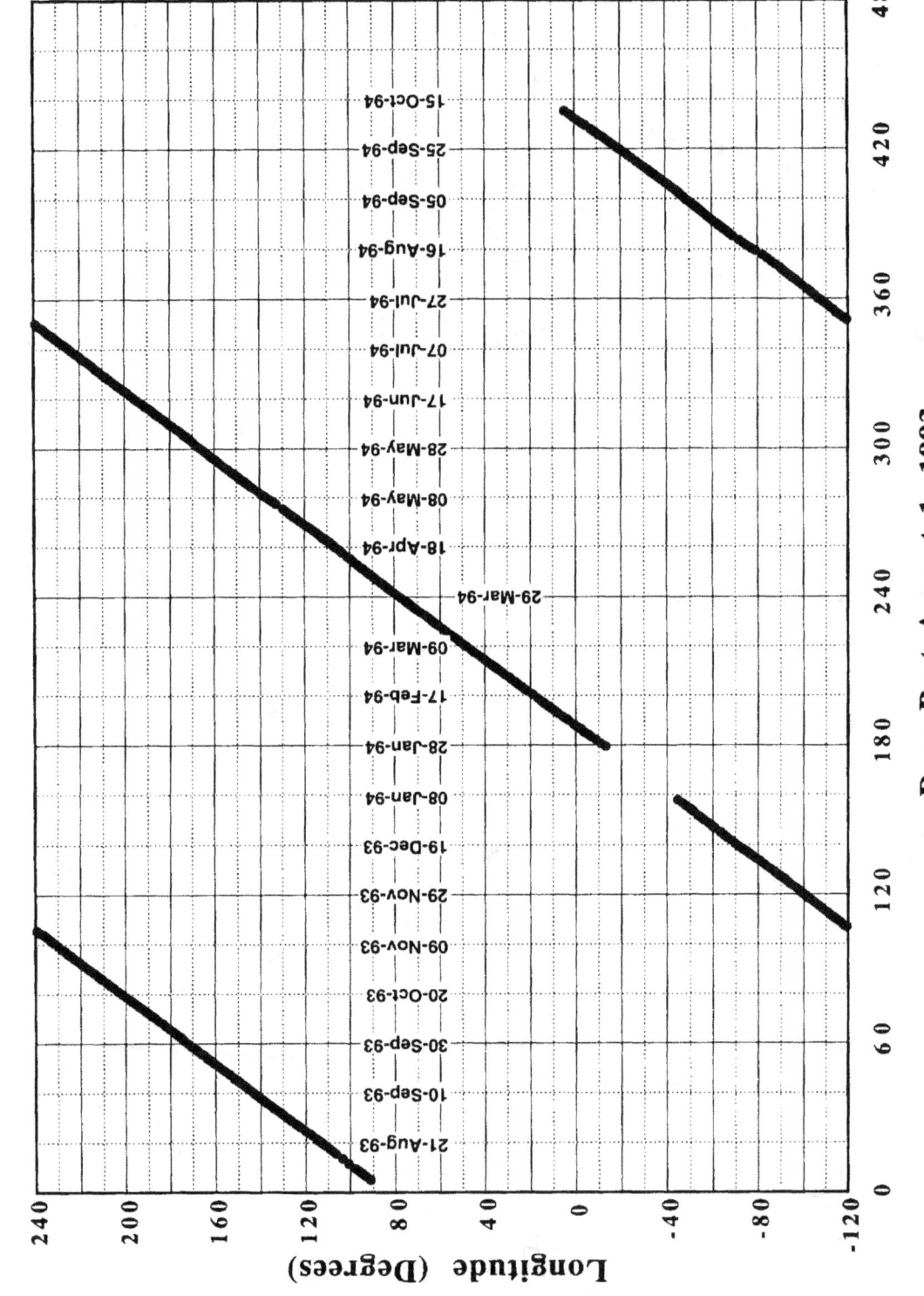

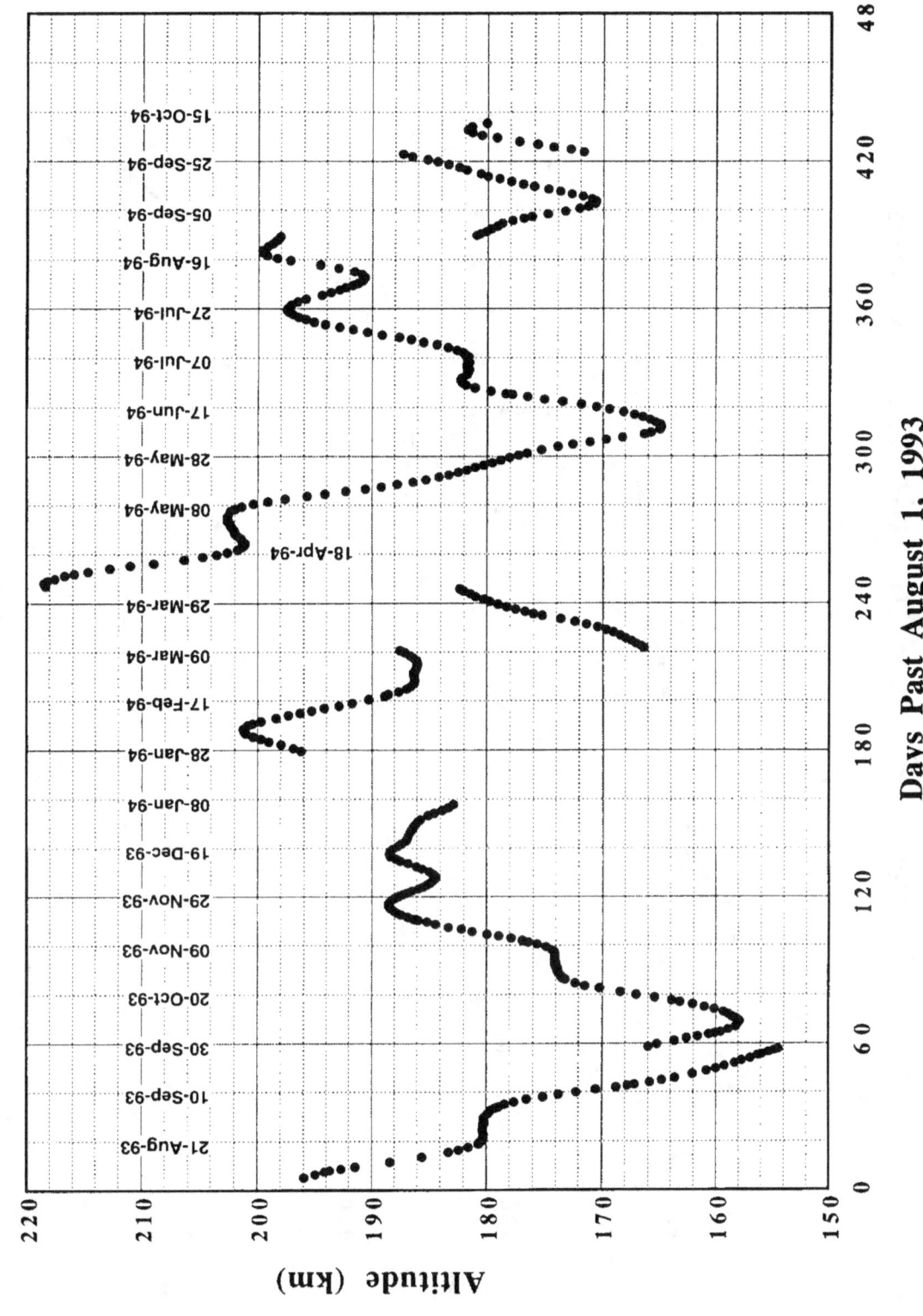

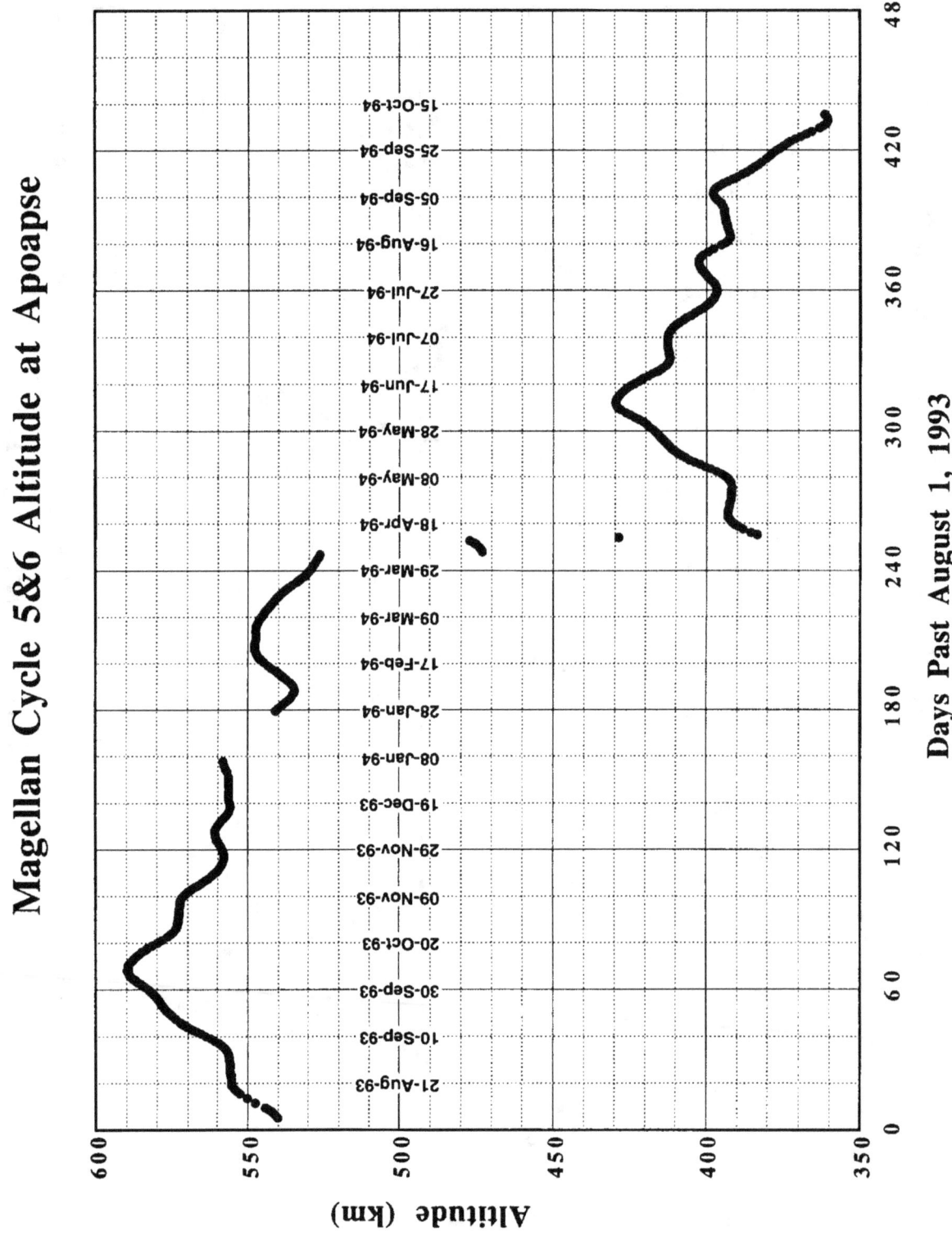

Magellan Cycle 5&6 Plane-of-Sky Inclination

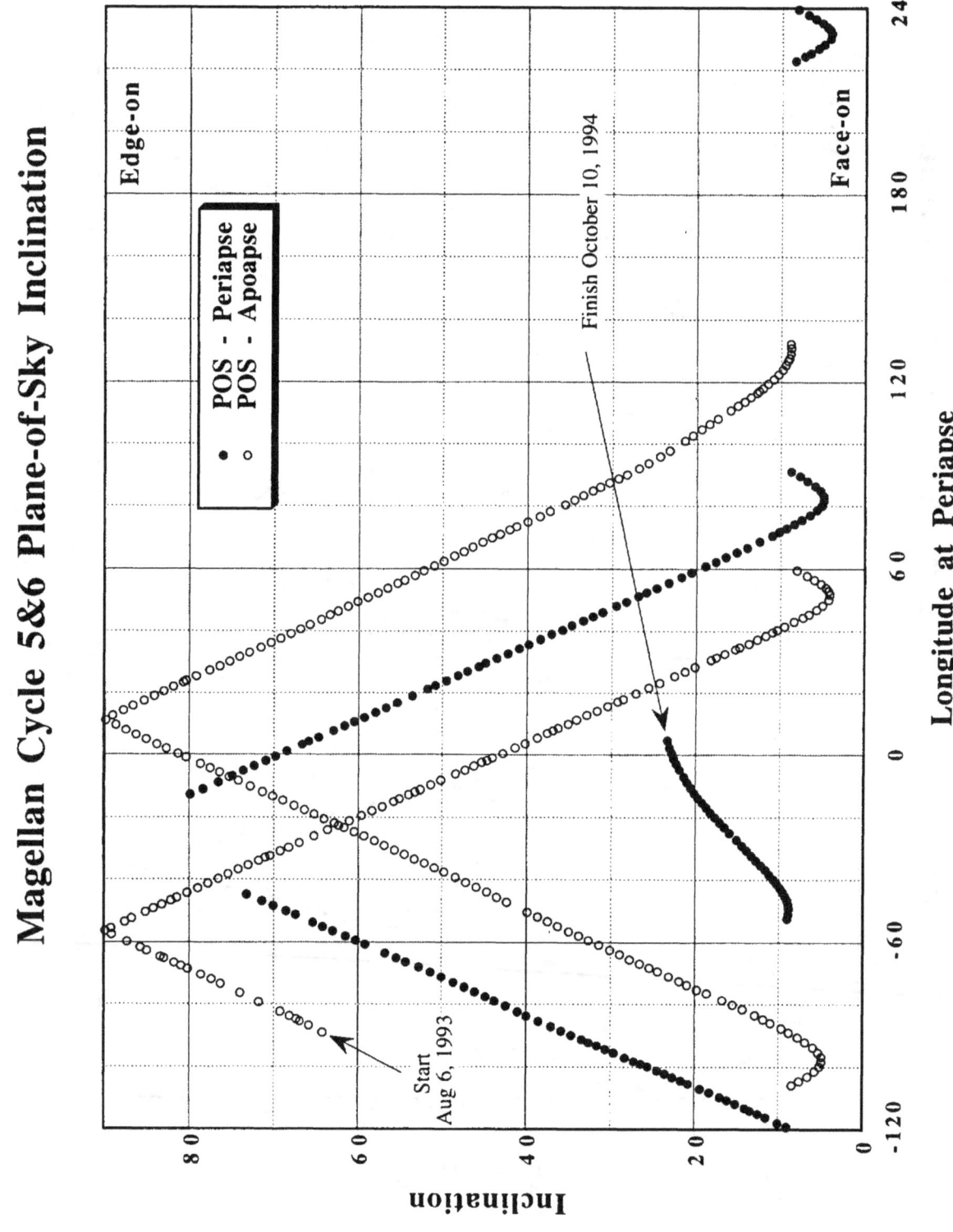

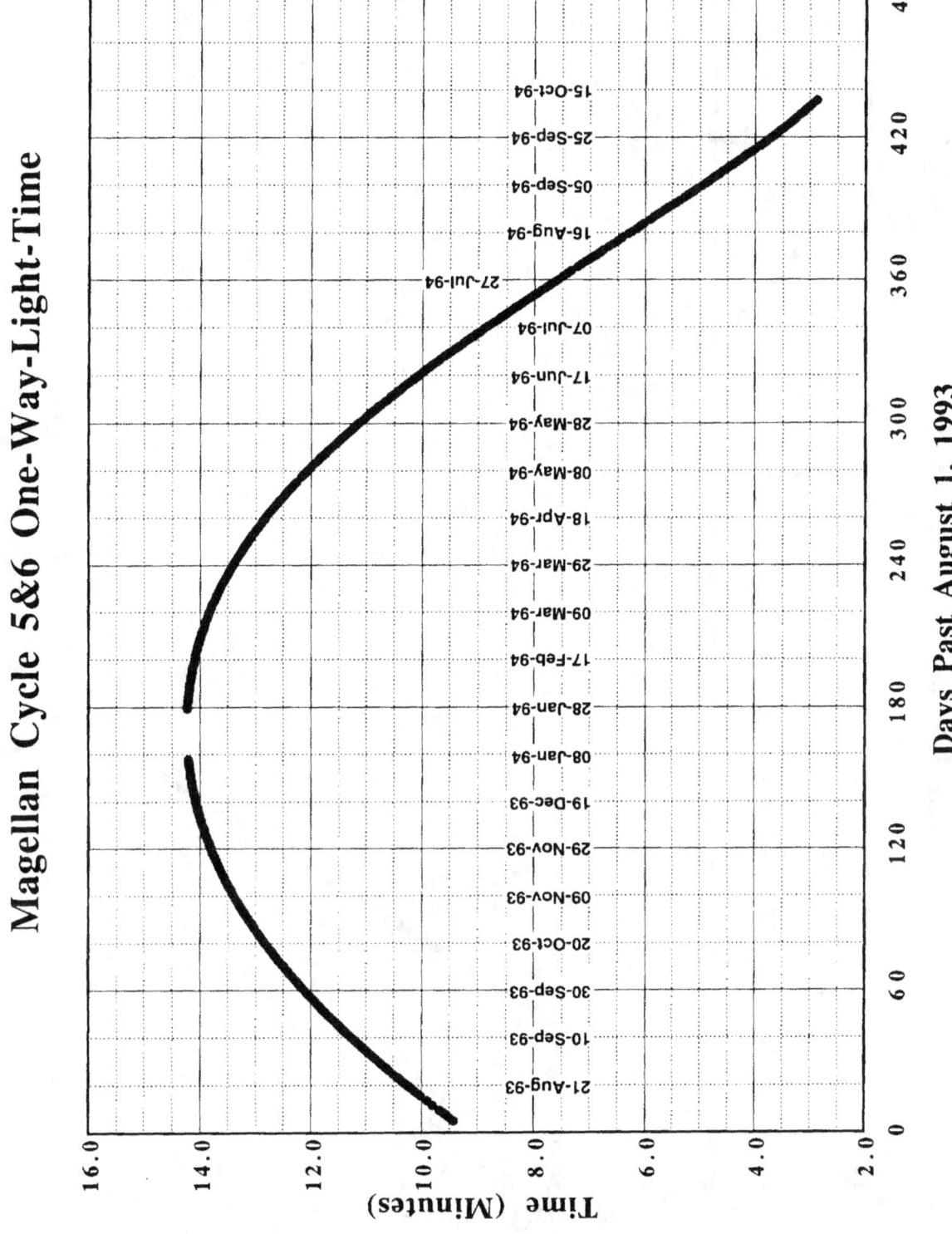

D-11

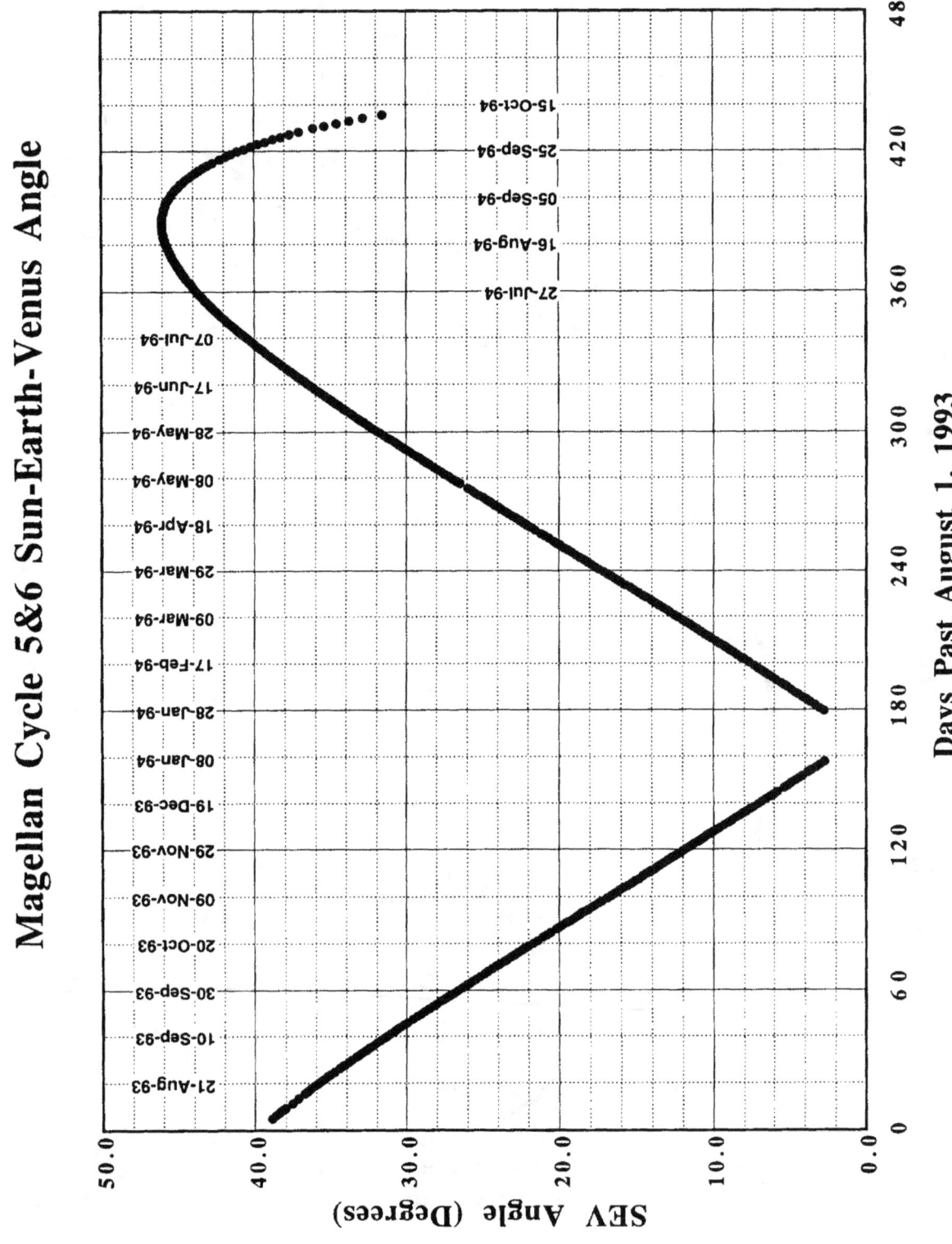

Magellan Cycle 5&6 Earth-Venus-Probe at Periapse Angle

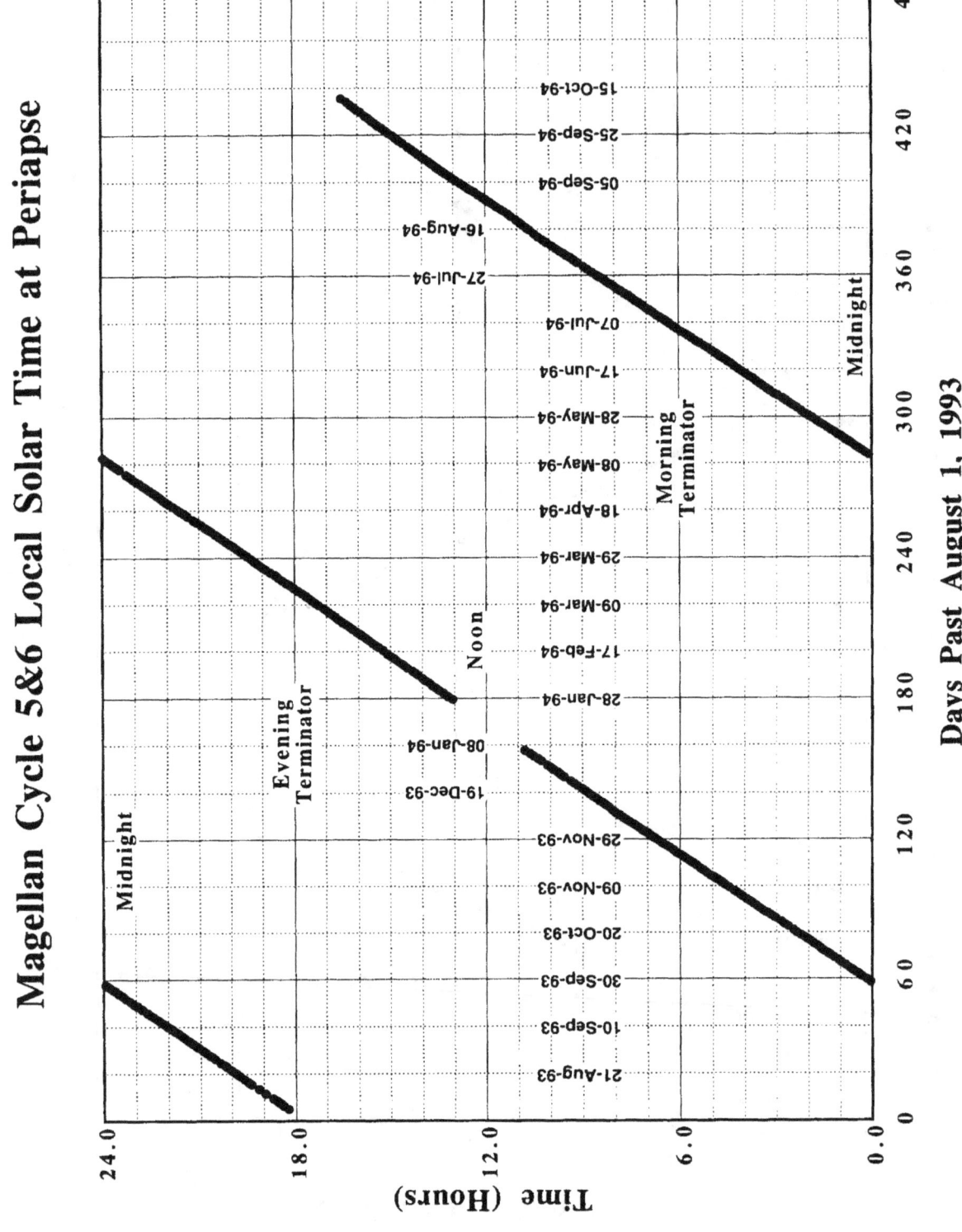

D-14

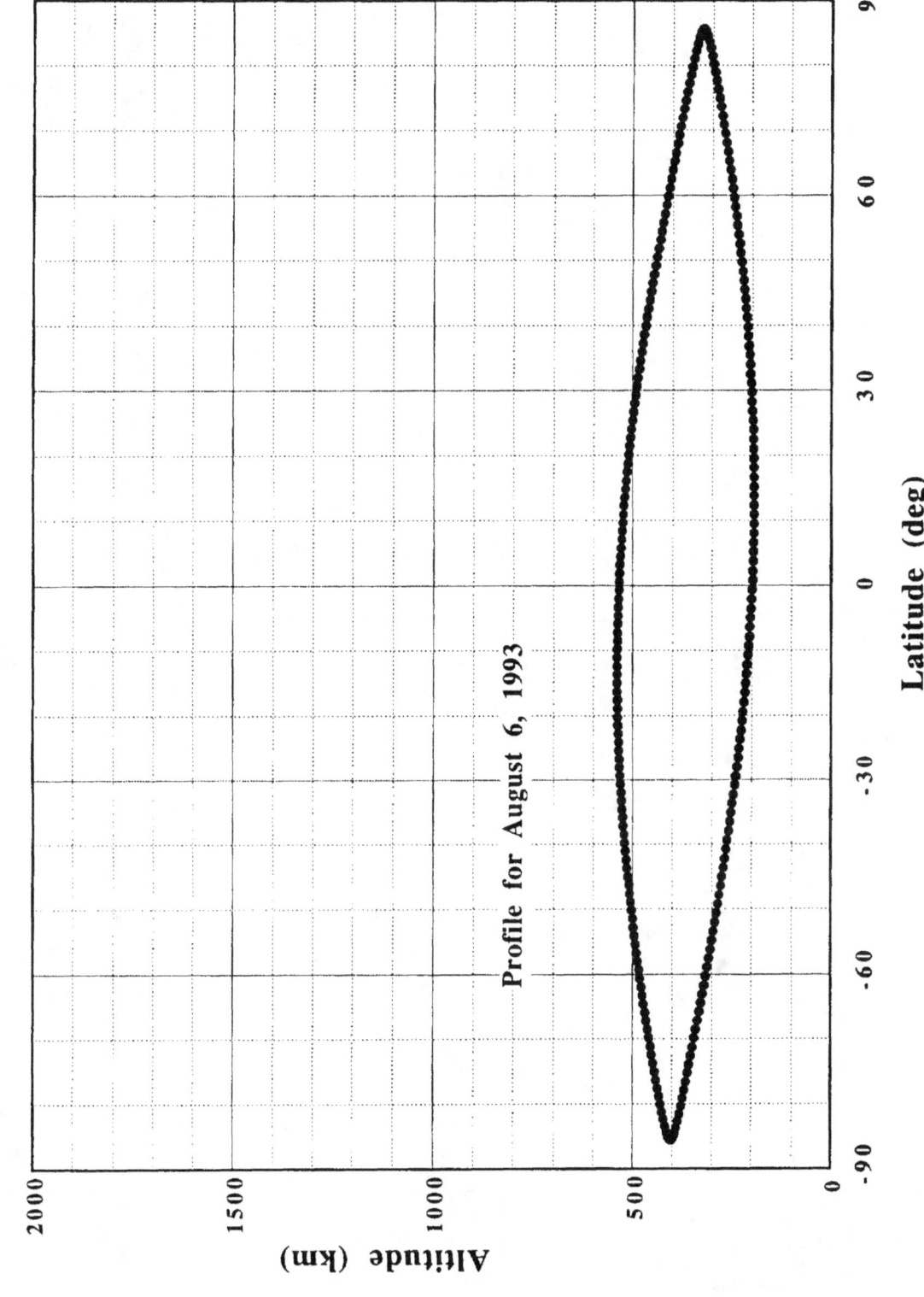

Appendix E

Magellan X-band Orbit Information

The columns contain the following information (in order):

1. Orbit number
2. Tracking data file name ("_2" is 2 second data and "_10" is 10 second data)
3. Number of Doppler observations
4. Tracking station number (15=Goldstone, 45=Canberra, 65=Madrid)
5. Time of first observation

5757	92258j261_10	878	65	15-SEP-1992 12:35:35	5830	92265j272_2	2	45	25-SEP-1992 09:14:30	5904	92279j286_2	51	45	5-OCT-1992 09:42:15
5758	92258j261_10	756	65	15-SEP-1992 15:50:05	5831	92265j272_2	1906	65	25-SEP-1992 13:21:35	5904	92279j286_2	1701	65	5-OCT-1992 10:21:59
5760	92258j261_10	836	45	15-SEP-1992 22:59:45	5833	92265j272_2	686	15	25-SEP-1992 20:14:08	5905	92279j286_2	2294	65	5-OCT-1992 12:56:45
5761	92258j261_10	872	45	16-SEP-1992 01:33:35	5834	92265j272_2	802	15	25-SEP-1992 22:50:30	5906	92279j286_2	90	15	5-OCT-1992 17:43:19
5762	92258j261_10	852	45	16-SEP-1992 04:48:05	5835	92265j272_2	1240	45	26-SEP-1992 02:09:30	5906	92279j286_2	1975	65	5-OCT-1992 16:11:15
5763	92258j261_10	268	45	16-SEP-1992 08:02:35	5836	92265j272_2	834	45	26-SEP-1992 05:19:30	5907	92279j286_2	1892	15	5-OCT-1992 19:27:30
5764	92258j261_10	338	65	16-SEP-1992 13:20:55	5837	92265j272_2	188	45	26-SEP-1992 08:34:30	5908	92279j286_2	1973	15	5-OCT-1992 22:41:30
5765	92258j261_10	707	65	16-SEP-1992 14:31:35	5837	92265j272_2	72	65	26-SEP-1992 10:29:30	5909	92279j286_2	2309	45	6-OCT-1992 01:54:35
5767	92258j261_10	2	45	17-SEP-1992 00:05:45	5838	92265j272_2	848	65	26-SEP-1992 11:48:30	5910	92279j286_2	2309	45	6-OCT-1992 05:09:05
5768	92258j261_10	10	45	17-SEP-1992 03:20:30	5839	92265j272_2	843	65	26-SEP-1992 15:03:30	5911	92279j286_2	1957	45	6-OCT-1992 08:23:35
5769	92258j261_10	847	45	17-SEP-1992 03:30:30	5840	92265j272_2	1001	15	26-SEP-1992 18:17:30	5911	92279j286_2	83	65	6-OCT-1992 10:24:35
5770	92258j261_10	740	45	17-SEP-1992 06:44:05	5841	92265j272_2	858	15	26-SEP-1992 21:32:30	5912	92279j286_2	2269	65	6-OCT-1992 11:38:05
5770	92258j261_10	2	65	17-SEP-1992 09:47:15	5842	92265j272_2	87	45	27-SEP-1992 02:25:30	5913	92279j286_2	2301	65	6-OCT-1992 14:52:35
5771	92261j262_10	830	65	17-SEP-1992 10:02:55	5843	92265j272_2	842	45	27-SEP-1992 04:01:30	5914	92279j286_2	50	65	6-OCT-1992 18:07:05
5772	92261j262_10	34	65	17-SEP-1992 13:13:05	5844	92265j272_2	770	45	27-SEP-1992 07:15:30	5916	92279j286_2	938	45	7-OCT-1992 01:40:55
5778	92261j262_10	59	65	18-SEP-1992 11:44:45	5845	92265j272_2	935	65	27-SEP-1992 10:32:47	5917	92279j286_2	2381	45	7-OCT-1992 03:50:25
5779	92262j266_2	872	65	18-SEP-1992 11:54:35	5846	92265j272_2	840	65	27-SEP-1992 13:44:30	5918	92279j286_2	2302	45	7-OCT-1992 07:04:55
5780	92262j266_2	36	65	18-SEP-1992 15:09:05	5847	92265j272_2	814	15	27-SEP-1992 17:24:30	5921	92279j286_2	1016	15	7-OCT-1992 17:58:27
5782	92262j266_2	60	45	19-SEP-1992 00:42:35	5848	92265j272_2	840	15	27-SEP-1992 20:13:30	5922	92279j286_2	2276	15	7-OCT-1992 20:02:45
5783	92262j266_2	879	45	19-SEP-1992 00:52:35	5849	92265j272_2	70	15	27-SEP-1992 23:27:30	5923	92279j286_2	2235	15	7-OCT-1992 23:17:15
5784	92262j266_2	288	45	19-SEP-1992 04:07:05	5851	92272j279_2	816	45	28-SEP-1992 06:20:30	5923	92279j286_2	35	45	8-OCT-1992 01:46:15
5785	92262j266_2	474	45	19-SEP-1992 07:21:35	5852	92272j279_2	7	45	28-SEP-1992 09:11:30	5924	92279j286_2	2294	45	8-OCT-1992 02:31:45
5785	92262j266_2	236	65	19-SEP-1992 09:43:45	5855	92272j279_2	997	15	28-SEP-1992 17:32:30	5925	92279j286_2	2301	45	8-OCT-1992 05:46:15
5786	92262j266_2	879	65	19-SEP-1992 10:36:05	5856	92272j279_2	816	15	28-SEP-1992 22:09:30	5926	92279j286_2	1063	45	8-OCT-1992 09:00:45
5787	92262j266_2	846	65	19-SEP-1992 13:50:35	5857	92272j279_2	21	15	29-SEP-1992 01:23:30	5926	92279j286_2	333	65	8-OCT-1992 10:30:15
5789	92262j266_2	570	15	19-SEP-1992 21:00:35	5859	92272j279_2	3	65	29-SEP-1992 10:22:30	5927	92279j286_2	2303	65	8-OCT-1992 12:15:05
5789	92262j266_2	261	45	19-SEP-1992 22:37:05	5860	92272j279_2	842	65	29-SEP-1992 11:07:30	5928	92279j286_2	2306	65	8-OCT-1992 15:29:35
5790	92262j266_2	866	45	19-SEP-1992 23:34:05	5861	92272j279_2	761	65	29-SEP-1992 14:21:30	5929	92279j286_2	2300	15	8-OCT-1992 18:44:05
5791	92262j266_2	866	45	20-SEP-1992 02:48:35	5862	92272j279_2	148	65	29-SEP-1992 17:36:30	5930	92279j286_2	974	15	8-OCT-1992 21:58:35
5792	92262j266_2	862	45	20-SEP-1992 06:03:05	5864	92272j279_2	9	45	30-SEP-1992 02:13:30	5934	92279j286_2	2304	65	9-OCT-1992 10:56:25
5793	92262j266_2	837	65	20-SEP-1992 09:58:35	5865	92272j279_2	824	45	30-SEP-1992 03:19:30	5935	92279j286_2	2305	65	9-OCT-1992 14:10:55
5794	92262j266_2	874	65	20-SEP-1992 12:32:05	5866	92272j279_2	914	45	30-SEP-1992 06:33:30	5936	92279j286_2	1964	15	9-OCT-1992 17:56:07
5795	92262j266_2	613	15	20-SEP-1992 17:06:05	5868	92272j279_2	804	65	30-SEP-1992 13:15:30	5936	92279j286_2	255	65	9-OCT-1992 17:25:25
5795	92262j266_2	265	65	20-SEP-1992 15:46:35	5869	92272j279_2	352	15	30-SEP-1992 17:36:01	5937	92279j286_2	2181	15	9-OCT-1992 20:02:55
5796	92262j266_2	881	15	20-SEP-1992 19:01:05	5870	92272j279_2	1907	15	30-SEP-1992 19:31:30	5938	92279j286_2	34	15	9-OCT-1992 23:17:25
5797	92262j266_2	37	15	20-SEP-1992 22:15:35	5871	92272j279_2	1823	15	30-SEP-1992 22:46:30	5938	92279j286_2	342	45	10-OCT-1992 01:24:05
5797	92262j266_2	775	45	20-SEP-1992 22:56:35	5872	92272j279_2	1894	45	1-OCT-1992 02:10:30	5939	92279j286_2	2162	45	10-OCT-1992 02:31:45
5799	92262j266_2	2068	45	21-SEP-1992 05:44:45	5873	92272j279_2	1899	45	1-OCT-1992 05:15:30	5940	92279j286_2	2192	45	10-OCT-1992 06:46:15
5800	92262j266_2	495	45	21-SEP-1992 07:59:05	5874	92272j279_2	1189	45	1-OCT-1992 08:29:30	5941	92279j286_2	41	45	10-OCT-1992 09:00:45
5800	92262j266_2	863	65	21-SEP-1992 09:48:09	5877	92272j279_2	2277	15	1-OCT-1992 18:27:21	5941	92279j286_2	1141	65	10-OCT-1992 10:35:27
5801	92262j266_2	1858	65	21-SEP-1992 11:13:35	5878	92272j279_2	2299	15	1-OCT-1992 21:27:30	5942	92279j286_2	2194	65	10-OCT-1992 12:15:15
5802	92262j266_2	1771	65	21-SEP-1992 14:28:05	5879	92272j279_2	1389	15	2-OCT-1992 01:04:05	5943	92279j286_2	5	65	10-OCT-1992 15:29:45
5804	92262j266_2	937	45	21-SEP-1992 22:39:23	5879	92272j279_2	540	45	2-OCT-1992 02:02:57	5946	92279j286_2	2154	45	11-OCT-1992 02:14:15
5805	92262j266_2	1603	45	22-SEP-1992 00:12:30	5880	92272j279_2	2300	45	2-OCT-1992 03:54:45	5947	92279j286_2	2169	45	11-OCT-1992 04:27:35
5806	92262j266_2	1907	45	22-SEP-1992 03:26:30	5881	92272j279_2	2267	45	2-OCT-1992 07:09:15	5948	92279j286_2	1742	45	11-OCT-1992 07:42:05
5807	92262j266_2	1856	45	22-SEP-1992 06:41:30	5882	92272j279_2	2304	65	2-OCT-1992 10:23:45	5948	92279j286_2	100	65	11-OCT-1992 10:39:55
5811	92265j272_2	1927	15	22-SEP-1992 20:26:30	5883	92272j279_2	2219	65	2-OCT-1992 13:38:15	5949	92279j286_2	2188	65	11-OCT-1992 10:56:35
5812	92265j272_2	1876	15	22-SEP-1992 22:53:30	5884	92272j279_2	1750	15	2-OCT-1992 17:31:43	5950	92279j286_2	453	65	11-OCT-1992 14:10:55
5815	92265j272_2	1644	65	23-SEP-1992 09:52:23	5885	92272j279_2	2108	15	2-OCT-1992 19:19:35	5951	92279j286_2	2132	15	11-OCT-1992 18:26:31
5816	92265j272_2	1936	65	23-SEP-1992 11:51:30	5886	92272j279_2	12	45	3-OCT-1992 01:54:35	5952	92279j286_2	2187	15	11-OCT-1992 20:39:55
5817	92265j272_2	78	15	23-SEP-1992 17:18:30	5887	92272j279_2	2296	45	3-OCT-1992 03:10:30	5953	92279j286_2	554	15	11-OCT-1992 23:54:25
5817	92265j272_2	1005	65	23-SEP-1992 15:06:30	5888	92272j279_2	2310	45	3-OCT-1992 05:50:45	5953	92279j286_2	1607	45	12-OCT-1992 01:13:27
5818	92265j272_2	1630	15	23-SEP-1992 18:20:30	5889	92272j279_2	449	45	3-OCT-1992 09:05:15	5954	92279j286_2	2183	45	12-OCT-1992 03:08:55
5819	92265j272_2	1661	15	23-SEP-1992 21:35:30	5889	92272j279_2	764	65	3-OCT-1992 10:16:41	5955	92279j293_2	2170	45	12-OCT-1992 06:23:15
5820	92265j272_2	6	15	24-SEP-1992 00:49:30	5890	92272j279_2	2297	65	3-OCT-1992 12:19:35	5956	92286j293_2	40	45	12-OCT-1992 09:37:45
5820	92265j272_2	581	45	24-SEP-1992 02:38:37	5891	92272j279_2	62	15	3-OCT-1992 17:44:35	5956	92286j293_2	2077	65	12-OCT-1992 10:40:53
5821	92265j272_2	1939	45	24-SEP-1992 04:04:30	5891	92272j279_2	127	65	3-OCT-1992 15:34:05	5957	92286j293_2	1934	65	12-OCT-1992 12:52:15
5822	92265j272_2	1568	45	24-SEP-1992 07:18:30	5892	92272j279_2	2300	15	3-OCT-1992 18:48:35	5958	92286j293_2	468	15	12-OCT-1992 18:03:37
5822	92265j272_2	18	65	24-SEP-1992 10:00:30	5893	92272j279_2	1923	15	3-OCT-1992 22:03:05	5959	92286j293_2	2174	15	12-OCT-1992 19:21:15
5823	92265j272_2	1799	65	24-SEP-1992 10:33:30	5894	92272j279_2	1692	45	4-OCT-1992 01:57:23	5960	92286j293_2	1992	15	12-OCT-1992 22:35:35
5824	92265j272_2	1914	65	24-SEP-1992 13:47:30	5895	92272j279_2	2302	45	4-OCT-1992 04:32:05	5960	92286j293_2	136	45	13-OCT-1992 01:13:15
5825	92265j272_2	1893	15	24-SEP-1992 17:48:30	5896	92272j279_2	2154	45	4-OCT-1992 07:46:25	5961	92286j293_2	2097	45	13-OCT-1992 01:50:05
5825	92265j272_2	8	65	24-SEP-1992 17:02:30	5896	92272j279_2	10	65	4-OCT-1992 10:19:35	5962	92286j293_2	2176	45	13-OCT-1992 05:04:35
5826	92265j272_2	1822	15	24-SEP-1992 20:16:30	5897	92272j279_2	2306	65	4-OCT-1992 11:00:55	5963	92286j293_2	1385	45	13-OCT-1992 08:19:05
5826	92265j272_2	108	45	24-SEP-1992 22:27:43	5898	92272j279_2	2040	65	4-OCT-1992 14:15:25	5963	92286j293_2	241	65	13-OCT-1992 10:43:55
5827	92265j272_2	1888	45	24-SEP-1992 23:31:30	5901	92272j279_2	241	45	5-OCT-1992 01:43:55	5964	92286j293_2	2149	65	13-OCT-1992 11:33:35
5828	92265j272_2	1902	45	25-SEP-1992 02:45:30	5902	92272j279_2	2300	45	5-OCT-1992 03:13:25	5965	92286j293_2	40	65	13-OCT-1992 14:47:55
5829	92265j272_2	1903	45	25-SEP-1992 05:59:30	5903	92272j286_2	2302	45	5-OCT-1992 06:27:45	5966	92286j293_2	2086	15	13-OCT-1992 19:03:25

5967	92286j293_2	877	15	13-OCT-1992 21:16:55	6093	92300j307_2	2054	45	30-OCT-1992 22:38:15	6254	92321j328_2	1825	65	21-NOV-1992 16:20:45
5970	92286j293_2	1899	45	14-OCT-1992 08:01:17	6094	92300j307_2	2198	45	31-OCT-1992 01:04:45	6256	92321j328_2	1835	45	21-NOV-1992 22:49:25
5971	92286j293_2	1885	65	14-OCT-1992 11:23:29	6095	92300j307_2	68	45	31-OCT-1992 04:19:15	6257	92321j328_2	1833	45	22-NOV-1992 02:03:45
5972	92286j293_2	1576	65	14-OCT-1992 13:29:15	6097	92300j307_2	2046	45	31-OCT-1992 11:35:55	6258	92321j328_2	1787	45	22-NOV-1992 05:18:05
5973	92286j293_2	1681	15	14-OCT-1992 17:59:45	6098	92300j307_2	2193	65	31-OCT-1992 14:02:35	6259	92321j328_2	1836	45	22-NOV-1992 08:32:35
5974	92286j293_2	2170	15	14-OCT-1992 19:58:05	6099	92300j307_2	1186	15	31-OCT-1992 18:37:47	6264	92321j328_2	1834	45	23-NOV-1992 00:44:25
5975	92286j293_2	39	15	14-OCT-1992 23:12:35	6099	92300j307_2	136	65	31-OCT-1992 17:16:55	6265	92321j335_2	1836	45	23-NOV-1992 03:58:45
5975	92286j293_2	100	45	15-OCT-1992 02:10:25	6100	92300j307_2	1866	15	31-OCT-1992 20:31:25	6268	92321j335_2	1825	65	23-NOV-1992 13:41:55
5976	92286j293_2	1838	45	15-OCT-1992 02:27:05	6101	92300j307_2	64	45	1-NOV-1992 01:31:23	6269	92321j335_2	1829	65	23-NOV-1992 16:56:15
5977	92286j293_2	1965	45	15-OCT-1992 05:41:35	6102	92300j307_2	2142	45	1-NOV-1992 03:00:15	6271	92328j335_2	1828	45	23-NOV-1992 23:24:55
5978	92286j293_2	39	45	15-OCT-1992 08:56:05	6104	92300j307_2	175	65	1-NOV-1992 11:32:15	6272	92328j335_2	1821	45	24-NOV-1992 02:39:25
5979	92286j293_2	2099	65	15-OCT-1992 13:11:25	6105	92300j307_2	1408	65	1-NOV-1992 12:43:25	6273	92328j335_2	1831	45	24-NOV-1992 05:53:45
5981	92286j293_2	101	15	15-OCT-1992 21:37:05	6112	92300j314_2	1866	65	2-NOV-1992 12:12:15	6274	92328j335_2	1828	45	24-NOV-1992 09:08:15
5982	92286j293_2	803	15	15-OCT-1992 21:53:55	6113	92300j314_2	1934	65	2-NOV-1992 14:38:15	6277	92328j335_2	1221	15	24-NOV-1992 19:12:45
5982	92286j293_2	59	45	16-OCT-1992 00:58:35	6116	92307j314_2	987	15	3-NOV-1992 00:22:15	6278	92328j335_2	550	45	24-NOV-1992 22:48:25
5983	92286j293_2	2134	45	16-OCT-1992 01:08:15	6124	92307j314_2	39	45	4-NOV-1992 04:16:35	6279	92328j335_2	1778	45	25-NOV-1992 01:19:55
5984	92286j293_2	2152	45	16-OCT-1992 04:22:45	6125	92307j314_2	1933	45	4-NOV-1992 05:31:55	6280	92328j335_2	1807	45	25-NOV-1992 04:34:25
5985	92286j293_2	1996	45	16-OCT-1992 07:37:15	6127	92307j314_2	30	65	4-NOV-1992 14:01:25	6283	92328j335_2	1830	65	25-NOV-1992 14:17:25
5986	92286j293_2	752	65	16-OCT-1992 12:38:05	6128	92307j314_2	1932	65	4-NOV-1992 15:15:05	6284	92328j335_2	1272	45	25-NOV-1992 17:31:55
5987	92286j293_2	92	65	16-OCT-1992 14:06:05	6130	92307j314_2	1886	45	4-NOV-1992 22:31:45	6286	92328j335_2	1819	45	26-NOV-1992 00:00:45
5990	92286j293_2	2032	45	17-OCT-1992 00:52:59	6131	92307j314_2	1937	45	5-NOV-1992 00:58:25	6287	92328j335_2	1832	45	26-NOV-1992 03:15:05
5991	92286j293_2	2148	45	17-OCT-1992 03:03:55	6132	92307j314_2	1940	45	5-NOV-1992 04:12:15	6290	92328j335_2	1674	65	26-NOV-1992 12:58:15
5992	92286j293_2	2150	45	17-OCT-1992 06:18:25	6133	92307j314_2	34	45	5-NOV-1992 07:27:25	6291	92328j335_2	1828	65	26-NOV-1992 16:12:35
5993	92286j293_2	39	45	17-OCT-1992 09:32:55	6136	92307j314_2	528	15	5-NOV-1992 18:47:33	6292	92328j335_2	1832	15	26-NOV-1992 19:26:45
5995	92286j293_2	361	15	17-OCT-1992 18:06:25	6137	92307j314_2	1266	15	5-NOV-1992 20:24:45	6293	92328j335_2	1829	15	26-NOV-1992 22:41:05
5996	92286j293_2	2145	15	17-OCT-1992 19:16:15	6137	92307j314_2	36	45	5-NOV-1992 22:24:45	6294	92328j335_2	728	15	27-NOV-1992 01:55:35
5997	92286j293_2	2134	15	17-OCT-1992 22:30:45	6138	92307j314_2	1934	45	5-NOV-1992 23:39:15	6295	92328j335_2	1806	45	27-NOV-1992 05:09:55
5998	92286j293_2	39	15	18-OCT-1992 01:45:15	6139	92307j314_2	1937	45	6-NOV-1992 02:53:45	6297	92328j335_2	1009	65	27-NOV-1992 12:06:53
6000	92286j293_2	101	65	18-OCT-1992 11:11:45	6140	92307j314_2	2009	45	6-NOV-1992 06:08:15	6298	92328j335_2	1824	65	27-NOV-1992 14:53:05
6001	92286j293_2	2142	65	18-OCT-1992 11:28:35	6142	92307j314_2	1941	65	6-NOV-1992 12:36:55	6299	92328j335_2	882	65	27-NOV-1992 18:07:25
6002	92286j293_2	2127	65	18-OCT-1992 14:43:05	6143	92307j314_2	1901	45	6-NOV-1992 15:51:15	6301	92328j335_2	1825	45	28-NOV-1992 00:36:15
6003	92286j293_2	2103	15	18-OCT-1992 18:58:13	6144	92307j314_2	1659	15	6-NOV-1992 19:05:45	6302	92328j335_2	1832	45	28-NOV-1992 03:50:35
6003	92286j293_2	38	65	18-OCT-1992 17:57:25	6145	92307j314_2	1510	45	6-NOV-1992 23:10:55	6303	92328j335_2	1830	45	28-NOV-1992 07:05:05
6004	92286j293_2	38	15	18-OCT-1992 21:11:55	6146	92307j314_2	1604	45	7-NOV-1992 02:25:25	6304	92328j335_2	1801	45	28-NOV-1992 10:19:25
6005	92286j293_2	2100	15	19-OCT-1992 01:27:07	6147	92307j314_2	1606	45	7-NOV-1992 05:39:45	6307	92328j335_2	1805	15	28-NOV-1992 20:02:15
6006	92286j293_2	2122	45	19-OCT-1992 03:40:57	6149	92307j314_2	32	65	7-NOV-1992 12:17:43	6308	92328j335_2	1598	45	28-NOV-1992 23:16:35
6007	92293j300_2	2103	45	19-OCT-1992 07:00:05	6150	92307j314_2	1284	65	7-NOV-1992 15:27:49	6309	92328j335_2	1834	45	29-NOV-1992 02:30:55
6008	92293j300_2	37	45	19-OCT-1992 10:09:45	6151	92307j314_2	1246	15	7-NOV-1992 18:51:15	6310	92328j335_2	1817	45	29-NOV-1992 05:45:25
6008	92293j300_2	2084	15	19-OCT-1992 11:10:27	6152	92307j314_2	198	15	7-NOV-1992 21:51:55	6311	92328j335_2	1833	45	29-NOV-1992 08:59:45
6009	92293j300_2	1741	65	19-OCT-1992 13:24:15	6152	92307j314_2	593	45	7-NOV-1992 22:27:09	6314	92328j335_2	890	15	29-NOV-1992 19:14:55
6010	92293j300_2	940	15	19-OCT-1992 18:17:43	6153	92307j314_2	1592	45	8-NOV-1992 01:06:25	6315	92328j335_2	79	45	29-NOV-1992 22:56:29
6011	92293j300_2	2129	15	19-OCT-1992 19:53:05	6154	92307j314_2	1606	45	8-NOV-1992 04:20:45	6316	92328j335_2	1837	45	30-NOV-1992 01:11:25
6012	92293j300_2	1996	45	19-OCT-1992 23:07:35	6155	92307j314_2	1601	45	8-NOV-1992 07:35:15	6320	92335j342_2	1539	65	30-NOV-1992 14:08:55
6013	92293j300_2	233	45	20-OCT-1992 04:47:45	6156	92307j314_2	15	45	8-NOV-1992 10:49:35	6321	92335j342_2	1829	65	30-NOV-1992 17:23:15
6014	92293j300_2	2069	45	20-OCT-1992 05:36:25	6159	92307j314_2	1609	15	8-NOV-1992 20:32:45	6322	92335j342_2	1824	15	30-NOV-1992 20:37:35
6015	92293j300_2	38	45	20-OCT-1992 08:50:55	6160	92307j314_2	1607	45	8-NOV-1992 23:47:05	6323	92335j342_2	1781	15	30-NOV-1992 23:52:05
6018	92293j300_2	2085	15	20-OCT-1992 19:34:53	6161	92307j314_2	1608	45	9-NOV-1992 03:01:35	6326	92335j342_2	1825	45	1-DEC-1992 09:35:25
6019	92293j300_2	546	15	20-OCT-1992 21:48:45	6162	92307j321_2	1829	45	9-NOV-1992 06:15:55	6329	92335j342_2	1810	15	1-DEC-1992 19:18:05
6019	92293j300_2	102	15	21-OCT-1992 00:46:15	6163	92307j321_2	1861	45	9-NOV-1992 09:30:25	6330	92335j342_2	1043	15	1-DEC-1992 22:59:29
6020	92293j300_2	1288	15	21-OCT-1992 01:03:15	6164	92307j321_2	1854	65	9-NOV-1992 12:44:55	6331	92335j342_2	1821	45	2-DEC-1992 01:46:45
6022	92293j300_2	1973	45	21-OCT-1992 08:32:45	6170	92314j321_2	1871	45	10-NOV-1992 08:11:15	6332	92335j342_2	1830	45	2-DEC-1992 05:01:15
6023	92293j300_2	2074	65	21-OCT-1992 11:47:09	6171	92314j321_2	1167	65	10-NOV-1992 11:50:19	6335	92335j342_2	1812	65	2-DEC-1992 14:44:15
6024	92293j300_2	30	65	21-OCT-1992 14:00:55	6172	92314j321_2	1857	65	10-NOV-1992 14:40:05	6336	92335j342_2	1478	65	2-DEC-1992 17:58:35
6025	92293j300_2	2033	15	21-OCT-1992 18:17:37	6173	92314j321_2	495	65	10-NOV-1992 17:54:25	6343	92335j342_2	1529	45	3-DEC-1992 16:39:05
6026	92293j300_2	2117	65	21-OCT-1992 20:29:55	6175	92314j321_2	1869	45	11-NOV-1992 00:23:15	6345	92335j342_2	1957	45	3-DEC-1992 23:07:45
6027	92293j300_2	1792	15	21-OCT-1992 23:44:15	6176	92314j321_2	1845	45	11-NOV-1992 03:37:45	6346	92335j342_2	823	15	4-DEC-1992 02:22:05
6028	92293j300_2	1048	45	22-OCT-1992 04:33:29	6177	92314j321_2	1866	45	11-NOV-1992 06:52:15	6349	92335j342_2	1768	65	4-DEC-1992 12:07:19
6029	92293j300_2	1982	45	22-OCT-1992 06:13:15	6178	92314j321_2	1705	45	11-NOV-1992 10:06:45	6350	92335j342_2	1820	65	4-DEC-1992 15:19:35
6030	92293j300_2	1079	45	22-OCT-1992 11:08:13	6181	92314j321_2	1858	15	11-NOV-1992 19:49:45	6352	92335j342_2	1770	15	4-DEC-1992 21:48:15
6031	92293j300_2	1249	65	22-OCT-1992 12:42:05	6182	92314j321_2	1867	45	11-NOV-1992 23:04:05	6353	92335j342_2	1833	45	5-DEC-1992 01:02:25
6034	92293j300_2	351	45	23-OCT-1992 00:31:55	6183	92314j321_2	1868	45	12-NOV-1992 02:18:25	6354	92335j342_2	1832	45	5-DEC-1992 04:16:55
6035	92293j300_2	2310	45	23-OCT-1992 01:39:55	6184	92314j321_2	1862	45	12-NOV-1992 05:32:55	6355	92335j342_2	1528	45	5-DEC-1992 07:31:15
6036	92293j300_2	2204	45	23-OCT-1992 04:54:45	6185	92314j321_2	1868	45	12-NOV-1992 08:47:15	6356	92335j342_2	1335	45	5-DEC-1992 10:45:35
6037	92293j300_2	1955	45	23-OCT-1992 08:08:45	6186	92314j321_2	1868	65	12-NOV-1992 12:01:35	6360	92335j342_2	1576	45	5-DEC-1992 23:52:45
6037	92293j300_2	78	65	23-OCT-1992 11:10:15	6187	92314j321_2	1863	65	12-NOV-1992 15:16:05	6361	92335j342_2	1831	45	6-DEC-1992 02:57:15
6038	92293j300_2	2215	65	23-OCT-1992 11:23:15	6189	92314j321_2	421	45	12-NOV-1992 22:34:47	6364	92335j342_2	1827	65	6-DEC-1992 12:40:25
6039	92293j300_2	2209	65	23-OCT-1992 14:37:45	6190	92314j321_2	1868	45	13-NOV-1992 00:59:15	6365	92335j342_2	1825	65	6-DEC-1992 15:55:05
6040	92293j300_2	37	65	23-OCT-1992 17:52:05	6191	92314j321_2	1838	45	13-NOV-1992 04:13:35	6366	92335j342_2	755	15	6-DEC-1992 19:15:07
6042	92293j300_2	35	45	24-OCT-1992 00:34:15	6192	92314j321_2	1865	45	13-NOV-1992 07:28:05	6370	92335j349_2	1831	45	7-DEC-1992 08:06:25
6043	92293j300_2	1538	45	24-OCT-1992 04:50:13	6193	92314j321_2	54	45	13-NOV-1992 10:42:25	6371	92335j349_2	247	15	7-DEC-1992 11:20:45
6044	92293j300_2	25	45	24-OCT-1992 08:56:15	6196	92314j321_2	877	15	13-NOV-1992 21:00:13	6371	92335j349_2	475	65	7-DEC-1992 12:07:29
6048	92293j300_2	1822	15	24-OCT-1992 21:00:55	6197	92314j321_2	1868	45	13-NOV-1992 23:40:05	6372	92335j349_2	1835	65	7-DEC-1992 14:35:15
6049	92293j300_2	140	15	24-OCT-1992 23:15:25	6198	92314j321_2	1868	45	14-NOV-1992 02:54:25	6373	92335j349_2	1829	65	7-DEC-1992 17:49:25
6049	92293j300_2	1978	45	25-OCT-1992 00:07:15	6199	92314j321_2	1865	45	14-NOV-1992 06:08:45	6375	92335j349_2	1812	45	8-DEC-1992 00:18:05
6052	92293j300_2	36	45	25-OCT-1992 10:17:23	6201	92314j321_2	1861	45	14-NOV-1992 12:37:35	6379	92335j349_2	1836	45	8-DEC-1992 13:15:25
6053	92293j300_2	1666	65	25-OCT-1992 13:17:35	6203	92314j321_2	1873	15	14-NOV-1992 19:06:25	6380	92335j349_2	1896	65	8-DEC-1992 16:29:45
6054	92293j300_2	2124	65	25-OCT-1992 15:27:35	6204	92314j321_2	87	15	14-NOV-1992 22:20:45	6382	92335j349_2	1432	45	8-DEC-1992 23:12:51
6055	92293j300_2	2148	15	25-OCT-1992 18:41:55	6204	92314j321_2	1484	45	14-NOV-1992 22:35:09	6383	92335j349_2	1830	45	9-DEC-1992 02:12:45
6056	92293j300_2	72	45	26-OCT-1992 00:17:25	6205	92314j321_2	1865	45	15-NOV-1992 01:35:15	6384	92335j349_2	830	45	9-DEC-1992 05:27:15
6057	92293j300_2	2203	45	26-OCT-1992 01:10:55	6206	92314j321_2	1870	45	15-NOV-1992 04:49:35	6387	92335j349_2	1055	65	9-DEC-1992 15:37:15
6058	92293j300_2	2197	45	26-OCT-1992 04:25:15	6207	92314j321_2	1840	45	15-NOV-1992 08:04:05	6388	92335j349_2	1166	65	9-DEC-1992 18:24:35
6059	92300j307_2	152	45	26-OCT-1992 07:39:45	6208	92314j321_2	709	65	15-NOV-1992 11:58:25	6389	92335j349_2	1834	15	9-DEC-1992 21:38:55
6060	92300j307_2	2058	65	26-OCT-1992 11:42:05	6216	92321j328_2	1864	65	16-NOV-1992 13:13:25	6390	92335j349_2	1584	15	10-DEC-1992 00:53:05
6061	92300j307_2	132	65	26-OCT-1992 14:08:35	6217	92321j328_2	1865	65	16-NOV-1992 16:27:45	6396	92335j349_2	1799	45	10-DEC-1992 20:19:05
6063	92300j307_2	2034	15	26-OCT-1992 21:29:05	6219	92321j328_2	1867	45	16-NOV-1992 22:56:35	6397	92335j349_2	1831	45	10-DEC-1992 23:33:35
6064	92300j307_2	1968	15	26-OCT-1992 23:51:55	6220	92321j328_2	1869	45	17-NOV-1992 02:10:55	6402	92335j349_2	1828	45	11-DEC-1992 02:47:45
6066	92300j307_2	2056	45	27-OCT-1992 07:08:45	6221	92321j328_2	1862	45	17-NOV-1992 05:25:25	6402	92335j349_2	1755	65	11-DEC-1992 15:45:15
6067	92300j307_2	387	45	27-OCT-1992 09:35:15	6222	92321j328_2	1865	45	17-NOV-1992 08:39:45	6404	92335j349_2	11	45	11-DEC-1992 23:16:15
6067	92300j307_2	459	65	27-OCT-1992 11:20:31	6223	92321j328_2	1731	45	17-NOV-1992 11:59:55	6405	92335j349_2	1833	45	12-DEC-1992 01:28:05
6068	92300j307_2	2187	45	27-OCT-1992 12:49:45	6224	92321j328_2	1867	65	17-NOV-1992 15:08:05	6408	92335j349_2	593	45	12-DEC-1992 11:11:15
6069	92300j307_2	58	15	27-OCT-1992 18:51:55	6226	92321j328_2	1838	15	17-NOV-1992 21:37:15	6408	92335j349_2	328	15	12-DEC-1992 12:02:43
6069	92300j307_2	2012	65	27-OCT-1992 16:04:05	6231	92321j328_2	1124	65	18-NOV-1992 14:15:21	6409	92335j349_2	1833	65	12-DEC-1992 14:25:25
6070	92300j307_2	1629	15	27-OCT-1992 19:18:35	6232	92321j328_2	1866	45	18-NOV-1992 17:03:25	6410	92335j349_2	1944	65	12-DEC-1992 17:39:45
6071	92300j307_2	789	45	28-OCT-1992 00:07:23	6234	92321j328_2	1864	45	18-NOV-1992 23:32:35	6412	92335j349_2	1829	45	13-DEC-1992 00:08:25
6072	92300j307_2	2063	45	28-OCT-1992 01:47:25	6235	92321j328_2	1866	45	19-NOV-1992 02:46:45	6413	92335j349_2	1835	45	13-DEC-1992 03:22:45
6079	92300j307_2	2197	45	29-OCT-1992 00:28:25	6236	92321j328_2	1541	65	19-NOV-1992 06:01:05	6416	92335j349_2	1836	45	13-DEC-1992 13:05:45
6080	92300j307_2	2203	45	29-OCT-1992 03:42:55	6241	92321j328_2	1865	15	19-NOV-1992 22:12:55	6417	92335j349_2	1829	65	13-DEC-1992 16:20:35
6081	92300j307_2	2203	45	29-OCT-1992 06:57:15	6242	92321j328_2	1521	15	20-NOV-1992 01:27:25	6423	92349j356_2	1422	45	14-DEC-1992 12:00:55
6082	92300j307_2	138	15	29-OCT-1992 10:11:45	6243	92321j328_2	1870	45	20-NOV-1992 04:41:45	6424	92349j356_2	1820	65	14-DEC-1992 15:00:25
6082	92300j307_2	1411	45	29-OCT-1992 11:25:27	6244	92321j328_2	1867	45	20-NOV-1992 07:56:05	6425	92349j356_2	1768	65	14-DEC-1992 18:14:45
6083	92300j307_2	2181	65	29-OCT-1992 13:26:05	6245	92321j328_2	195	45	20-NOV-1992 11:10:35	6426	92349j356_2	1804	15	14-DEC-1992 21:29:05
6084	92300j307_2	1102	65	29-OCT-1992 16:40:35	6245	92321j328_2	330	65	20-NOV-1992 12:03:23	6427	92349j356_2	1782	15	15-DEC-1992 00:43:25
6086	92300j307_2	2055	45	29-OCT-1992 23:57:25	6246	92321j328_2	1850	65	20-NOV-1992 14:24:25	6431	92349j356_2	1829	45	15-DEC-1992 13:40:45
6087	92300j307_2	2179	45	30-OCT-1992 02:23:55	6247	92321j328_2	1262	65	20-NOV-1992 17:39:15	6432	92349j356_2	1830	65	15-DEC-1992 16:55:05
6088	92300j307_2	2199	45	30-OCT-1992 05:38:15	6249	92321j328_2	1795	45	21-NOV-1992 00:08:45	6433	92349j356_2	1829	45	15-DEC-1992 20:09:15
6089	92300j307_2	1916	45	30-OCT-1992 08:52:45	6250	92321j328_2	1832	45	21-NOV-1992 03:23:05	6434	92349j356_2	1268	15	15-DEC-1992 23:23:35
6089	92300j307_2	60	15	30-OCT-1992 11:40:25	6251	92321j328_2	800	45	21-NOV-1992 06:37:35	6435	92349j356_2	893	15	16-DEC-1992 02:38:05
6090	92300j307_2	1780	65	30-OCT-1992 12:07:05	6253	92321j328_2	1829	65	21-NOV-1992 13:06:15	6438	92349j356_2	360	65	16-DEC-1992 13:11:31

6439	92349j356_2	1157	65	16-DEC-1992 15:35:15	6641	93011j018_2	1830	15	12-JAN-1993 21:45:05	6769	93025j032_2	4	45	30-JAN-1993 06:10:55
6442	92349j356_2	1779	45	17-DEC-1992 01:18:15	6642	93011j018_2	1702	15	13-JAN-1993 00:59:15	6770	93025j032_2	1559	45	30-JAN-1993 06:11:35
6443	92349j356_2	1795	45	17-DEC-1992 04:32:35	6643	93011j018_2	1814	45	13-JAN-1993 04:15:53	6771	93025j032_2	54	15	30-JAN-1993 09:25:55
6444	92349j356_2	1829	45	17-DEC-1992 07:46:55	6644	93011j018_2	1834	45	13-JAN-1993 07:27:45	6773	93025j032_2	166	15	30-JAN-1993 18:01:33
6445	92349j356_2	731	45	17-DEC-1992 11:01:15	6647	93011j018_2	1822	65	13-JAN-1993 17:10:55	6774	93025j032_2	2018	45	30-JAN-1993 19:08:35
6445	92349j356_2	179	65	17-DEC-1992 11:57:51	6648	93011j018_2	597	15	13-JAN-1993 21:08:05	6775	93025j032_2	1892	15	30-JAN-1993 22:22:45
6446	92349j356_2	1823	65	17-DEC-1992 14:15:35	6649	93011j018_2	1801	15	13-JAN-1993 23:39:25	6775	93025j032_2	2	45	31-JAN-1993 01:36:35
6447	92349j356_2	1828	65	17-DEC-1992 17:29:55	6650	93011j018_2	1894	45	14-JAN-1993 02:53:35	6776	93025j032_2	2014	45	31-JAN-1993 01:36:55
6449	92349j356_2	1790	45	17-DEC-1992 23:58:25	6654	93011j018_2	1830	65	14-JAN-1993 15:50:55	6777	93025j032_2	1894	45	31-JAN-1993 04:51:15
6450	92349j356_2	1832	45	18-DEC-1992 03:12:45	6655	93011j018_2	1814	15	14-JAN-1993 19:05:05	6778	93025j032_2	181	45	31-JAN-1993 08:05:25
6451	92349j356_2	1829	45	18-DEC-1992 06:27:05	6656	93011j018_2	1821	15	14-JAN-1993 22:19:15	6782	93025j032_2	2017	15	31-JAN-1993 21:02:15
6452	92349j356_2	1831	45	18-DEC-1992 09:41:25	6657	93011j018_2	1820	15	15-JAN-1993 01:33:45	6783	93025j032_2	20	15	1-FEB-1993 00:16:35
6453	92349j356_2	1739	65	18-DEC-1992 12:55:45	6658	93011j018_2	1799	45	15-JAN-1993 04:47:55	6783	93025j032_2	1878	45	1-FEB-1993 00:38:35
6454	92349j356_2	1832	65	18-DEC-1992 16:10:05	6659	93011j018_2	1828	45	15-JAN-1993 08:02:15	6784	93025j032_2	1987	45	1-FEB-1993 03:30:45
6456	92349j356_2	442	45	18-DEC-1992 23:26:19	6661	93011j018_2	42	65	15-JAN-1993 15:32:17	6789	93032j039_2	1706	15	1-FEB-1993 21:06:49
6457	92349j356_2	1833	45	19-DEC-1992 01:53:15	6662	93011j018_2	1833	65	15-JAN-1993 17:45:25	6790	93032j039_2	1590	15	1-FEB-1993 23:00:15
6458	92349j356_2	494	45	19-DEC-1992 05:07:25	6663	93011j018_2	1841	15	15-JAN-1993 21:00:35	6790	93032j039_2	386	45	2-FEB-1993 04:04:59
6460	92349j356_2	1269	65	19-DEC-1992 11:55:37	6663	93011j018_2	50	65	15-JAN-1993 20:03:25	6791	93032j039_2	1893	45	2-FEB-1993 02:10:55
6461	92349j356_2	1832	65	19-DEC-1992 14:50:25	6664	93011j018_2	1688	15	15-JAN-1993 23:02:45	6792	93032j039_2	2017	45	2-FEB-1993 05:25:15
6462	92349j356_2	1820	65	19-DEC-1992 18:04:45	6664	93011j018_2	259	45	16-JAN-1993 01:08:29	6793	93032j039_2	20	45	2-FEB-1993 08:39:25
6464	92349j356_2	1835	45	20-DEC-1992 00:33:25	6665	93011j018_2	771	45	16-JAN-1993 02:31:55	6793	93032j039_2	967	65	2-FEB-1993 10:28:43
6465	92349j356_2	1833	45	20-DEC-1992 03:47:45	6666	93011j018_2	1824	45	16-JAN-1993 05:48:25	6794	93032j039_2	161	65	2-FEB-1993 11:53:45
6468	92349j356_2	1833	65	20-DEC-1992 13:30:35	6667	93011j018_2	1580	45	16-JAN-1993 09:00:35	6797	93032j039_2	1898	15	2-FEB-1993 21:36:35
6469	92349j356_2	1825	65	20-DEC-1992 16:45:05	6669	93011j018_2	4	15	16-JAN-1993 18:27:45	6798	93032j039_2	182	15	3-FEB-1993 00:50:55
6475	92356j363_2	1823	65	21-DEC-1992 12:10:55	6670	93011j018_2	1948	15	16-JAN-1993 18:28:15	6799	93032j039_2	1897	45	3-FEB-1993 04:06:05
6476	92356j363_2	739	65	21-DEC-1992 15:25:05	6671	93011j018_2	1896	15	16-JAN-1993 22:19:05	6800	93032j039_2	647	45	3-FEB-1993 07:19:15
6479	92356j363_2	1822	15	22-DEC-1992 01:08:05	6671	93011j018_2	22	45	17-JAN-1993 00:53:15	6801	93032j039_2	1828	65	3-FEB-1993 11:53:05
6481	92356j363_2	1823	45	22-DEC-1992 07:36:45	6672	93011j018_2	1912	45	17-JAN-1993 00:56:55	6804	93032j039_2	1852	15	3-FEB-1993 21:35:55
6482	92356j363_2	549	45	22-DEC-1992 10:51:05	6673	93011j018_2	1910	45	17-JAN-1993 04:26:25	6805	93032j039_2	1261	15	3-FEB-1993 23:53:05
6484	92356j363_2	1822	65	22-DEC-1992 17:19:45	6674	93011j018_2	1922	45	17-JAN-1993 07:25:25	6806	93032j039_2	241	45	4-FEB-1993 04:40:45
6486	92356j363_2	1831	45	22-DEC-1992 23:48:15	6675	93011j018_2	32	45	17-JAN-1993 18:26:45	6807	93032j039_2	1868	45	4-FEB-1993 05:59:25
6487	92356j363_2	1828	45	23-DEC-1992 03:02:35	6677	93011j018_2	1687	15	17-JAN-1993 18:26:45	6810	93032j039_2	530	15	4-FEB-1993 17:46:11
6491	92356j363_2	1825	65	23-DEC-1992 15:59:55	6678	93011j018_2	1919	15	17-JAN-1993 20:22:25	6811	93032j039_2	1902	15	4-FEB-1993 18:56:35
6492	92356j363_2	468	65	23-DEC-1992 19:14:15	6679	93011j018_2	47	15	17-JAN-1993 23:52:05	6812	93032j039_2	1786	15	4-FEB-1993 22:10:55
6494	92356j363_2	1828	45	24-DEC-1992 01:42:45	6679	93011j018_2	1857	45	18-JAN-1993 00:49:25	6813	93032j039_2	1889	45	5-FEB-1993 01:25:25
6495	92356j363_2	960	45	24-DEC-1992 04:57:15	6680	93011j018_2	1964	45	18-JAN-1993 02:50:55	6814	93032j039_2	2010	45	5-FEB-1993 04:39:35
6499	92356j363_2	1830	65	24-DEC-1992 17:54:25	6681	93011j018_2	1902	15	18-JAN-1993 06:20:45	6818	93032j039_2	1961	15	5-FEB-1993 17:45:15
6501	92356j363_2	1823	15	25-DEC-1992 00:23:05	6682	93018j025_2	34	45	18-JAN-1993 09:19:25	6819	93032j039_2	1886	15	5-FEB-1993 21:13:05
6509	92356j363_2	1827	45	26-DEC-1992 02:17:25	6685	93018j025_2	1640	15	18-JAN-1993 20:23:21	6820	93032j039_2	182	15	6-FEB-1993 00:05:15
6512	92356j363_2	1829	65	26-DEC-1992 12:00:25	6686	93018j025_2	1596	15	19-JAN-1993 00:22:05	6820	93032j039_2	1834	45	6-FEB-1993 00:22:35
6513	92356j363_2	1830	65	26-DEC-1992 15:14:45	6686	93018j025_2	261	45	19-JAN-1993 00:22:51	6821	93032j039_2	1881	45	6-FEB-1993 03:41:35
6514	92356j363_2	1830	65	26-DEC-1992 18:29:05	6687	93018j025_2	1912	15	19-JAN-1993 01:46:25	6822	93032j039_2	1701	45	6-FEB-1993 06:33:55
6516	92356j363_2	1827	45	27-DEC-1992 00:57:35	6688	93018j025_2	1912	45	19-JAN-1993 04:45:05	6825	93032j039_2	1499	15	6-FEB-1993 17:45:09
6517	92356j363_2	1828	45	27-DEC-1992 04:11:55	6689	93018j025_2	690	45	19-JAN-1993 08:15:05	6826	93032j039_2	2017	15	6-FEB-1993 19:30:55
6523	92356j363_2	1663	45	27-DEC-1992 23:44:35	6691	93018j025_2	1914	65	19-JAN-1993 14:43:45	6827	93032j039_2	1798	15	6-FEB-1993 23:07:15
6524	92356j363_2	1724	45	28-DEC-1992 02:52:05	6692	93018j025_2	335	65	19-JAN-1993 17:42:05	6827	93032j039_2	53	45	7-FEB-1993 01:04:59
6525	92363j004_2	1834	45	28-DEC-1992 06:06:25	6695	93018j025_2	247	45	20-JAN-1993 05:08:45	6828	93032j039_2	2008	45	7-FEB-1993 01:59:35
6526	92363j004_2	1832	45	28-DEC-1992 09:20:45	6696	93018j025_2	1926	45	20-JAN-1993 06:39:05	6829	93032j039_2	1884	45	7-FEB-1993 05:35:55
6527	92363j004_2	1830	65	28-DEC-1992 12:35:05	6697	93018j025_2	48	45	20-JAN-1993 10:09:35	6830	93032j039_2	103	45	7-FEB-1993 08:28:15
6528	92363j004_2	1833	65	28-DEC-1992 15:49:15	6701	93018j025_2	1867	15	20-JAN-1993 23:07:45	6833	93032j039_2	1879	15	7-FEB-1993 18:33:05
6529	92363j004_2	1254	65	28-DEC-1992 19:03:45	6705	93018j025_2	1900	65	21-JAN-1993 12:03:45	6834	93032j039_2	1971	15	7-FEB-1993 21:25:25
6530	92363j004_2	752	15	28-DEC-1992 22:55:03	6706	93018j025_2	1973	65	21-JAN-1993 15:01:15	6835	93032j039_2	11	15	8-FEB-1993 01:01:45
6531	92363j004_2	1824	45	29-DEC-1992 01:32:15	6707	93018j025_2	389	15	21-JAN-1993 20:19:41	6835	93032j039_2	1864	15	8-FEB-1993 01:05:05
6532	92363j004_2	70	45	29-DEC-1992 04:46:55	6707	93018j025_2	50	65	21-JAN-1993 21:04:15	6836	93032j039_2	2007	45	8-FEB-1993 03:54:25
6534	92363j004_2	195	65	29-DEC-1992 12:10:55	6708	93018j025_2	1983	15	21-JAN-1993 21:30:05	6837	93032j039_2	499	45	8-FEB-1993 07:30:15
6535	92363j004_2	1831	65	29-DEC-1992 14:29:25	6709	93018j025_2	1415	45	22-JAN-1993 01:00:55	6840	93039j046_2	1823	15	8-FEB-1993 18:10:15
6536	92363j004_2	1803	45	29-DEC-1992 17:43:45	6712	93018j025_2	1669	65	22-JAN-1993 11:43:15	6841	93039j046_2	1732	15	8-FEB-1993 20:27:25
6537	92363j004_2	1830	15	29-DEC-1992 20:58:05	6714	93018j025_2	19	15	22-JAN-1993 20:04:45	6842	93039j046_2	178	15	8-FEB-1993 23:19:35
6538	92363j004_2	1831	45	30-DEC-1992 00:12:25	6715	93018j025_2	1880	15	22-JAN-1993 20:26:45	6845	93039j046_2	1259	65	9-FEB-1993 10:44:37
6539	92363j004_2	1824	45	30-DEC-1992 03:26:45	6716	93018j025_2	911	15	23-JAN-1993 01:10:09	6846	93039j046_2	93	65	9-FEB-1993 12:30:55
6543	92363j004_2	1402	65	30-DEC-1992 16:23:55	6719	93018j025_2	1077	45	23-JAN-1993 10:48:17	6848	93039j046_2	1824	15	9-FEB-1993 20:04:35
6544	92363j004_2	406	65	30-DEC-1992 19:38:05	6720	93018j025_2	1997	15	23-JAN-1993 12:20:55	6849	93039j046_2	1883	15	9-FEB-1993 22:24:05
6553	92363j004_2	1252	15	1-JAN-1993 00:46:55	6721	93018j025_2	1791	15	23-JAN-1993 15:52:35	6851	93039j046_2	1911	45	10-FEB-1993 05:47:35
6554	92363j004_2	1575	45	1-JAN-1993 04:01:15	6722	93018j025_2	19	15	23-JAN-1993 21:59:05	6852	93039j046_2	92	45	10-FEB-1993 07:56:45
6560	92363j004_2	951	45	1-JAN-1993 23:57:53	6722	93018j025_2	771	65	23-JAN-1993 18:49:25	6855	93039j046_2	86	15	10-FEB-1993 20:39:35
6561	92363j004_2	1819	45	2-JAN-1993 02:41:25	6723	93018j025_2	1738	15	23-JAN-1993 19:43:05	6856	93039j046_2	1742	15	10-FEB-1993 20:51:55
6564	92363j004_2	1822	65	2-JAN-1993 12:24:05	6723	93018j025_2	142	45	24-JAN-1993 00:16:27	6857	93039j046_2	1961	15	11-FEB-1993 00:15:55
6566	92363j004_2	1569	15	2-JAN-1993 19:02:57	6724	93018j025_2	2010	45	24-JAN-1993 01:16:55	6858	93039j046_2	90	15	11-FEB-1993 03:22:35
6567	92363j004_2	1826	15	2-JAN-1993 22:07:05	6725	93018j025_2	1873	45	24-JAN-1993 04:49:45	6863	93039j046_2	1911	15	11-FEB-1993 20:38:55
6568	92363j004_2	1835	15	3-JAN-1993 01:21:25	6726	93018j025_2	1980	15	24-JAN-1993 07:45:25	6864	93039j046_2	1928	15	11-FEB-1993 22:48:15
6569	92363j004_2	811	45	3-JAN-1993 04:35:45	6728	93018j025_2	1992	65	24-JAN-1993 14:16:55	6867	93039j046_2	1372	65	12-FEB-1993 09:55:07
6570	92363j004_2	1828	45	3-JAN-1993 07:50:05	6729	93018j025_2	1875	65	24-JAN-1993 17:46:55	6868	93039j046_2	1925	65	12-FEB-1993 11:45:25
6575	92363j004_2	1816	45	4-JAN-1993 00:01:35	6730	93018j025_2	1853	15	24-JAN-1993 21:50:35	6873	93039j046_2	1899	45	13-FEB-1993 05:02:05
6576	92363j004_2	1823	45	4-JAN-1993 03:15:45	6731	93018j025_2	3	15	24-JAN-1993 23:56:35	6874	93039j046_2	1380	45	13-FEB-1993 07:11:15
6577	93004j011_2	1830	45	4-JAN-1993 06:30:15	6731	93018j025_2	1881	45	25-JAN-1993 00:08:15	6875	93039j046_2	2	65	13-FEB-1993 13:39:35
6578	93004j011_2	890	45	4-JAN-1993 09:44:25	6732	93018j025_2	2009	45	25-JAN-1993 03:10:55	6876	93039j046_2	1915	45	13-FEB-1993 13:39:55
6579	93004j011_2	1821	65	4-JAN-1993 12:58:45	6733	93018j025_2	1848	45	25-JAN-1993 06:25:05	6877	93039j046_2	1919	15	13-FEB-1993 17:59:05
6580	93004j011_2	9	65	4-JAN-1993 15:08:05	6737	93025j032_2	1011	15	25-JAN-1993 21:07:19	6877	93039j046_2	53	65	13-FEB-1993 17:00:05
6583	93004j011_2	1924	45	5-JAN-1993 01:55:55	6738	93025j032_2	766	15	25-JAN-1993 22:36:15	6878	93039j046_2	1930	15	13-FEB-1993 20:08:25
6586	93004j011_2	1761	45	5-JAN-1993 11:38:55	6738	93025j032_2	1190	45	26-JAN-1993 00:15:43	6879	93039j046_2	66	15	13-FEB-1993 23:28:35
6589	93004j011_2	1826	15	5-JAN-1993 21:22:05	6739	93025j032_2	1895	15	26-JAN-1993 01:50:35	6879	93039j046_2	1819	45	14-FEB-1993 00:31:55
6590	93004j011_2	1821	15	6-JAN-1993 00:36:05	6740	93025j032_2	2000	45	26-JAN-1993 05:04:45	6880	93039j046_2	1874	45	14-FEB-1993 02:37:05
6592	93004j011_2	1833	45	6-JAN-1993 07:04:35	6741	93025j032_2	51	45	26-JAN-1993 08:19:05	6881	93039j046_2	1884	15	14-FEB-1993 05:57:15
6595	93004j011_2	1772	65	6-JAN-1993 16:47:45	6741	93025j032_2	16	65	26-JAN-1993 10:41:35	6883	93039j046_2	1830	65	14-FEB-1993 13:24:55
6597	93004j011_2	1819	15	6-JAN-1993 23:17:53	6742	93025j032_2	2010	45	26-JAN-1993 11:33:15	6887	93039j046_2	1787	45	15-FEB-1993 01:22:55
6598	93004j011_2	1919	45	7-JAN-1993 02:30:15	6743	93025j032_2	53	65	26-JAN-1993 14:47:25	6888	93039j046_2	1571	45	15-FEB-1993 05:43:21
6601	93004j011_2	1802	65	7-JAN-1993 12:13:05	6745	93025j032_2	14	15	27-JAN-1993 00:27:55	6892	93046j053_2	1819	65	15-FEB-1993 18:33:35
6604	93004j011_2	1634	15	7-JAN-1993 22:03:39	6746	93025j032_2	167	15	27-JAN-1993 00:30:15	6895	93046j053_2	1439	45	16-FEB-1993 04:33:21
6605	93004j011_2	1821	45	8-JAN-1993 01:10:15	6746	93025j032_2	1847	45	27-JAN-1993 01:47:05	6896	93046j053_2	1924	45	16-FEB-1993 06:25:45
6606	93004j011_2	1826	45	8-JAN-1993 04:24:35	6747	93025j032_2	1876	45	27-JAN-1993 03:44:25	6900	93046j053_2	862	15	16-FEB-1993 21:03:53
6607	93004j011_2	1819	45	8-JAN-1993 07:38:55	6748	93025j032_2	1979	45	27-JAN-1993 06:58:35	6901	93046j053_2	1892	15	16-FEB-1993 22:42:05
6608	93004j011_2	973	65	8-JAN-1993 11:22:05	6749	93025j032_2	33	45	27-JAN-1993 10:12:55	6903	93046j053_2	1975	45	17-FEB-1993 05:11:35
6609	93004j011_2	781	65	8-JAN-1993 14:07:45	6751	93025j032_2	1657	15	27-JAN-1993 18:05:51	6904	93046j053_2	40	45	17-FEB-1993 08:20:05
6612	93004j011_2	1498	45	9-JAN-1993 00:03:15	6752	93025j032_2	1991	15	27-JAN-1993 19:55:35	6905	93046j053_2	1916	15	17-FEB-1993 12:39:15
6613	93004j011_2	1831	45	9-JAN-1993 03:04:45	6753	93025j032_2	1051	15	27-JAN-1993 23:09:45	6906	93046j053_2	2028	15	17-FEB-1993 14:48:45
6616	93004j011_2	1235	15	9-JAN-1993 12:47:25	6756	93025j032_2	1057	15	28-JAN-1993 10:36:35	6907	93046j053_2	2277	15	17-FEB-1993 18:08:25
6618	93004j011_2	1827	15	9-JAN-1993 19:16:05	6757	93025j032_2	1894	15	28-JAN-1993 12:06:45	6910	93046j053_2	1438	45	18-FEB-1993 05:04:37
6619	93004j011_2	1835	15	9-JAN-1993 22:30:25	6758	93025j032_2	170	65	28-JAN-1993 15:20:55	6911	93046j053_2	1892	45	18-FEB-1993 07:05:55
6621	93004j011_2	1828	45	10-JAN-1993 04:59:15	6759	93025j032_2	9	15	28-JAN-1993 21:47:55	6913	93046j053_2	1915	15	18-FEB-1993 14:33:25
6622	93004j011_2	1832	45	10-JAN-1993 08:13:35	6760	93025j032_2	2017	15	28-JAN-1993 21:49:25	6914	93046j053_2	1826	15	18-FEB-1993 17:47:45
6627	93004j011_2	1833	45	11-JAN-1993 00:24:45	6761	93025j032_2	1891	15	29-JAN-1993 01:03:35	6914	93046j053_2	93	65	18-FEB-1993 16:43:05
6628	93004j011_2	627	45	11-JAN-1993 03:39:05	6762	93025j032_2	1961	45	29-JAN-1993 04:26:15	6915	93046j053_2	148	15	18-FEB-1993 20:02:55
6631	93011j018_2	1826	65	11-JAN-1993 13:21:55	6763	93025j032_2	1892	45	29-JAN-1993 07:32:05	6916	93046j053_2	1458	15	23-FEB-1993 11:05:35
6634	93011j018_2	430	15	11-JAN-1993 23:53:21	6765	93025j032_2	1885	15	29-JAN-1993 14:21:25	6916	93046j053_2	179	45	19-FEB-1993 01:12:31
6635	93011j018_2	1830	15	12-JAN-1993 02:19:15	6766	93025j032_2	1158	15	29-JAN-1993 18:56:43	6917	93046j053_2	1666	45	19-FEB-1993 02:31:35
6637	93011j018_2	1805	45	12-JAN-1993 08:47:55	6766	93025j032_2	176	65	29-JAN-1993 17:14:45	6920	93046j053_2	1867	15	19-FEB-1993 12:18:45
6639	93011j018_2	1224	65	12-JAN-1993 15:38:09	6767	93025j032_2	1901	15	29-JAN-1993 20:28:55	6921	93046j053_2	791	15	19-FEB-1993 17:06:03
6640	93011j018_2	1874	65	12-JAN-1993 18:30:35	6768	93025j032_2	1319	15	29-JAN-1993 23:43:15	6921	93046j053_2	1161	65	19-FEB-1993 15:28:45

E-3

6922	93046j053_2	1967	15	19-FEB-1993	18:37:25	7038	93060j067_2	419	65	7-MAR-1993	12:06:45	7202	93081j095_2	1833	15	29-MAR-1993	14:20:45
6923	93046j053_2	1895	15	19-FEB-1993	21:57:15	7039	93060j067_2	87	15	7-MAR-1993	16:28:15	7203	93081j095_2	1977	15	29-MAR-1993	16:35:55
6923	93046j053_2	87	45	20-FEB-1993	00:51:25	7039	93060j067_2	1862	65	7-MAR-1993	13:34:15	7204	93081j095_2	1880	15	29-MAR-1993	19:44:05
6924	93046j053_2	1929	45	20-FEB-1993	01:05:55	7040	93060j067_2	1926	15	7-MAR-1993	16:42:45	7204	93081j095_2	40	45	29-MAR-1993	21:49:57
6925	93046j053_2	555	45	20-FEB-1993	04:25:45	7041	93060j067_2	1963	15	7-MAR-1993	20:02:45	7205	93081j095_2	1976	45	29-MAR-1993	23:04:25
6927	93046j053_2	1177	65	20-FEB-1993	12:19:05	7042	93060j067_2	1925	15	7-MAR-1993	23:11:15	7206	93081j095_2	1930	45	30-MAR-1993	02:12:35
6928	93046j053_2	1925	65	20-FEB-1993	14:03:05	7045	93060j067_2	11	65	8-MAR-1993	12:06:35	7207	93081j095_2	17	45	30-MAR-1993	05:32:55
6929	93046j053_2	1982	15	20-FEB-1993	17:22:55	7046	93067j074_2	1931	65	8-MAR-1993	12:07:25	7209	93081j095_2	82	15	30-MAR-1993	14:55:35
6930	93046j053_2	1923	15	20-FEB-1993	20:31:45	7047	93067j074_2	1920	15	8-MAR-1993	16:27:25	7210	93081j095_2	1917	15	30-MAR-1993	15:09:45
6931	93046j053_2	61	15	20-FEB-1993	23:51:35	7047	93067j074_2	59	65	8-MAR-1993	15:28:35	7211	93081j095_2	61	15	30-MAR-1993	18:30:05
6931	93046j053_2	1923	45	21-FEB-1993	00:50:35	7048	93067j074_2	1928	15	8-MAR-1993	18:37:05	7213	93081j095_2	85	45	31-MAR-1993	03:52:35
6932	93046j053_2	1924	45	21-FEB-1993	03:00:15	7049	93067j074_2	1534	15	8-MAR-1993	21:57:05	7214	93081j095_2	1950	45	31-MAR-1993	04:06:45
6933	93046j053_2	1722	45	21-FEB-1993	06:20:05	7049	93067j074_2	409	45	8-MAR-1993	23:45:59	7218	93081j095_2	1829	15	31-MAR-1993	18:09:05
6935	93046j053_2	1986	65	21-FEB-1993	12:48:35	7050	93067j074_2	1927	45	9-MAR-1993	01:05:25	7219	93081j095_2	1891	15	31-MAR-1993	20:24:15
6936	93046j053_2	1834	15	21-FEB-1993	17:02:05	7051	93067j074_2	1958	45	9-MAR-1993	04:25:35	7221	93081j095_2	1422	45	1-APR-1993	04:03:07
6936	93046j053_2	94	45	21-FEB-1993	15:57:25	7053	93067j074_2	1	65	9-MAR-1993	14:02:30	7222	93081j095_2	98	45	1-APR-1993	06:00:55
6937	93046j053_2	1986	15	21-FEB-1993	19:17:15	7054	93067j074_2	250	15	9-MAR-1993	16:00:53	7224	93081j095_2	762	15	1-APR-1993	14:11:17
6938	93046j053_2	1475	15	21-FEB-1993	22:26:05	7054	93067j074_2	1608	65	9-MAR-1993	14:03:30	7225	93081j095_2	1924	15	1-APR-1993	15:50:05
6938	93046j053_2	438	45	22-FEB-1993	00:17:57	7055	93067j074_2	2036	15	9-MAR-1993	17:22:45	7226	93081j095_2	1891	15	1-APR-1993	18:58:05
6939	93046j053_2	1975	45	22-FEB-1993	01:45:45	7056	93067j074_2	1927	15	9-MAR-1993	20:31:25	7227	93081j095_2	1917	45	1-APR-1993	22:20:25
6940	93046j053_2	1324	45	22-FEB-1993	04:54:35	7057	93067j074_2	1921	45	10-MAR-1993	00:50:25	7228	93081j095_2	1928	45	2-APR-1993	01:26:35
6942	93053j060_2	1824	65	22-FEB-1993	12:27:45	7058	93067j074_2	1929	45	10-MAR-1993	02:59:55	7229	93081j095_2	56	45	2-APR-1993	04:47:05
6943	93053j060_2	1974	15	22-FEB-1993	14:42:55	7059	93067j074_2	653	45	10-MAR-1993	06:19:55	7231	93081j095_2	224	15	2-APR-1993	14:09:23
6944	93053j060_2	1911	65	22-FEB-1993	17:51:45	7061	93067j074_2	5	15	10-MAR-1993	15:56:15	7232	93081j095_2	1922	15	2-APR-1993	14:23:45
6945	93053j060_2	1957	15	22-FEB-1993	21:11:35	7062	93067j074_2	1924	15	10-MAR-1993	15:57:05	7233	93081j095_2	61	15	2-APR-1993	17:44:05
6945	93053j060_2	18	45	23-FEB-1993	00:17:25	7063	93067j074_2	61	15	10-MAR-1993	19:16:55	7234	93081j095_2	1827	45	2-APR-1993	21:57:25
6946	93053j060_2	1132	45	23-FEB-1993	00:20:25	7066	93067j074_2	1896	45	11-MAR-1993	04:58:25	7235	93081j095_2	1884	45	3-APR-1993	00:12:45
6947	93053j060_2	1971	45	23-FEB-1993	03:40:05	7069	93067j074_2	1712	15	11-MAR-1993	15:49:19	7239	93081j095_2	1913	15	3-APR-1993	14:08:55
6948	93053j060_2	579	45	23-FEB-1993	06:48:55	7070	93067j074_2	1857	15	11-MAR-1993	17:51:25	7240	93081j095_2	1926	15	3-APR-1993	16:17:55
6949	93053j060_2	1891	65	23-FEB-1993	10:08:35	7071	93067j074_2	1973	15	11-MAR-1993	21:11:15	7241	93081j095_2	1965	15	3-APR-1993	19:38:25
6951	93053j060_2	1920	15	23-FEB-1993	17:36:15	7072	93067j074_2	94	15	12-MAR-1993	00:19:55	7243	93081j095_2	1899	45	4-APR-1993	03:06:05
6952	93053j060_2	1924	15	23-FEB-1993	19:46:05	7074	93067j074_2	1921	45	12-MAR-1993	04:38:55	7244	93081j095_2	97	45	4-APR-1993	05:15:05
6953	93053j060_2	1566	45	24-FEB-1993	00:17:19	7074	93067j074_2	454	45	12-MAR-1993	06:48:35	7246	93081j095_2	28	15	4-APR-1993	13:50:01
6954	93053j060_2	1915	45	24-FEB-1993	02:14:45	7074	93067j074_2	1060	65	12-MAR-1993	08:11:55	7247	93081j095_2	1961	15	4-APR-1993	15:04:05
6955	93053j060_2	1948	45	24-FEB-1993	05:34:25	7075	93067j074_2	188	65	12-MAR-1993	10:08:25	7248	93081j095_2	1919	15	4-APR-1993	18:12:05
6956	93053j060_2	92	45	24-FEB-1993	08:43:15	7077	93067j074_2	1923	15	12-MAR-1993	16:36:55	7249	93081j095_2	1961	45	4-APR-1993	21:32:35
6956	93053j060_2	1820	65	24-FEB-1993	09:47:45	7079	93067j074_2	1913	45	13-MAR-1993	00:04:25	7250	93081j095_2	1927	45	5-APR-1993	00:40:45
6957	93053j060_2	1968	15	24-FEB-1993	12:02:55	7080	93067j074_2	1925	45	13-MAR-1993	02:14:05	7251	93081j095_2	477	45	5-APR-1993	04:01:05
6958	93053j060_2	1907	65	24-FEB-1993	15:11:55	7081	93067j074_2	1862	15	13-MAR-1993	15:05:55	7254	93081j095_2	1926	15	5-APR-1993	13:45:15
6959	93053j060_2	1914	15	24-FEB-1993	19:30:35	7084	93067j074_2	1828	15	13-MAR-1993	16:15:55	7255	93095j102_2	1981	15	5-APR-1993	16:58:15
6959	93053j060_2	59	65	24-FEB-1993	18:31:35	7085	93067j074_2	1969	15	13-MAR-1993	18:31:05	7256	93095j102_2	97	15	5-APR-1993	20:06:15
6960	93053j060_2	1913	45	24-FEB-1993	21:40:25	7086	93067j074_2	1746	15	13-MAR-1993	21:39:55	7256	93095j102_2	1816	45	5-APR-1993	21:11:35
6964	93053j060_2	1364	65	25-FEB-1993	11:58:19	7089	93067j074_2	1917	65	14-MAR-1993	08:27:15	7257	93095j102_2	1965	45	5-APR-1993	23:26:45
6965	93053j060_2	1958	65	25-FEB-1993	14:01:05	7090	93067j074_2	91	65	14-MAR-1993	10:36:55	7258	93095j102_2	1928	45	6-APR-1993	02:34:55
6966	93053j060_2	91	65	25-FEB-1993	17:06:15	7097	93074j081_2	1830	65	15-MAR-1993	10:21:25	7261	93095j102_2	1041	15	6-APR-1993	13:53:11
6967	93053j060_2	1976	15	25-FEB-1993	20:25:45	7099	93074j081_2	1950	15	15-MAR-1993	15:51:05	7262	93095j102_2	1924	15	6-APR-1993	15:31:55
6968	93053j060_2	91	15	25-FEB-1993	23:34:45	7100	93074j081_2	1910	15	15-MAR-1993	18:59:55	7263	93095j102_2	1590	15	6-APR-1993	18:52:25
6968	93053j060_2	1582	45	26-FEB-1993	00:47:35	7101	93074j081_2	1453	15	15-MAR-1993	22:19:35	7265	93095j102_2	84	45	7-APR-1993	04:15:05
6969	93053j060_2	1980	45	26-FEB-1993	02:54:15	7102	93074j081_2	135	45	16-MAR-1993	03:16:15	7266	93095j102_2	972	45	7-APR-1993	04:29:05
6970	93053j060_2	1926	45	26-FEB-1993	06:02:25	7103	93074j081_2	1892	45	16-MAR-1993	04:48:05	7269	93095j102_2	1920	15	7-APR-1993	15:17:15
6972	93053j060_2	1047	65	26-FEB-1993	14:02:09	7107	93074j081_2	1289	15	16-MAR-1993	19:05:47	7270	93095j102_2	1920	15	7-APR-1993	17:26:05
6973	93053j060_2	1907	15	26-FEB-1993	16:50:35	7108	93074j081_2	1902	15	16-MAR-1993	20:54:05	7271	93095j102_2	1882	15	7-APR-1993	20:46:35
6973	93053j060_2	60	65	26-FEB-1993	15:51:35	7109	93074j081_2	1860	45	17-MAR-1993	01:12:55	7276	93095j102_2	1830	15	8-APR-1993	13:57:05
6974	93053j060_2	1918	15	26-FEB-1993	18:59:35	7110	93074j081_2	1921	45	17-MAR-1993	03:22:45	7277	93095j102_2	1963	15	8-APR-1993	16:12:15
6975	93053j060_2	1546	15	27-FEB-1993	22:20:05	7111	93074j081_2	60	45	17-MAR-1993	06:42:25	7278	93095j102_2	612	15	8-APR-1993	19:20:25
6975	93053j060_2	433	45	27-FEB-1993	00:12:13	7113	93074j081_2	89	15	17-MAR-1993	16:04:55	7278	93095j102_2	1280	45	8-APR-1993	20:44:47
6976	93053j060_2	1926	45	27-FEB-1993	01:28:05	7114	93074j081_2	1922	15	17-MAR-1993	17:51:25	7279	93095j102_2	1966	45	8-APR-1993	22:40:45
6977	93053j060_2	1975	45	27-FEB-1993	04:48:35	7115	93074j081_2	1979	15	17-MAR-1993	19:39:25	7280	93095j102_2	1926	45	9-APR-1993	01:48:55
6978	93053j060_2	405	45	27-FEB-1993	07:56:45	7116	93074j081_2	315	15	17-MAR-1993	22:48:25	7281	93095j102_2	34	45	9-APR-1993	05:09:45
6981	93053j060_2	1983	15	27-FEB-1993	17:45:45	7118	93074j081_2	1851	45	18-MAR-1993	05:30:35	7281	93095j102_2	1916	65	9-APR-1993	06:08:35
6982	93053j060_2	1930	15	27-FEB-1993	20:53:55	7122	93074j081_2	1820	15	18-MAR-1993	19:18:25	7282	93095j102_2	193	65	9-APR-1993	08:17:25
6983	93053j060_2	1887	45	28-FEB-1993	00:14:15	7123	93074j081_2	1887	15	18-MAR-1993	21:33:45	7285	93095j102_2	55	45	9-APR-1993	21:05:25
6984	93053j060_2	1932	45	28-FEB-1993	03:22:25	7126	93074j081_2	1830	65	19-MAR-1993	08:15:35	7286	93095j102_2	1927	45	9-APR-1993	21:14:35
6985	93053j060_2	1887	45	28-FEB-1993	06:42:55	7127	93074j081_2	61	65	19-MAR-1993	10:30:45	7287	93095j102_2	1977	45	10-APR-1993	00:34:55
6987	93053j060_2	1953	65	28-FEB-1993	13:11:35	7131	93074j081_2	89	15	19-MAR-1993	23:29:15	7288	93095j102_2	730	45	10-APR-1993	03:43:05
6988	93053j060_2	1832	15	28-FEB-1993	16:35:15	7134	93074j081_2	524	65	20-MAR-1993	10:21:09	7291	93095j102_2	1965	15	10-APR-1993	13:32:05
6988	93053j060_2	89	65	28-FEB-1993	16:19:35	7135	93074j081_2	90	15	20-MAR-1993	15:19:05	7292	93095j102_2	1922	15	10-APR-1993	16:40:05
6989	93053j060_2	1974	15	28-FEB-1993	19:40:05	7136	93074j081_2	1920	15	20-MAR-1993	15:34:05	7293	93095j102_2	60	15	10-APR-1993	20:00:35
6990	93053j060_2	474	15	28-FEB-1993	22:48:15	7137	93074j081_2	1016	15	20-MAR-1993	18:53:35	7293	93095j102_2	1912	45	10-APR-1993	20:59:45
6990	93053j060_2	1439	45	1-MAR-1993	00:07:11	7138	93074j081_2	551	45	20-MAR-1993	23:50:33	7294	93095j102_2	2020	45	10-APR-1993	23:08:35
6991	93053j060_2	1979	45	1-MAR-1993	02:08:35	7139	93074j081_2	1889	45	21-MAR-1993	01:22:05	7295	93095j102_2	1983	45	11-APR-1993	02:29:05
6992	93053j060_2	1381	45	1-MAR-1993	05:16:45	7145	93074j081_2	23	15	21-MAR-1993	23:56:55	7298	93095j102_2	1745	15	11-APR-1993	13:14:47
6993	93060j067_2	58	45	1-MAR-1993	08:37:15	7146	93074j081_2	1923	45	21-MAR-1993	23:56:55	7299	93095j102_2	1975	15	11-APR-1993	15:26:15
6994	93060j067_2	1810	65	1-MAR-1993	12:50:25	7147	93074j081_2	1892	45	22-MAR-1993	03:16:25	7300	93095j102_2	1564	15	11-APR-1993	18:34:15
6995	93060j067_2	233	15	1-MAR-1993	17:01:37	7150	93074j088_2	161	15	22-MAR-1993	14:55:03	7300	93095j102_2	353	45	11-APR-1993	20:29:55
6995	93060j067_2	1722	65	1-MAR-1993	15:05:45	7151	93074j088_2	1965	15	22-MAR-1993	16:13:25	7301	93095j102_2	1968	15	11-APR-1993	21:54:45
6996	93060j067_2	1915	15	1-MAR-1993	18:13:55	7152	93074j088_2	1930	15	22-MAR-1993	19:21:35	7302	93095j102_2	2037	45	12-APR-1993	01:02:45
6997	93060j067_2	1886	15	1-MAR-1993	21:34:15	7155	93074j088_2	85	45	23-MAR-1993	08:04:35	7303	93095j102_2	265	45	12-APR-1993	04:23:15
6999	93060j067_2	73	45	2-MAR-1993	06:58:05	7156	93074j088_2	1930	65	23-MAR-1993	08:18:35	7304	93102j109_2	1127	65	12-APR-1993	09:01:01
7000	93060j067_2	1240	45	2-MAR-1993	07:11:05	7159	93074j088_2	84	15	23-MAR-1993	21:01:45	7305	93102j109_2	84	15	12-APR-1993	13:45:55
7000	93060j067_2	665	65	2-MAR-1993	08:55:49	7160	93074j088_2	1874	15	23-MAR-1993	21:15:45	7305	93102j109_2	2159	65	12-APR-1993	10:51:37
7001	93060j067_2	1971	65	2-MAR-1993	10:31:25	7161	93074j088_2	1157	45	24-MAR-1993	02:01:43	7306	93102j109_2	1922	15	12-APR-1993	13:59:55
7002	93060j067_2	1924	65	2-MAR-1993	13:39:45	7162	93074j088_2	1722	45	24-MAR-1993	03:44:25	7307	93102j109_2	1875	15	12-APR-1993	17:20:25
7003	93060j067_2	1970	65	2-MAR-1993	17:00:05	7165	93074j088_2	1567	15	24-MAR-1993	14:45:05	7308	93102j109_2	1925	15	12-APR-1993	20:28:25
7004	93060j067_2	92	65	2-MAR-1993	20:08:15	7166	93074j088_2	229	15	24-MAR-1993	16:41:25	7309	93102j109_2	1788	15	13-APR-1993	00:48:05
7005	93060j067_2	1901	45	3-MAR-1993	00:27:35	7167	93074j088_2	1914	15	24-MAR-1993	21:01:05	7310	93102j109_2	1911	45	13-APR-1993	02:57:55
7006	93060j067_2	1922	45	3-MAR-1993	02:36:55	7168	93074j088_2	62	15	24-MAR-1993	14:25:21	7313	93102j109_2	745	15	13-APR-1993	14:25:21
7009	93060j067_2	1959	65	3-MAR-1993	12:25:45	7173	93074j088_2	1782	15	25-MAR-1993	15:27:35	7314	93102j109_2	1909	15	13-APR-1993	15:54:55
7010	93060j067_2	1916	65	3-MAR-1993	15:34:05	7176	93074j088_2	1812	45	26-MAR-1993	02:09:35	7318	93102j109_2	1291	65	14-APR-1993	05:56:25
7011	93060j067_2	938	65	3-MAR-1993	18:54:15	7177	93074j088_2	1892	45	26-MAR-1993	04:24:45	7319	93102j109_2	1977	15	14-APR-1993	08:11:35
7016	93060j067_2	1834	65	4-MAR-1993	12:04:45	7180	93074j088_2	1812	15	26-MAR-1993	15:06:35	7320	93102j109_2	657	15	14-APR-1993	13:04:55
7017	93060j067_2	1966	65	4-MAR-1993	14:19:55	7181	93074j088_2	61	15	26-MAR-1993	17:21:45	7320	93102j109_2	1253	15	14-APR-1993	11:20:35
7018	93060j067_2	95	65	4-MAR-1993	17:28:25	7182	93074j088_2	16	45	26-MAR-1993	23:41:35	7321	93102j109_2	2085	15	14-APR-1993	14:40:05
7020	93060j067_2	1869	45	4-MAR-1993	23:57:45	7183	93074j088_2	405	45	26-MAR-1993	23:44:15	7322	93102j109_2	1920	15	14-APR-1993	17:49:05
7021	93060j067_2	1979	45	5-MAR-1993	03:17:05	7184	93074j088_2	1907	45	27-MAR-1993	02:58:25	7323	93102j109_2	1780	15	14-APR-1993	21:08:35
7022	93060j067_2	96	45	5-MAR-1993	06:25:35	7187	93074j088_2	632	15	27-MAR-1993	14:30:07	7327	93102j109_2	1976	65	15-APR-1993	10:05:45
7025	93060j067_2	16	15	5-MAR-1993	16:16:05	7188	93074j088_2	1929	15	27-MAR-1993	15:55:15	7328	93102j109_2	1910	45	15-APR-1993	13:14:25
7025	93060j067_2	1182	65	5-MAR-1993	17:29:55	7189	93074j088_2	1889	15	27-MAR-1993	19:16:05	7330	93102j109_2	1830	15	15-APR-1993	20:47:35
7027	93060j067_2	1594	45	5-MAR-1993	23:52:35	7189	93074j088_2	64	45	27-MAR-1993	22:13:25	7331	93102j109_2	22	15	15-APR-1993	23:02:45
7028	93060j067_2	1928	45	6-MAR-1993	01:51:15	7190	93074j088_2	1932	45	27-MAR-1993	22:24:05	7333	93102j109_2	1968	65	16-APR-1993	05:31:15
7031	93060j067_2	1927	65	6-MAR-1993	12:38:53	7191	93074j088_2	1977	45	28-MAR-1993	01:44:35	7334	93102j109_2	1917	65	16-APR-1993	08:40:15
7032	93060j067_2	1253	15	6-MAR-1993	16:11:27	7192	93074j088_2	97	45	28-MAR-1993	04:52:45	7335	93102j109_2	1979	65	16-APR-1993	11:59:55
7032	93060j067_2	607	65	6-MAR-1993	14:48:25	7195	93074j088_2	1975	15	28-MAR-1993	14:41:45	7336	93102j109_2	1826	15	16-APR-1993	16:13:25
7033	93060j067_2	1979	15	6-MAR-1993	18:08:25	7196	93074j088_2	1931	15	28-MAR-1993	17:49:45	7336	93102j109_2	91	15	16-APR-1993	15:08:45
7034	93060j067_2	1931	15	6-MAR-1993	21:16:55	7197	93074j088_2	61	15	28-MAR-1993	21:10:15	7338	93102j109_2	1734	45	16-APR-1993	22:46:45
7035	93060j067_2	1978	45	7-MAR-1993	00:37:05	7197	93074j088_2	2022	45	28-MAR-1993	22:09:15	7339	93102j109_2	59	45	17-APR-1993	00:56:55
7036	93060j067_2	1926	45	7-MAR-1993	04:35:05	7198	93074j088_2	1925	45	29-MAR-1993	00:18:25	7341	93102j109_2	1294	65	17-APR-1993	08:46:35
7037	93060j067_2	61	45	7-MAR-1993	07:05:35	7199	93074j088_2	1770	45	29-MAR-1993	03:38:45	7342	93102j109_2	1918	65	17-APR-1993	10:34:15

7343	93102j109_2	1974	15	17-APR-1993	13:53:55	7449	93116j123_2	1957	45	1-MAY-1993	21:02:55	7573	93137j144_2	1825 15 18-MAY-1993 15:16:35
7344	93102j109_2	1920	15	17-APR-1993	17:02:45	7450	93116j123_2	1370	45	2-MAY-1993	00:10:45	7574	93137j144_2	1850 15 18-MAY-1993 17:25:45
7345	93102j109_2	1973	45	17-APR-1993	20:22:25	7455	93116j123_2	84	45	2-MAY-1993	19:21:55	7575	93137j144_2	949 15 18-MAY-1993 21:44:55
7346	93102j109_2	1913	45	17-APR-1993	23:31:15	7456	93116j123_2	1967	45	2-MAY-1993	19:35:55	7578	93137j144_2	1830 65 19-MAY-1993 07:27:15
7347	93102j109_2	1882	45	18-APR-1993	02:50:55	7457	93116j123_2	1939	45	2-MAY-1993	22:56:35	7579	93137j144_2	1829 65 19-MAY-1993 10:41:25
7349	93102j109_2	1968	65	18-APR-1993	09:20:15	7458	93116j123_2	94	45	3-MAY-1993	02:04:15	7583	93137j144_2	924 45 20-MAY-1993 00:11:37
7350	93102j109_2	1829	15	18-APR-1993	13:32:45	7459	93116j130_2	1192	65	3-MAY-1993	06:47:47	7587	93137j144_2	1833 15 20-MAY-1993 12:34:15
7350	93102j109_2	92	65	18-APR-1993	12:28:15	7460	93123j130_2	1311	65	3-MAY-1993	08:32:45	7588	93137j144_2	1844 15 20-MAY-1993 14:43:05
7351	93102j109_2	1969	15	18-APR-1993	15:47:55	7461	93123j130_2	1905	15	3-MAY-1993	12:52:35	7591	93137j144_2	1833 45 21-MAY-1993 01:30:35
7352	93102j109_2	91	15	18-APR-1993	18:56:45	7462	93123j130_2	1265	15	3-MAY-1993	15:01:05	7592	93137j144_2	13 45 21-MAY-1993 03:39:35
7352	93102j109_2	1922	45	18-APR-1993	20:01:25	7463	93123j130_2	60	15	3-MAY-1993	18:21:35	7593	93137j144_2	1829 65 21-MAY-1993 07:58:45
7353	93102j109_2	1978	45	18-APR-1993	22:16:35	7463	93123j130_2	1247	45	3-MAY-1993	19:20:45	7594	93137j144_2	1820 65 21-MAY-1993 11:14:35
7354	93102j109_2	1914	45	19-APR-1993	01:25:15	7464	93123j130_2	1262	45	3-MAY-1993	21:29:25	7595	93137j144_2	8 15 21-MAY-1993 15:14:45
7355	93102j109_2	58	45	19-APR-1993	04:45:05	7465	93123j130_2	1965	45	4-MAY-1993	00:50:05	7596	93137j144_2	1833 15 21-MAY-1993 16:35:55
7356	93109j116_2	1830	65	19-APR-1993	08:58:25	7466	93123j130_2	99	45	4-MAY-1993	03:57:45	7597	93137j144_2	1228 15 21-MAY-1993 20:55:05
7357	93109j116_2	853	15	19-APR-1993	12:49:35	7467	93123j130_2	1917	65	4-MAY-1993	08:17:25	7600	93137j144_2	1145 65 22-MAY-1993 07:04:09
7357	93109j116_2	955	65	19-APR-1993	11:13:35	7468	93123j130_2	2005	45	4-MAY-1993	10:26:05	7601	93137j144_2	1826 45 22-MAY-1993 09:51:25
7358	93109j116_2	94	15	19-APR-1993	14:22:15	7469	93123j130_2	865	65	4-MAY-1993	13:46:55	7603	93137j144_2	1831 15 22-MAY-1993 16:19:45
7359	93109j116_2	1880	15	19-APR-1993	17:42:05	7471	93123j130_2	1855	45	4-MAY-1993	21:17:49	7604	93137j144_2	1848 15 22-MAY-1993 18:28:35
7359	93109j116_2	88	45	19-APR-1993	20:36:05	7472	93123j130_2	1175	45	4-MAY-1993	23:22:55	7605	93137j144_2	1 45 23-MAY-1993 00:56:35
7360	93109j116_2	1926	45	19-APR-1993	20:50:45	7474	93123j130_2	1796	65	5-MAY-1993	06:26:35	7606	93137j144_2	1851 45 23-MAY-1993 06:46:35
7361	93109j116_2	2137	45	20-APR-1993	00:10:17	7475	93123j130_2	61	65	5-MAY-1993	09:11:45	7608	93137j144_2	1725 65 23-MAY-1993 07:24:55
7363	93109j116_2	515	65	20-APR-1993	08:26:17	7478	93123j130_2	1826	45	5-MAY-1993	19:00:15	7609	93137j144_2	1821 15 23-MAY-1993 11:44:15
7364	93109j116_2	1896	65	20-APR-1993	09:47:45	7479	93123j130_2	1919	45	5-MAY-1993	22:08:35	7610	93137j144_2	1848 15 23-MAY-1993 13:53:05
7365	93109j116_2	1972	15	20-APR-1993	13:07:45	7483	93123j130_2	1910	15	6-MAY-1993	12:04:25	7611	93137j144_2	1834 15 23-MAY-1993 18:12:25
7366	93109j116_2	1911	15	20-APR-1993	16:16:15	7484	93123j130_2	1926	15	6-MAY-1993	14:13:55	7612	93137j144_2	1681 15 23-MAY-1993 20:21:15
7367	93109j116_2	1977	15	20-APR-1993	19:36:05	7485	93123j130_2	61	15	6-MAY-1993	17:33:45	7615	93144j151_2	1835 65 24-MAY-1993 07:08:55
7368	93109j116_2	44	15	20-APR-1993	22:44:45	7485	93123j130_2	1374	45	6-MAY-1993	18:52:45	7616	93144j151_2	1847 65 24-MAY-1993 09:17:35
7369	93109j116_2	1891	45	21-APR-1993	02:04:35	7486	93123j130_2	1906	45	6-MAY-1993	20:42:15	7617	93144j151_2	1 15 24-MAY-1993 15:45:45
7371	93109j116_2	1905	65	21-APR-1993	09:32:05	7487	93123j130_2	1661	45	7-MAY-1993	00:02:05	7617	93144j151_2	1832 65 24-MAY-1993 13:36:55
7372	93109j116_2	1824	15	21-APR-1993	12:46:25	7488	93123j130_2	644	45	7-MAY-1993	03:10:45	7618	93144j151_2	1851 15 24-MAY-1993 15:45:35
7372	93109j116_2	94	65	21-APR-1993	11:41:45	7489	93123j130_2	259	65	7-MAY-1993	08:06:25	7619	93144j151_2	1833 15 24-MAY-1993 20:05:15
7373	93109j116_2	432	15	21-APR-1993	15:01:45	7490	93123j130_2	1888	65	7-MAY-1993	09:18:25	7620	93144j151_2	18 15 24-MAY-1993 22:13:45
7374	93109j116_2	1100	45	21-APR-1993	19:40:41	7491	93123j130_2	1972	15	7-MAY-1993	12:58:45	8559	93228j235_2	508 45 16-AUG-1993 20:33:57
7375	93109j116_2	1891	45	21-APR-1993	21:30:05	7492	93123j130_2	94	15	7-MAY-1993	16:07:25	8574	93228j235_2	640 45 17-AUG-1993 20:46:51
7378	93109j116_2	1529	65	22-APR-1993	08:21:27	7493	93123j130_2	1838	45	7-MAY-1993	20:25:15	8575	93228j235_2	507 45 17-AUG-1993 21:46:13
7379	93109j116_2	87	15	22-APR-1993	13:21:05	7494	93123j130_2	1841	45	7-MAY-1993	22:35:45	8576	93228j235_2	651 45 17-AUG-1993 23:55:53
7379	93109j116_2	9	65	22-APR-1993	10:27:05	7495	93123j130_2	1943	45	8-MAY-1993	02:53:35	8584	93228j235_2	644 15 18-AUG-1993 12:32:01
7380	93109j116_2	1925	15	22-APR-1993	13:35:35	7497	93123j130_2	1827	65	8-MAY-1993	09:21:55	8585	93228j235_2	509 15 18-AUG-1993 13:31:21
7381	93109j116_2	1916	15	22-APR-1993	17:15:55	7498	93123j130_2	1815	15	8-MAY-1993	12:36:05	8589	93228j235_2	71 45 18-AUG-1993 20:03:53
7382	93109j116_2	1916	15	22-APR-1993	20:04:05	7498	93123j130_2	11	65	8-MAY-1993	11:31:55	8601	93228j235_2	505 15 19-AUG-1993 14:43:35
7383	93109j116_2	1916	45	23-APR-1993	00:23:05	7504	93123j130_2	1817	65	9-MAY-1993	08:01:05	8604	93228j235_2	499 45 19-AUG-1993 20:06:11
7384	93109j116_2	1915	45	23-APR-1993	02:32:35	7505	93123j130_2	1822	65	9-MAY-1993	11:15:15	8605	93228j235_2	505 45 19-AUG-1993 21:01:39
7385	93109j116_2	1971	65	23-APR-1993	05:52:35	7507	93123j130_2	7	45	9-MAY-1993	19:52:15	8606	93228j235_2	1678 45 19-AUG-1993 22:36:09
7386	93109j116_2	1919	65	23-APR-1993	09:01:05	7508	93123j130_2	1845	45	9-MAY-1993	19:53:25	8607	93228j235_2	420 45 20-AUG-1993 00:13:35
7387	93109j116_2	1917	15	23-APR-1993	13:20:05	7509	93123j130_2	1829	45	10-MAY-1993	00:11:55	8620	93228j235_2	1690 45 20-AUG-1993 20:39:23
7387	93109j116_2	59	65	23-APR-1993	12:20:55	7510	93123j130_2	1827	45	10-MAY-1993	02:21:45	8621	93228j235_2	501 45 20-AUG-1993 22:13:53
7388	93109j116_2	1922	15	23-APR-1993	15:29:25	7511	93130j137_2	576	65	10-MAY-1993	07:23:35	8622	93228j235_2	1689 45 20-AUG-1993 23:48:25
7389	93109j116_2	59	15	23-APR-1993	18:49:35	7512	93130j137_2	548	65	10-MAY-1993	08:49:55	8625	93228j235_2	501 65 21-AUG-1993 04:31:59
7389	93109j116_2	1915	45	23-APR-1993	19:48:35	7513	93130j137_2	1825	15	10-MAY-1993	13:08:35	8626	93228j235_2	1690 65 21-AUG-1993 06:06:27
7390	93109j116_2	1924	45	23-APR-1993	21:57:55	7514	93130j137_2	1815	15	10-MAY-1993	15:18:15	8627	93228j235_2	513 65 21-AUG-1993 07:40:59
7391	93109j116_2	1976	45	24-APR-1993	01:17:55	7515	93130j137_2	1819	45	10-MAY-1993	19:36:55	8630	93228j235_2	1689 15 21-AUG-1993 12:24:33
7392	93109j116_2	93	45	24-APR-1993	04:26:25	7516	93130j137_2	1848	45	10-MAY-1993	21:46:35	8631	93228j235_2	502 15 21-AUG-1993 13:59:03
7393	93109j116_2	1580	65	24-APR-1993	08:57:49	7517	93130j137_2	1828	45	11-MAY-1993	01:00:35	8632	93228j235_2	798 15 21-AUG-1993 15:33:33
7394	93109j116_2	789	15	24-APR-1993	12:35:53	7518	93130j137_2	13	45	11-MAY-1993	04:14:45	8635	93228j235_2	499 45 21-AUG-1993 20:17:07
7394	93109j116_2	964	65	24-APR-1993	10:54:55	7520	93130j137_2	1671	15	11-MAY-1993	11:53:51	8636	93228j235_2	1613 45 21-AUG-1993 21:51:37
7395	93109j116_2	1968	15	24-APR-1993	14:14:55	7521	93130j137_2	1840	15	11-MAY-1993	15:01:45	8640	93228j235_2	1682 65 22-AUG-1993 04:09:43
7396	93109j116_2	1926	15	24-APR-1993	17:23:15	7522	93130j137_2	11	15	11-MAY-1993	17:11:25	8641	93228j235_2	499 65 22-AUG-1993 05:44:11
7397	93109j116_2	1979	45	24-APR-1993	20:43:25	7523	93130j137_2	1812	15	11-MAY-1993	21:30:15	8642	93228j235_2	1306 65 22-AUG-1993 07:18:41
7398	93109j116_2	1911	45	24-APR-1993	23:51:45	7524	93130j137_2	1832	45	11-MAY-1993	23:39:45	8645	93228j235_2	500 15 22-AUG-1993 12:02:17
7399	93109j116_2	1541	45	25-APR-1993	03:11:45	7525	93130j137_2	770	45	12-MAY-1993	05:13:55	8646	93228j235_2	1643 15 22-AUG-1993 13:36:47
7400	93109j116_2	916	65	25-APR-1993	07:56:43	7526	93130j137_2	1818	65	12-MAY-1993	07:13:25	8647	93228j235_2	495 15 22-AUG-1993 15:11:19
7401	93109j116_2	87	15	25-APR-1993	12:34:15	7527	93130j137_2	6	15	12-MAY-1993	12:35:13	8648	93228j235_2	359 15 22-AUG-1993 16:45:47
7401	93109j116_2	1733	65	25-APR-1993	09:40:25	7527	93130j137_2	1817	65	12-MAY-1993	10:26:35	8650	93228j235_2	1338 45 22-AUG-1993 20:06:23
7402	93109j116_2	1898	15	25-APR-1993	12:48:45	7528	93130j137_2	1824	15	12-MAY-1993	12:36:15	8651	93228j235_2	491 45 22-AUG-1993 21:29:21
7403	93109j116_2	1971	15	25-APR-1993	16:08:45	7529	93130j137_2	1834	15	12-MAY-1993	16:54:55	8652	93228j235_2	1683 45 22-AUG-1993 23:03:53
7404	93109j116_2	22	15	25-APR-1993	19:17:05	7529	93130j137_2	6	45	12-MAY-1993	19:03:35	8653	93228j235_2	492 15 23-AUG-1993 00:38:21
7404	93109j116_2	1894	45	25-APR-1993	19:21:05	7530	93130j137_2	1836	45	12-MAY-1993	19:04:25	8655	93228j235_2	53 15 23-AUG-1993 04:02:13
7405	93109j116_2	1968	45	25-APR-1993	22:37:15	7533	93130j137_2	6	65	13-MAY-1993	08:00:05	8656	93228j235_2	1683 65 23-AUG-1993 05:21:55
7406	93109j116_2	1839	45	26-APR-1993	01:45:35	7534	93130j137_2	1844	45	13-MAY-1993	08:01:05	8657	93228j235_2	53 65 23-AUG-1993 06:56:25
7408	93116j123_2	1829	65	26-APR-1993	09:19:05	7535	93130j137_2	1822	15	13-MAY-1993	12:21:31	8666	93235j242_2	853 45 23-AUG-1993 21:07:05
7409	93116j123_2	1903	15	26-APR-1993	12:33:25	7536	93130j137_2	1846	15	13-MAY-1993	14:29:15	8670	93235j242_2	2 65 24-AUG-1993 04:03:17
7409	93116j123_2	58	65	26-APR-1993	11:34:15	7537	93130j137_2	1687	45	13-MAY-1993	18:48:05	8671	93235j242_2	509 65 24-AUG-1993 04:59:37
7410	93116j123_2	1927	15	26-APR-1993	14:42:25	7538	93130j137_2	1845	45	13-MAY-1993	20:57:35	8672	93235j242_2	1697 65 24-AUG-1993 06:34:09
7411	93116j123_2	448	15	26-APR-1993	18:02:35	7539	93130j137_2	1835	45	14-MAY-1993	00:06:45	8681	93235j242_2	508 45 24-AUG-1993 20:44:47
7411	93116j123_2	1623	45	26-APR-1993	19:16:17	7540	93130j137_2	14	45	14-MAY-1993	03:25:45	8682	93235j242_2	1692 45 24-AUG-1993 22:19:23
7412	93116j123_2	1927	45	26-APR-1993	21:10:55	7541	93130j137_2	1840	65	14-MAY-1993	09:54:05	8683	93235j242_2	505 45 24-AUG-1993 23:53:49
7413	93116j123_2	1880	45	27-APR-1993	00:31:05	7542	93130j137_2	413	15	14-MAY-1993	11:48:49	8690	93235j242_2	195 15 25-AUG-1993 11:45:19
7417	93116j123_2	1947	15	27-APR-1993	13:28:05	7542	93130j137_2	13	65	14-MAY-1993	09:54:05	8691	93235j242_2	504 15 25-AUG-1993 12:29:55
7418	93116j123_2	94	15	27-APR-1993	16:36:15	7543	93130j137_2	1839	15	14-MAY-1993	14:12:55	8692	93235j242_2	1694 15 25-AUG-1993 14:04:25
7419	93116j123_2	1976	45	27-APR-1993	19:56:15	7544	93130j137_2	1845	15	14-MAY-1993	16:22:15	8693	93235j242_2	502 15 25-AUG-1993 15:38:57
7420	93116j123_2	1917	45	27-APR-1993	23:04:35	7545	93130j137_2	1840	45	14-MAY-1993	20:41:05	8696	93235j242_2	1684 45 25-AUG-1993 20:22:37
7421	93116j123_2	1871	45	28-APR-1993	02:24:55	7546	93130j137_2	1848	45	14-MAY-1993	22:50:35	8697	93235j242_2	501 45 25-AUG-1993 21:57:01
7424	93116j123_2	1863	15	28-APR-1993	13:06:27	7549	93130j137_2	1841	65	15-MAY-1993	09:37:35	8698	93235j242_2	1691 45 25-AUG-1993 23:31:31
7425	93116j123_2	1915	15	28-APR-1993	15:21:45	7550	93130j137_2	1835	15	15-MAY-1993	11:48:35	8699	93235j242_2	426 45 26-AUG-1993 01:06:03
7426	93116j123_2	86	15	28-APR-1993	18:29:55	7550	93130j137_2	7	65	15-MAY-1993	14:47:05	8707	93235j242_2	502 15 26-AUG-1993 13:42:09
7426	93116j123_2	1880	45	28-APR-1993	19:34:39	7551	93130j137_2	1836	15	15-MAY-1993	16:05:55	8708	93235j242_2	1684 15 26-AUG-1993 15:16:41
7427	93116j123_2	59	45	28-APR-1993	21:50:05	7552	93130j137_2	14	15	15-MAY-1993	18:15:15	8716	93235j242_2	1175 65 27-AUG-1993 04:09:05
7430	93116j123_2	1732	65	29-APR-1993	08:36:23	7552	93130j137_2	1824	15	15-MAY-1993	19:20:05	8717	93235j242_2	1687 65 27-AUG-1993 05:27:35
7431	93116j123_2	59	65	29-APR-1993	10:46:55	7553	93130j137_2	1840	45	15-MAY-1993	22:34:05	8718	93235j242_2	1329 65 27-AUG-1993 07:02:05
7433	93116j123_2	1553	15	29-APR-1993	17:16:55	7554	93130j137_2	1844	45	16-MAY-1993	07:11:15	8726	93235j242_2	814 45 27-AUG-1993 20:07:05
7433	93116j123_2	404	45	29-APR-1993	19:06:27	7555	93130j137_2	3	65	16-MAY-1993	07:11:15	8727	93235j242_2	1683 45 27-AUG-1993 21:12:43
7434	93116j123_2	1238	45	29-APR-1993	20:23:35	7556	93130j137_2	1493	15	16-MAY-1993	11:45:23	8728	93235j242_2	1685 45 27-AUG-1993 22:47:13
7435	93116j123_2	1312	45	29-APR-1993	23:43:55	7557	93130j137_2	312	65	16-MAY-1993	11:30:35	8731	93235j242_2	470 65 28-AUG-1993 04:11:09
7436	93116j123_2	95	45	30-APR-1993	02:51:55	7557	93130j137_2	1234	15	16-MAY-1993	17:58:55	8732	93235j242_2	1681 65 28-AUG-1993 05:05:19
7436	93116j123_2	403	65	30-APR-1993	04:46:23	7558	93130j137_2	1845	15	16-MAY-1993	13:39:55	8733	93235j242_2	1674 65 28-AUG-1993 06:39:49
7437	93116j123_2	1968	65	30-APR-1993	06:12:25	7559	93130j137_2	552	45	16-MAY-1993	18:45:05	8736	93235j242_2	17 15 28-AUG-1993 11:49:51
7438	93116j123_2	94	65	30-APR-1993	09:20:25	7560	93130j137_2	1846	45	16-MAY-1993	20:08:15	8737	93235j242_2	1669 15 28-AUG-1993 12:57:53
7441	93116j123_2	1961	45	30-APR-1993	19:09:25	7561	93130j137_2	1826	45	17-MAY-1993	19:20:45	8738	93235j242_2	1681 15 28-AUG-1993 14:32:25
7442	93116j123_2	1921	45	30-APR-1993	22:17:15	7562	93130j137_2	1003	45	17-MAY-1993	02:36:25	8739	93235j242_2	1160 15 28-AUG-1993 16:06:55
7443	93116j123_2	1966	45	1-MAY-1993	01:37:35	7563	93130j137_2	231	45	17-MAY-1993	07:23:55	8742	93235j242_2	1681 45 28-AUG-1993 20:50:27
7444	93116j123_2	11	45	1-MAY-1993	04:45:35	7564	93137j144_2	1851	65	17-MAY-1993	09:04:35	8744	93235j242_2	1681 45 28-AUG-1993 23:59:31
7445	93116j123_2	1907	65	1-MAY-1993	09:05:15	7565	93137j144_2	1839	15	17-MAY-1993	13:23:35	8747	93235j242_2	1679 65 29-AUG-1993 04:43:03
7446	93116j123_2	795	15	1-MAY-1993	12:55:23	7566	93137j144_2	1847	15	17-MAY-1993	11:13:55	8748	93235j242_2	1679 65 29-AUG-1993 06:17:37
7446	93116j123_2	1103	65	1-MAY-1993	11:13:55	7569	93137j144_2	3	65	18-MAY-1993	04:28:45	8749	93235j242_2	265 65 29-AUG-1993 07:52:07
7447	93116j123_2	1952	15	1-MAY-1993	14:34:15	7570	93137j144_2	1849	65	18-MAY-1993	04:29:15	8751	93235j242_2	166 15 29-AUG-1993 11:51:35
7448	93116j123_2	448	15	1-MAY-1993	17:42:25	7571	93137j144_2	1830	65	18-MAY-1993	08:48:25	8752	93235j242_2	1678 45 29-AUG-1993 12:35:39
7448	93116j123_2	1443	45	1-MAY-1993	19:01:39							8753	93235j242_2	1611 15 29-AUG-1993 14:10:09

E-5

```
8754 93235j242_2  1225 15 29-AUG-1993 15:44:39    9026 93256j263_2  1049 15 16-SEP-1993 12:30:19    9366 93277j284_2  1746 45  8-OCT-1993 20:09:17
8762 93235j242_2  1671 65 30-AUG-1993 04:20:49    9027 93256j263_2  1710 15 16-SEP-1993 13:46:37    9367 93277j284_2   368 45  8-OCT-1993 21:43:21
8763 93235j242_2  1675 65 30-AUG-1993 05:55:21    9028 93256j263_2  1286 15 16-SEP-1993 15:21:27    9372 93277j284_2  1584 65  9-OCT-1993 05:41:49
8764 93242j249_2   789 65 30-AUG-1993 07:29:51    9031 93256j263_2  1426 45 16-SEP-1993 20:08:51    9373 93277j284_2  1162 65  9-OCT-1993 07:11:21
8772 93242j249_2  1562 45 30-AUG-1993 20:10:17    9032 93256j263_2  1694 45 16-SEP-1993 21:39:29    9377 93277j284_2   340 15  9-OCT-1993 13:29:29
8773 93242j249_2  1694 45 30-AUG-1993 21:40:29    9033 93256j263_2   155 15 16-SEP-1993 23:13:59    9378 93277j284_2  1746 15  9-OCT-1993 15:04:43
8774 93242j249_2  1692 45 30-AUG-1993 23:14:59    9034 93256j263_2   287 45 17-SEP-1993 00:48:25    9382 93277j284_2  1071 45  9-OCT-1993 21:23:41
8775 93242j249_2   808 45 31-AUG-1993 00:49:31    9041 93256j263_2   541 15 17-SEP-1993 12:28:43    9387 93277j284_2   552 65 10-OCT-1993 05:56:37
8782 93242j249_2   465 15 31-AUG-1993 12:32:11    9042 93256j263_2  1673 15 17-SEP-1993 13:24:35    9388 93277j284_2  1378 45 10-OCT-1993 06:51:03
8783 93242j249_2    32 15 31-AUG-1993 13:50:43    9043 93256j263_2  1702 15 17-SEP-1993 14:59:01    9392 93277j284_2  1667 15 10-OCT-1993 13:12:49
8785 93242j249_2    73 15 31-AUG-1993 16:37:29    9044 93256j263_2   740 15 17-SEP-1993 16:33:17    9393 93277j284_2  1764 15 10-OCT-1993 14:44:17
8787 93242j249_2   853 45 31-AUG-1993 20:11:43    9046 93256j263_2   736 45 17-SEP-1993 20:14:47    9397 93277j284_2  1269 45 10-OCT-1993 21:02:53
8788 93242j249_2  1694 45 31-AUG-1993 21:18:15    9047 93256j263_2   576 45 17-SEP-1993 21:16:51    9402 93277j284_2   214 65 11-OCT-1993 05:47:37
8789 93242j249_2  1688 45 31-AUG-1993 22:52:45    9052 93256j263_2  1692 65 18-SEP-1993 05:09:41    9403 93277j284_2  1768 65 11-OCT-1993 06:30:47
8790 93242j249_2  1002 45  1-SEP-1993 00:27:15    9053 93256j263_2  1705 65 18-SEP-1993 06:43:55    9407 93284j291_2   389 15 11-OCT-1993 13:34:59
8792 93242j249_2   421 65  1-SEP-1993 04:18:31    9057 93256j263_2   827 15 18-SEP-1993 13:02:13    9412 93284j291_2  1613 45 11-OCT-1993 20:42:19
8793 93242j249_2  1691 65  1-SEP-1993 05:10:51    9058 93256j263_2   191 15 18-SEP-1993 14:37:01    9418 93284j291_2  1122 65 12-OCT-1993 06:13:39
8794 93242j249_2  1687 65  1-SEP-1993 06:45:21    9059 93256j263_2   100 15 18-SEP-1993 16:11:29    9422 93284j291_2   319 15 12-OCT-1993 13:17:11
8805 93242j249_2  1679 45  2-SEP-1993 00:05:17    9067 93256j263_2   154 65 19-SEP-1993 04:55:13    9423 93284j291_2  1765 15 12-OCT-1993 14:03:39
8813 93242j249_2  1393 15  2-SEP-1993 12:41:09    9068 93256j263_2  1697 65 19-SEP-1993 06:21:49    9424 93284j291_2   665 15 12-OCT-1993 15:38:17
8814 93242j249_2  1692 15  2-SEP-1993 14:15:39    9072 93256j263_2  1703 15 19-SEP-1993 12:39:47    9427 93284j291_2  1642 45 12-OCT-1993 20:26:13
8815 93242j249_2  1687 15  2-SEP-1993 15:50:11    9073 93256j263_2  1163 15 19-SEP-1993 14:14:37    9428 93284j291_2  1781 45 12-OCT-1993 21:56:41
8816 93242j249_2    47 15  2-SEP-1993 17:24:41    9082 93256j263_2   725 65 20-SEP-1993 04:57:25    9429 93284j291_2   446 45 12-OCT-1993 23:31:17
8823 93242j249_2  1703 65  3-SEP-1993 04:26:19    9083 93256j263_2   760 65 20-SEP-1993 05:59:27    9438 93284j291_2  1775 15 13-OCT-1993 13:42:43
8824 93242j249_2  1703 65  3-SEP-1993 06:00:49    9084 93256j263_2   146 65 20-SEP-1993 07:34:13    9439 93284j291_2  1186 15 13-OCT-1993 15:17:21
8825 93242j249_2   484 65  3-SEP-1993 07:35:21    9092 93263j270_2    57 45 20-SEP-1993 20:10:23    9448 93284j291_2  1099 65 14-OCT-1993 05:52:29
8834 93242j249_2  1655 45  3-SEP-1993 21:46:59    9093 93263j270_2  1717 45 20-SEP-1993 21:44:43    9449 93284j291_2   846 65 14-OCT-1993 07:02:41
8835 93242j249_2  1497 45  3-SEP-1993 23:20:31    9103 93263j270_2  1720 15 21-SEP-1993 13:29:47    9453 93284j291_2  1795 15 14-OCT-1993 13:22:23
8838 93242j249_2  1062 65  4-SEP-1993 04:24:41    9107 93263j270_2  1037 45 21-SEP-1993 20:10:35    9454 93284j291_2  1795 45 14-OCT-1993 14:57:01
8839 93242j249_2  1683 65  4-SEP-1993 05:38:35    9108 93263j270_2  1736 45 21-SEP-1993 21:22:03    9457 93284j291_2  1198 45 14-OCT-1993 20:01:03
8840 93242j249_2  1048 65  4-SEP-1993 07:13:05    9109 93263j270_2  1297 15 21-SEP-1993 22:56:27    9458 93284j291_2   454 45 14-OCT-1993 21:15:35
8843 93242j249_2  1076 15  4-SEP-1993 12:16:59    9118 93263j270_2  1675 15 22-SEP-1993 13:07:41    9468 93284j291_2  1149 15 15-OCT-1993 13:23:09
8844 93242j249_2  1651 15  4-SEP-1993 13:31:11    9119 93263j270_2    74 15 22-SEP-1993 14:42:37    9469 93284j291_2  1807 15 15-OCT-1993 14:36:39
8845 93242j249_2  1673 15  4-SEP-1993 15:05:41    9122 93263j270_2   379 45 22-SEP-1993 20:10:15    9473 93284j291_2   206 45 15-OCT-1993 20:55:11
8846 93242j249_2   217 15  4-SEP-1993 16:40:17    9124 93263j270_2    79 45 22-SEP-1993 22:34:25    9479 93284j291_2  1814 65 16-OCT-1993 06:23:03
8848 93242j249_2  1067 45  4-SEP-1993 20:09:43    9128 93263j270_2   782 45 23-SEP-1993 05:04:27    9483 93284j291_2   499 15 16-OCT-1993 13:25:29
8849 93242j249_2  1679 45  4-SEP-1993 21:23:43    9138 93263j270_2   891 45 23-SEP-1993 21:05:23    9487 93284j291_2   157 45 16-OCT-1993 19:55:29
8850 93242j249_2  1684 45  4-SEP-1993 22:58:15    9139 93263j270_2   364 45 23-SEP-1993 22:12:21    9488 93284j291_2  1392 45 16-OCT-1993 20:33:37
8853 93242j249_2   359 65  5-SEP-1993 04:26:35    9143 93263j270_2   630 65 24-SEP-1993 05:06:37    9494 93284j291_2  1813 65 17-OCT-1993 06:02:33
8854 93242j249_2  1701 65  5-SEP-1993 05:16:21    9144 93263j270_2  1643 65 24-SEP-1993 06:05:13    9500 93291j298_2   651 15 17-OCT-1993 15:30:23
8855 93242j249_2  1689 65  5-SEP-1993 06:50:51    9145 93263j270_2   294 65 24-SEP-1993 07:39:03    9503 93291j298_2  1799 45 17-OCT-1993 20:15:19
8858 93242j249_2   789 15  5-SEP-1993 12:04:11    9149 93263j270_2   128 15 24-SEP-1993 13:57:23    9509 93291j298_2  1232 65 18-OCT-1993 06:02:05
8859 93242j249_2  1682 15  5-SEP-1993 13:08:55    9150 93263j270_2   646 15 24-SEP-1993 15:31:25    9510 93291j298_2   466 65 18-OCT-1993 07:16:45
8860 93242j249_2  1686 15  5-SEP-1993 14:43:25    9153 93263j270_2  1678 45 24-SEP-1993 20:16:33    9518 93291j298_2  1778 45 18-OCT-1993 19:55:57
8861 93242j249_2  1336 15  5-SEP-1993 16:17:59    9154 93263j270_2  1735 45 24-SEP-1993 21:51:05    9519 93291j298_2  1563 45 18-OCT-1993 21:28:11
8869 93242j249_2  1703 65  6-SEP-1993 04:54:05    9159 93263j270_2  1735 65 25-SEP-1993 05:43:35    9530 93291j298_2  1496 15 19-OCT-1993 15:01:17
8870 93242j249_2  1703 65  6-SEP-1993 06:28:35    9160 93263j270_2   770 65 25-SEP-1993 07:18:07    9531 93291j298_2   478 15 19-OCT-1993 16:24:03
8874 93249j256_2  1674 15  6-SEP-1993 12:46:39    9163 93263j270_2     1 15 25-SEP-1993 12:44:07    9536 93291j298_2   194 45 20-OCT-1993 01:12:19
8875 93249j256_2  1681 15  6-SEP-1993 14:21:11    9164 93263j270_2  1710 15 25-SEP-1993 13:37:05    9537 93291j298_2  1592 45 20-OCT-1993 01:52:07
8876 93249j256_2    16 15  6-SEP-1993 15:55:41    9165 93263j270_2  1736 15 25-SEP-1993 15:10:35    9540 93291j298_2  1634 65 20-OCT-1993 06:35:59
8889 93249j256_2   752 15  7-SEP-1993 12:05:19    9168 93263j270_2  1235 45 25-SEP-1993 20:10:57    9544 93291j298_2   641 15 20-OCT-1993 13:35:15
8890 93249j256_2  1676 15  7-SEP-1993 13:58:55    9169 93263j270_2  1740 45 25-SEP-1993 21:28:37    9545 93291j298_2  1866 15 20-OCT-1993 14:29:13
8891 93249j256_2   527 15  7-SEP-1993 15:33:27    9174 93263j270_2  1748 65 26-SEP-1993 05:20:33    9555 93291j298_2  1871 65 21-OCT-1993 06:15:27
8894 93249j256_2  1676 45  7-SEP-1993 20:16:57    9175 93263j270_2  1382 65 26-SEP-1993 06:55:39    9559 93291j298_2     5 15 21-OCT-1993 13:36:29
8895 93249j256_2  1676 45  7-SEP-1993 21:51:29    9181 93263j270_2   935 15 26-SEP-1993 16:49:31    9560 93291j298_2  1874 15 21-OCT-1993 14:08:33
8896 93249j256_2   430 45  7-SEP-1993 23:26:01    9182 93263j270_2  1736 15 26-SEP-1993 17:57:23    9561 93291j298_2  1871 15 21-OCT-1993 15:43:11
8899 93249j256_2   974 65  8-SEP-1993 04:32:59    9189 93263j270_2  1257 15 27-SEP-1993 05:15:01    9562 93291j298_2  1836 15 21-OCT-1993 17:17:49
8900 93249j256_2  1678 65  8-SEP-1993 05:44:05    9190 93263j270_2  1749 65 27-SEP-1993 06:33:11    9563 93291j298_2   830 15 21-OCT-1993 18:47:03
8901 93249j256_2  1024 65  8-SEP-1993 07:18:37    9199 93263j270_2  1751 45 27-SEP-1993 20:43:39    9564 93291j298_2  2005 45 21-OCT-1993 20:21:37
8909 93249j256_2   128 45  8-SEP-1993 20:09:29    9200 93270j277_2  1753 45 27-SEP-1993 22:18:07    9575 93291j298_2  2010 15 22-OCT-1993 13:42:41
8910 93249j256_2  1669 45  8-SEP-1993 21:29:13    9201 93270j277_2   593 45 27-SEP-1993 23:52:39    9579 93291j298_2  2009 45 22-OCT-1993 20:01:13
8911 93249j256_2  1045 45  8-SEP-1993 23:03:45    9209 93270j277_2  1070 15 28-SEP-1993 12:49:45    9580 93291j298_2  1905 45 22-OCT-1993 21:37:19
8914 93249j256_2   212 65  9-SEP-1993 04:36:01    9210 93270j277_2  1754 15 28-SEP-1993 14:03:07    9581 93291j298_2  1844 45 22-OCT-1993 23:10:45
8915 93249j256_2  1677 65  9-SEP-1993 05:21:49    9211 93270j277_2   669 15 28-SEP-1993 15:37:37    9582 93291j298_2  1983 45 23-OCT-1993 00:45:45
8916 93249j256_2  1690 65  9-SEP-1993 06:56:19    9214 93270j277_2  1639 45 28-SEP-1993 20:25:29    9583 93291j298_2  1236 45 23-OCT-1993 02:20:45
8921 93249j256_2   572 15  9-SEP-1993 14:49:07    9215 93270j277_2  1761 45 28-SEP-1993 21:56:01    9585 93291j298_2   685 65 23-OCT-1993 06:13:21
8922 93249j256_2  1651 15  9-SEP-1993 16:23:23    9216 93270j277_2  1224 45 28-SEP-1993 23:30:39    9586 93291j298_2  1018 65 23-OCT-1993 07:02:57
8924 93249j256_2   511 45  9-SEP-1993 20:10:39    9224 93270j277_2     2 15 29-SEP-1993 12:52:35    9590 93291j298_2  1439 15 23-OCT-1993 13:41:33
8925 93249j256_2  1673 45  9-SEP-1993 21:06:27    9225 93270j277_2  1727 15 29-SEP-1993 13:42:27    9591 93291j298_2  1309 15 23-OCT-1993 14:56:51
8926 93249j256_2  1672 45  9-SEP-1993 22:40:59    9226 93270j277_2  1760 15 29-SEP-1993 15:17:07    9594 93291j298_2  1849 45 23-OCT-1993 19:46:31
8935 93249j256_2  1671 15 10-SEP-1993 12:51:35    9229 93270j277_2  1502 45 29-SEP-1993 20:09:41    9595 93291j298_2   297 45 23-OCT-1993 21:15:21
8936 93249j256_2  1674 15 10-SEP-1993 14:26:05    9230 93270j277_2  1633 45 29-SEP-1993 21:35:43    9601 93291j298_2  1973 65 24-OCT-1993 06:43:11
8937 93249j256_2  1674 15 10-SEP-1993 16:00:37    9240 93270j277_2  1768 15 30-SEP-1993 13:22:17    9605 93291j298_2   785 15 24-OCT-1993 13:43:17
8940 93249j256_2  1671 45 10-SEP-1993 20:44:07    9241 93270j277_2  1733 15 30-SEP-1993 14:58:01    9606 93291j298_2  2076 15 24-OCT-1993 14:34:59
8941 93249j256_2  1672 45 10-SEP-1993 22:18:39    9250 93270j277_2  1297 65  1-OCT-1993 05:23:07    9607 93291j298_2  2062 15 24-OCT-1993 16:10:01
8945 93249j256_2  1606 65 11-SEP-1993 04:38:51    9251 93270j277_2  1547 65  1-OCT-1993 06:43:47    9608 93291j298_2   638 15 24-OCT-1993 17:44:23
8946 93249j256_2  1666 65 11-SEP-1993 06:11:21    9255 93270j277_2  1771 15  1-OCT-1993 13:01:57    9616 93291j298_2  1000 65 25-OCT-1993 06:20:55
8947 93249j256_2   142 65 11-SEP-1993 07:45:51    9256 93270j277_2   707 15  1-OCT-1993 14:36:35    9681 93298j305_2   362 15 29-OCT-1993 13:53:41
8950 93249j256_2  1672 15 11-SEP-1993 12:29:15    9259 93270j277_2   272 45  1-OCT-1993 20:10:35    9682 93298j305_2  2023 15 29-OCT-1993 14:22:35
8951 93249j256_2  1672 15 11-SEP-1993 14:03:45    9260 93270j277_2  1769 45  1-OCT-1993 20:55:17    9685 93298j305_2   322 45 29-OCT-1993 19:57:08
8952 93249j256_2  1666 15 11-SEP-1993 15:38:17    9265 93270j277_2   590 65  2-OCT-1993 05:27:25    9686 93298j305_2   369 45 29-OCT-1993 21:25:48
8955 93249j256_2  1643 45 11-SEP-1993 20:22:49    9266 93270j277_2  1751 65  2-OCT-1993 06:23:41    9687 93298j305_2   436 45 29-OCT-1993 22:23:58
8956 93249j256_2  1670 45 11-SEP-1993 21:56:19    9271 93270j277_2  1754 15  2-OCT-1993 14:16:29    9688 93298j305_2  2366 15 29-OCT-1993 23:49:51
8960 93249j256_2   883 65 12-SEP-1993 04:41:55    9272 93270j277_2   853 15  2-OCT-1993 15:51:09    9689 93298j305_2  2111 45 30-OCT-1993 01:33:07
8961 93249j256_2  1705 65 12-SEP-1993 05:49:01    9275 93270j277_2  1765 45  2-OCT-1993 20:35:09    9690 93298j305_2  2342 45 30-OCT-1993 02:58:55
8962 93249j256_2   809 65 12-SEP-1993 07:23:31    9276 93270j277_2   173 45  2-OCT-1993 22:09:49    9691 93298j305_2  1893 45 30-OCT-1993 04:42:25
8966 93249j256_2  1669 15 12-SEP-1993 13:41:29    9281 93270j277_2  1750 65  3-OCT-1993 06:03:39    9692 93298j305_2  1696 65 30-OCT-1993 06:31:07
8967 93249j256_2  1669 15 12-SEP-1993 15:15:59    9282 93270j277_2   327 65  3-OCT-1993 07:38:17    9693 93298j305_2  2123 65 30-OCT-1993 07:51:35
8968 93249j256_2   461 15 12-SEP-1993 16:50:29    9285 93270j277_2   538 15  3-OCT-1993 13:02:27    9694 93298j305_2  2405 65 30-OCT-1993 09:16:57
8970 93249j256_2  1284 45 12-SEP-1993 20:12:27    9286 93270j277_2  1764 15  3-OCT-1993 13:56:17    9695 93298j305_2  2138 15 30-OCT-1993 11:00:35
8971 93249j256_2  1669 45 12-SEP-1993 21:33:59    9287 93270j277_2  1476 15  3-OCT-1993 15:30:57    9696 93298j305_2  2419 15 30-OCT-1993 12:25:59
8972 93249j256_2  1665 45 12-SEP-1993 23:08:31    9291 93270j277_2   774 45  3-OCT-1993 21:49:35    9697 93298j305_2  2146 15 30-OCT-1993 14:09:51
8973 93249j256_2    83 45 13-SEP-1993 00:43:01    9296 93270j277_2  1755 65  4-OCT-1993 05:43:23    9700 93298j305_2   541 45 30-OCT-1993 19:47:35
8976 93249j256_2  1667 65 13-SEP-1993 05:26:35    9297 93277j284_2   592 65  4-OCT-1993 07:18:05    9701 93298j305_2  2156 45 30-OCT-1993 20:28:27
8977 93256j263_2  1501 65 13-SEP-1993 07:01:05    9305 93277j284_2  1300 45  4-OCT-1993 20:10:07    9702 93298j305_2  2464 45 30-OCT-1993 21:53:01
8985 93256j263_2   764 45 13-SEP-1993 20:08:27    9306 93277j284_2  1766 45  4-OCT-1993 21:29:23    9703 93298j305_2  2164 45 30-OCT-1993 23:37:45
8986 93256j263_2  1712 45 13-SEP-1993 21:11:39    9316 93277j284_2  1758 15  5-OCT-1993 13:15:49    9704 93298j305_2  2478 15 31-OCT-1993 01:02:01
8987 93256j263_2  1706 45 13-SEP-1993 22:46:11    9317 93277j284_2  1716 15  5-OCT-1993 14:50:27    9705 93298j305_2  2173 45 31-OCT-1993 02:46:59
8988 93256j263_2   767 45 14-SEP-1993 00:20:41    9320 93277j284_2    85 45  5-OCT-1993 20:10:11    9706 93298j305_2  2497 45 31-OCT-1993 04:10:57
8996 93256j263_2  1680 15 14-SEP-1993 12:56:55    9321 93277j284_2  1769 45  5-OCT-1993 21:09:05    9707 93298j305_2   456 45 31-OCT-1993 05:56:31
8997 93256j263_2  1404 15 14-SEP-1993 14:31:19    9326 93277j284_2   839 65  6-OCT-1993 05:33:25    9708 93298j305_2  1048 65 31-OCT-1993 06:34:01
8998 93256j263_2  1677 15 14-SEP-1993 16:05:57    9327 93277j284_2  1767 65  6-OCT-1993 07:37:05    9709 93298j305_2  2517 65 31-OCT-1993 07:19:51
9006 93256j263_2  1561 65 15-SEP-1993 04:46:45    9332 93277j284_2   325 15  6-OCT-1993 15:18:27    9712 93298j305_2   175 65 31-OCT-1993 09:05:29
9007 93256j263_2  1698 15 15-SEP-1993 06:16:35    9333 93277j284_2   297 15  6-OCT-1993 16:04:59    9713 93298j305_2  1820 15 31-OCT-1993 14:02:01
9013 93256j263_2  1679 15 15-SEP-1993 15:43:35    9353 93277j284_2   428 15  7-OCT-1993 23:38:55    9714 93298j305_2  2213 15 31-OCT-1993 15:23:59
9021 93256j263_2   806 65 16-SEP-1993 04:48:45    9361 93277j284_2   116 15  8-OCT-1993 13:10:07    9715 93298j305_2  2587 65 31-OCT-1993 16:46:25
9022 93256j263_2  1701 65 16-SEP-1993 05:54:15    9362 93277j284_2   326 15  8-OCT-1993 13:49:43    9716 93298j305_2  2217 15 31-OCT-1993 18:33:17
9023 93256j263_2   696 65 16-SEP-1993 07:28:45    9363 93277j284_2  1515 15  8-OCT-1993 15:24:43                                                   
```

E-6

9722	93298j312_2	1197	65	1-NOV-1993 06:36:07	9869	93312j319_2	2830	45	10-NOV-1993 21:52:39	10011	93319j326_2	199	45	20-NOV-1993 06:39:15
9723	93305j312_2	2840	65	1-NOV-1993 07:35:41	9870	93312j319_2	2237	45	10-NOV-1993 23:27:17	10012	93319j326_2	2070	65	20-NOV-1993 07:27:37
9724	93305j312_2	2259	65	1-NOV-1993 09:10:19	9871	93312j319_2	2839	45	11-NOV-1993 01:01:55	10013	93319j326_2	1367	65	20-NOV-1993 09:48:27
9725	93305j312_2	2819	65	1-NOV-1993 10:44:57	9872	93312j319_2	2253	45	11-NOV-1993 02:36:33	10014	93319j326_2	2157	65	20-NOV-1993 10:33:59
9726	93305j312_2	2257	65	1-NOV-1993 12:19:35	9873	93312j319_2	2839	45	11-NOV-1993 04:11:11	10016	93319j326_2	308	15	20-NOV-1993 14:44:59
9727	93305j312_2	212	65	1-NOV-1993 13:54:13	9874	93312j319_2	1330	45	11-NOV-1993 05:45:47	10019	93319j326_2	819	45	20-NOV-1993 19:34:27
9730	93305j312_2	756	45	1-NOV-1993 19:47:31	9874	93312j319_2	529	65	11-NOV-1993 07:02:49	10020	93319j326_2	2132	45	20-NOV-1993 20:01:45
9731	93305j312_2	2539	45	1-NOV-1993 20:12:43	9875	93312j319_2	2840	65	11-NOV-1993 07:20:25	10021	93319j326_2	1361	45	20-NOV-1993 22:25:39
9732	93305j312_2	2256	45	1-NOV-1993 21:47:21	9876	93312j319_2	2251	65	11-NOV-1993 08:55:03	10022	93319j326_2	2161	45	20-NOV-1993 23:10:59
9733	93305j312_2	2829	45	1-NOV-1993 23:21:59	9877	93312j319_2	2730	65	11-NOV-1993 10:29:41	10023	93319j326_2	1361	45	21-NOV-1993 01:34:53
9734	93305j312_2	2258	45	2-NOV-1993 00:56:35	9878	93312j319_2	2034	45	11-NOV-1993 12:04:19	10024	93319j326_2	2180	45	21-NOV-1993 02:20:15
9735	93305j312_2	2712	45	2-NOV-1993 02:31:13	9879	93312j319_2	1491	15	11-NOV-1993 14:23:53	10025	93319j326_2	1361	45	21-NOV-1993 04:44:09
9736	93305j312_2	2258	45	2-NOV-1993 04:05:51	9880	93312j319_2	666	15	11-NOV-1993 15:13:35	10026	93319j326_2	2161	45	21-NOV-1993 05:29:29
9737	93305j312_2	176	45	2-NOV-1993 05:40:29	9883	93312j319_2	1383	45	11-NOV-1993 20:45:59	10027	93319j326_2	1294	65	21-NOV-1993 07:53:23
9737	93305j312_2	1051	65	2-NOV-1993 06:39:31	9884	93312j319_2	2141	45	11-NOV-1993 21:32:05	10031	93319j326_2	193	15	21-NOV-1993 14:47:59
9738	93305j312_2	2260	65	2-NOV-1993 07:15:07	9885	93312j319_2	1382	45	11-NOV-1993 23:55:17	10032	93319j326_2	1128	15	21-NOV-1993 14:57:15
9739	93305j312_2	475	65	2-NOV-1993 08:50:17	9886	93312j319_2	2124	45	12-NOV-1993 00:41:21	10033	93319j326_2	1361	15	21-NOV-1993 17:21:09
9742	93305j312_2	1395	15	2-NOV-1993 14:02:57	9887	93312j319_2	1383	45	12-NOV-1993 03:04:31	10034	93319j326_2	2156	15	21-NOV-1993 18:06:29
9743	93305j312_2	2839	15	2-NOV-1993 15:08:49	9888	93312j319_2	2137	45	12-NOV-1993 03:50:35	10035	93319j326_2	1334	45	21-NOV-1993 20:30:57
9744	93305j312_2	2259	15	2-NOV-1993 16:43:27	9889	93312j319_2	638	45	12-NOV-1993 06:13:49	10036	93319j326_2	2141	45	21-NOV-1993 21:15:45
9745	93305j312_2	1683	15	2-NOV-1993 18:18:05	9890	93312j319_2	1975	65	12-NOV-1993 07:05:31	10037	93319j326_2	1357	45	21-NOV-1993 23:39:47
9746	93305j312_2	1462	45	2-NOV-1993 20:19:13	9894	93312j319_2	129	15	12-NOV-1993 14:25:33	10038	93319j326_2	1071	45	22-NOV-1993 00:24:59
9747	93305j312_2	2839	45	2-NOV-1993 21:27:21	9895	93312j319_2	1383	15	12-NOV-1993 15:41:33	10042	93319j333_2	677	65	22-NOV-1993 07:33:17
9748	93305j312_2	2257	45	2-NOV-1993 23:01:59	9896	93312j319_2	2134	15	12-NOV-1993 16:27:37	10043	93326j333_2	1360	65	22-NOV-1993 09:07:31
9749	93305j312_2	2829	45	3-NOV-1993 00:36:37	9897	93312j319_2	1382	15	12-NOV-1993 18:50:49	10044	93326j333_2	2170	45	22-NOV-1993 09:52:49
9750	93305j312_2	2256	45	3-NOV-1993 02:11:15	9898	93312j319_2	2	15	12-NOV-1993 19:36:53	10045	93326j333_2	1357	65	22-NOV-1993 12:16:47
9751	93305j312_2	2829	45	3-NOV-1993 03:45:53	9898	93312j319_2	2135	45	12-NOV-1993 19:37:09	10046	93326j333_2	2164	65	22-NOV-1993 13:01:59
9752	93305j312_2	1478	45	3-NOV-1993 05:20:31	9899	93312j319_2	1379	45	12-NOV-1993 22:29:13	10050	93326j333_2	1709	45	22-NOV-1993 19:35:19
9752	93305j312_2	403	65	3-NOV-1993 06:41:43	9900	93312j319_2	2125	15	12-NOV-1993 22:46:09	10051	93326j333_2	1353	45	22-NOV-1993 21:44:39
9753	93305j312_2	2839	65	3-NOV-1993 06:55:09	9905	93312j319_2	1381	65	13-NOV-1993 07:27:55	10052	93326j333_2	469	45	22-NOV-1993 22:29:45
9754	93305j312_2	726	65	3-NOV-1993 08:29:47	9906	93312j319_2	2135	45	13-NOV-1993 08:13:57	10057	93326j333_2	569	65	23-NOV-1993 07:38:33
9757	93305j312_2	1236	15	3-NOV-1993 14:06:47	9907	93312j319_2	1376	65	13-NOV-1993 10:22:19	10058	93326j333_2	2174	65	23-NOV-1993 07:57:29
9758	93305j312_2	2258	15	3-NOV-1993 14:48:19	9908	93312j319_2	2145	65	13-NOV-1993 11:23:11	10059	93326j333_2	1351	65	23-NOV-1993 10:21:43
9759	93305j312_2	2839	15	3-NOV-1993 16:22:57	9909	93312j319_2	135	15	13-NOV-1993 14:27:59	10060	93326j333_2	2178	65	23-NOV-1993 11:06:43
9760	93305j312_2	2260	15	3-NOV-1993 17:57:35	9909	93312j319_2	1242	65	13-NOV-1993 13:46:27	10061	93326j333_2	1350	65	23-NOV-1993 13:30:59
9761	93305j312_2	279	15	3-NOV-1993 19:32:13	9910	93312j319_2	2141	15	13-NOV-1993 14:32:25	10062	93326j333_2	2115	65	23-NOV-1993 14:15:59
9761	93305j312_2	2542	45	3-NOV-1993 19:41:49	9911	93312j319_2	1379	15	13-NOV-1993 16:55:43	10065	93326j333_2	890	45	23-NOV-1993 20:04:49
9762	93305j312_2	1488	45	3-NOV-1993 21:06:51	9912	93312j319_2	2144	15	13-NOV-1993 17:41:41	10066	93326j333_2	2172	45	23-NOV-1993 20:34:29
9768	93305j312_2	1965	65	4-NOV-1993 06:44:23	9913	93312j319_2	1373	45	13-NOV-1993 20:05:07	10067	93326j333_2	1348	45	23-NOV-1993 22:58:49
9769	93305j312_2	2830	65	4-NOV-1993 08:09:17	9914	93312j319_2	2149	45	13-NOV-1993 20:50:55	10068	93326j333_2	2181	45	23-NOV-1993 23:43:43
9770	93305j312_2	1487	65	4-NOV-1993 09:43:55	9915	93312j319_2	1377	45	13-NOV-1993 23:14:19	10069	93326j333_2	1347	45	24-NOV-1993 02:08:05
9772	93305j312_2	608	15	4-NOV-1993 14:07:03	9916	93312j319_2	2148	45	14-NOV-1993 00:00:11	10070	93326j333_2	2177	45	24-NOV-1993 02:52:59
9773	93305j312_2	2322	15	4-NOV-1993 14:27:51	9917	93312j319_2	1366	45	14-NOV-1993 02:23:35	10071	93326j333_2	1346	45	24-NOV-1993 05:17:21
9776	93305j312_2	1241	45	4-NOV-1993 19:42:33	9918	93312j319_2	2145	45	14-NOV-1993 03:09:27	10072	93326j333_2	1743	45	24-NOV-1993 06:02:13
9777	93305j312_2	2839	45	4-NOV-1993 20:46:23	9919	93312j319_2	1376	45	14-NOV-1993 05:32:51	10073	93326j333_2	1342	65	24-NOV-1993 08:26:45
9778	93305j312_2	1950	45	4-NOV-1993 22:21:01	9920	93312j319_2	649	45	14-NOV-1993 06:18:41	10074	93326j333_2	1092	65	24-NOV-1993 09:11:27
9779	93305j312_2	1492	45	4-NOV-1993 23:55:39	9920	93312j319_2	563	65	14-NOV-1993 07:11:25	10077	93326j333_2	798	15	24-NOV-1993 15:03:15
9783	93305j312_2	1764	65	5-NOV-1993 06:50:03	9921	93312j319_2	1378	65	14-NOV-1993 08:42:03	10078	93326j333_2	1358	15	24-NOV-1993 15:29:57
9784	93305j312_2	2086	65	5-NOV-1993 07:48:49	9922	93312j319_2	1991	65	14-NOV-1993 09:27:57	10081	93326j333_2	1333	45	24-NOV-1993 20:03:41
9785	93305j312_2	2757	65	5-NOV-1993 09:23:27	9923	93312j319_2	1377	65	14-NOV-1993 11:51:21	10082	93326j333_2	2185	45	24-NOV-1993 21:48:27
9786	93305j312_2	2225	65	5-NOV-1993 10:58:05	9924	93312j319_2	2148	45	14-NOV-1993 12:37:13	10088	93326j333_2	441	65	25-NOV-1993 08:14:17
9787	93305j312_2	2218	65	5-NOV-1993 12:32:43	9925	93312j319_2	936	65	14-NOV-1993 15:00:37	10089	93326j333_2	1337	65	25-NOV-1993 09:40:53
9791	93305j312_2	1426	45	5-NOV-1993 19:38:23	9928	93312j319_2	433	45	14-NOV-1993 20:03:05	10090	93326j333_2	2184	65	25-NOV-1993 10:25:25
9792	93305j312_2	2243	45	5-NOV-1993 20:25:53	9929	93312j319_2	1374	45	14-NOV-1993 21:19:13	10093	93326j333_2	1337	15	25-NOV-1993 15:59:21
9793	93305j312_2	2840	45	5-NOV-1993 22:00:31	9930	93312j319_2	1210	45	14-NOV-1993 22:04:59	10094	93326j333_2	2188	15	25-NOV-1993 16:43:55
9794	93305j312_2	2213	45	5-NOV-1993 23:35:09	9936	93312j319_2	1333	45	15-NOV-1993 08:00:11	10095	93326j333_2	718	15	25-NOV-1993 19:08:35
9795	93305j312_2	2840	45	6-NOV-1993 01:09:47	9941	93319j326_2	682	15	15-NOV-1993 16:37:59	10095	93326j333_2	4	45	25-NOV-1993 19:35:51
9796	93305j312_2	2253	45	6-NOV-1993 02:44:25	9942	93319j326_2	1562	15	15-NOV-1993 17:00:43	10096	93326j333_2	1563	45	25-NOV-1993 23:13:57
9797	93305j312_2	2575	45	6-NOV-1993 04:19:03	9943	93319j326_2	391	15	15-NOV-1993 19:24:03	10099	93326j333_2	1218	45	26-NOV-1993 01:31:01
9798	93305j312_2	741	65	6-NOV-1993 07:03:39	9943	93319j326_2	966	45	15-NOV-1993 19:37:47	10100	93326j333_2	1932	45	26-NOV-1993 02:11:39
9799	93305j312_2	2839	65	6-NOV-1993 07:28:21	9944	93319j326_2	2143	45	15-NOV-1993 20:09:57	10101	93326j333_2	1326	45	26-NOV-1993 04:36:39
9800	93305j312_2	53	65	6-NOV-1993 09:02:59	9945	93319j326_2	1377	45	15-NOV-1993 22:33:21	10102	93326j333_2	2031	45	26-NOV-1993 05:20:53
9803	93305j312_2	2097	15	6-NOV-1993 14:11:27	9946	93319j326_2	2121	45	15-NOV-1993 23:19:13	10103	93326j333_2	1326	65	26-NOV-1993 07:45:53
9804	93305j312_2	2254	15	6-NOV-1993 15:21:31	9951	93319j326_2	697	65	16-NOV-1993 08:01:07	10104	93326j333_2	1964	65	26-NOV-1993 08:30:07
9805	93305j312_2	2831	15	6-NOV-1993 16:56:09	9952	93319j326_2	2152	65	16-NOV-1993 11:10:23	10111	93326j333_2	1286	45	26-NOV-1993 22:07:01
9806	93305j312_2	1483	15	6-NOV-1993 18:30:47	9953	93319j326_2	1377	65	16-NOV-1993 11:10:23	10112	93326j333_2	1887	45	26-NOV-1993 21:17:11
9806	93305j312_2	770	15	6-NOV-1993 19:39:45	9954	93319j326_2	2151	65	16-NOV-1993 11:46:05	10113	93326j333_2	1276	45	26-NOV-1993 23:32:15
9807	93305j312_2	2839	45	6-NOV-1993 20:05:25	9955	93319j326_2	906	15	16-NOV-1993 14:35:21	10114	93326j333_2	1861	45	27-NOV-1993 00:27:21
9808	93305j312_2	2252	45	6-NOV-1993 21:40:03	9955	93319j326_2	468	15	16-NOV-1993 14:19:39	10115	93326j333_2	1266	45	27-NOV-1993 02:41:31
9809	93305j312_2	2839	45	6-NOV-1993 23:14:41	9956	93319j326_2	1212	15	16-NOV-1993 15:05:31	10116	93326j333_2	1852	45	27-NOV-1993 03:36:59
9810	93305j312_2	513	45	7-NOV-1993 00:49:19	9959	93319j326_2	1374	45	16-NOV-1993 20:38:15	10119	93326j333_2	472	65	27-NOV-1993 08:59:47
9813	93305j312_2	453	65	7-NOV-1993 06:52:25	9960	93319j326_2	1952	45	16-NOV-1993 21:24:03	10123	93326j333_2	1241	15	27-NOV-1993 15:18:19
9814	93305j312_2	2255	65	7-NOV-1993 07:07:49	9961	93319j326_2	1374	45	16-NOV-1993 23:47:31	10124	93326j333_2	1572	15	27-NOV-1993 16:23:27
9815	93305j312_2	2830	65	7-NOV-1993 08:42:27	9962	93319j326_2	2154	45	17-NOV-1993 00:33:19	10125	93326j333_2	1230	15	27-NOV-1993 18:27:35
9816	93305j312_2	2253	65	7-NOV-1993 10:17:05	9963	93319j326_2	1374	45	17-NOV-1993 02:56:47	10126	93326j333_2	252	15	27-NOV-1993 19:24:11
9817	93305j312_2	2840	65	7-NOV-1993 11:51:43	9964	93319j326_2	2132	45	17-NOV-1993 03:42:31	10126	93326j333_2	1567	45	27-NOV-1993 19:32:37
9818	93305j312_2	817	15	7-NOV-1993 14:14:15	9965	93319j326_2	1183	45	17-NOV-1993 06:06:03	10127	93326j333_2	1212	45	27-NOV-1993 21:37:09
9818	93305j312_2	1396	65	7-NOV-1993 13:26:21	9966	93319j326_2	1326	65	17-NOV-1993 07:19:13	10128	93326j333_2	348	45	27-NOV-1993 22:33:47
9819	93305j312_2	2840	15	7-NOV-1993 15:00:59	9967	93319j326_2	1374	65	17-NOV-1993 09:15:17	10134	93326j333_2	1197	45	28-NOV-1993 08:02:09
9820	93305j312_2	2254	15	7-NOV-1993 16:35:37	9968	93319j326_2	1513	65	17-NOV-1993 10:01:05	10135	93326j333_2	1197	45	28-NOV-1993 10:13:55
9821	93305j312_2	2563	15	7-NOV-1993 18:10:15	9974	93319j326_2	1950	45	17-NOV-1993 21:52:25	10136	93326j333_2	1782	65	28-NOV-1993 11:11:37
9821	93305j312_2	242	45	7-NOV-1993 19:36:51	9975	93319j326_2	1371	45	17-NOV-1993 21:52:25	10137	93326j333_2	1189	65	28-NOV-1993 13:23:11
9822	93305j312_2	2249	45	7-NOV-1993 14:54:53	9976	93319j326_2	2156	45	17-NOV-1993 22:38:07	10138	93326j333_2	472	45	28-NOV-1993 15:04:51
9823	93305j312_2	2592	45	7-NOV-1993 21:19:31	9977	93319j326_2	1370	45	18-NOV-1993 01:01:43	10138	93326j333_2	1297	65	28-NOV-1993 14:21:11
9829	93305j319_2	2612	45	8-NOV-1993 06:54:29	9978	93319j326_2	1616	45	18-NOV-1993 01:47:23	10139	93326j333_2	1183	45	28-NOV-1993 16:32:29
9830	93312j319_2	2259	65	8-NOV-1993 08:22:07	9979	93319j326_2	1371	45	18-NOV-1993 03:22:29	10140	93326j333_2	1770	15	28-NOV-1993 17:30:39
9831	93312j319_2	2839	65	8-NOV-1993 09:56:45	9980	93319j326_2	1746	45	18-NOV-1993 04:56:37	10141	93326j333_2	1172	45	28-NOV-1993 19:42:01
9832	93312j319_2	1982	65	8-NOV-1993 11:31:23	9981	93319j326_2	1316	45	18-NOV-1993 07:02:07	10142	93326j333_2	1748	45	28-NOV-1993 20:40:33
9833	93312j319_2	2840	65	8-NOV-1993 13:05:59	9982	93319j326_2	1177	65	18-NOV-1993 08:05:53	10149	93333j340_2	1163	65	29-NOV-1993 08:18:47
9834	93312j319_2	1487	65	8-NOV-1993 14:40:37	9986	93319j326_2	1648	15	18-NOV-1993 14:41:47	10150	93333j340_2	1758	65	29-NOV-1993 09:17:49
9837	93312j319_2	1125	45	8-NOV-1993 20:21:41	9987	93319j326_2	1372	45	18-NOV-1993 16:47:57	10151	93333j340_2	1159	45	29-NOV-1993 11:28:05
9838	93312j319_2	1765	45	8-NOV-1993 20:59:09	9988	93319j326_2	2156	45	18-NOV-1993 17:33:39	10157	93333j340_2	1092	45	29-NOV-1993 20:56:29
9839	93312j319_2	2459	45	8-NOV-1993 22:46:27	9989	93319j326_2	1346	45	18-NOV-1993 19:12:49	10158	93333j340_2	1740	45	29-NOV-1993 21:55:25
9840	93312j319_2	2256	45	9-NOV-1993 00:08:25	9990	93319j326_2	2124	45	18-NOV-1993 20:42:55	10159	93333j340_2	1093	45	30-NOV-1993 00:05:09
9841	93312j319_2	1778	45	9-NOV-1993 01:43:03	9991	93319j326_2	1369	45	18-NOV-1993 21:29:03	10160	93333j340_2	1718	45	30-NOV-1993 01:04:49
9842	93312j319_2	2257	45	9-NOV-1993 03:17:41	9992	93319j326_2	1150	45	18-NOV-1993 23:52:11	10161	93333j340_2	1126	45	30-NOV-1993 03:14:25
9843	93312j319_2	1591	45	9-NOV-1993 04:52:19	9997	93319j326_2	688	45	19-NOV-1993 08:34:09	10162	93333j340_2	1713	45	30-NOV-1993 04:14:13
9848	93312j319_2	49	15	9-NOV-1993 14:18:27	9998	93319j326_2	2145	65	19-NOV-1993 09:19:57	10164	93333j340_2	747	65	30-NOV-1993 07:55:27
9849	93312j319_2	2560	15	9-NOV-1993 14:20:05	9999	93319j326_2	1359	65	19-NOV-1993 11:43:35	10169	93333j340_2	1003	15	30-NOV-1993 15:55:23
9852	93312j319_2	1182	45	9-NOV-1993 19:39:47	10000	93319j326_2	702	65	19-NOV-1993 12:29:13	10170	93333j340_2	1719	15	30-NOV-1993 16:51:47
9853	93312j319_2	1742	45	9-NOV-1993 20:38:35	10001	93319j326_2	325	15	19-NOV-1993 15:27:39	10171	93333j340_2	1042	15	30-NOV-1993 19:00:33
9854	93312j319_2	2198	45	9-NOV-1993 22:13:13	10002	93319j326_2	229	15	19-NOV-1993 19:53:47	10172	93333j340_2	1712	45	30-NOV-1993 20:01:09
9855	93312j319_2	2770	45	9-NOV-1993 23:47:51	10005	93319j326_2	1366	45	19-NOV-1993 21:11:27	10173	93333j340_2	1107	15	30-NOV-1993 22:10:03
9864	93312j319_2	771	15	10-NOV-1993 18:08:27	10006	93319j326_2	2156	45	19-NOV-1993 22:02:07	10174	93333j340_2	1693	45	30-NOV-1993 23:10:31
9865	93312j319_2	2829	15	10-NOV-1993 15:34:09	10007	93319j326_2	1366	45	20-NOV-1993 00:20:43	10179	93333j340_2	457	15	1-DEC-1993 07:58:09
9866	93312j319_2	2257	15	10-NOV-1993 17:08:47	10008	93319j326_2	2148	45	20-NOV-1993 01:06:13	10180	93333j340_2	1683	65	1-DEC-1993 08:38:41
9867	93312j319_2	1436	15	10-NOV-1993 18:43:25	10009	93319j326_2	1365	45	20-NOV-1993 03:29:59	10181	93333j340_2	1081	65	1-DEC-1993 10:47:03
9868	93312j319_2	1758	45	10-NOV-1993 20:34:01	10010	93319j326_2	2162	45	20-NOV-1993 04:15:29	10182	93333j340_2	1663	65	1-DEC-1993 11:49:05

E-7

10183	93333j340_2	1082	65	1-DEC-1993 13:56:21	10401	93347j354_2	958	45	15-DEC-1993 21:41:09	10831	94010j017_2	1411	45	13-JAN-1994 03:11:39
10184	93333j340_2	1205	15	1-DEC-1993 15:13:47	10402	93347j354_2	1442	45	15-DEC-1993 22:49:31	10835	94010j017_2	1048	65	13-JAN-1994 09:30:01
10186	93333j340_2	450	65	1-DEC-1993 14:57:21	10403	93347j354_2	965	45	16-DEC-1993 00:50:17	10837	94010j017_2	1359	65	13-JAN-1994 12:39:07
10187	93333j340_2	1083	45	1-DEC-1993 20:14:55	10404	93347j354_2	494	45	16-DEC-1993 01:58:45	10951	94017j024_2	981	45	21-JAN-1994 00:34:29
10188	93333j340_2	1663	45	1-DEC-1993 21:16:13	10408	93347j354_2	788	65	16-DEC-1993 08:38:35	10967	94017j024_2	1452	45	22-JAN-1994 01:32:23
10189	93333j340_2	1080	45	1-DEC-1993 23:24:09	10409	93347j354_2	939	65	16-DEC-1993 10:17:43	10969	94017j024_2	1441	45	22-JAN-1994 04:41:21
10190	93333j340_2	1640	45	2-DEC-1993 00:26:03	10413	93347j354_2	1429	15	16-DEC-1993 16:21:45	10971	94017j024_2	1453	45	22-JAN-1994 07:50:27
10191	93333j340_2	1056	45	2-DEC-1993 02:33:23	10415	93347j354_2	26	45	16-DEC-1993 20:17:23	10973	94017j024_2	293	65	22-JAN-1994 11:05:09
10192	93333j340_2	1674	45	2-DEC-1993 03:34:51	10417	93347j354_2	1734	45	16-DEC-1993 22:30:11	10983	94017j024_2	1438	45	23-JAN-1994 02:45:17
10195	93333j340_2	1058	65	2-DEC-1993 08:51:51	10423	93347j354_2	210	65	17-DEC-1993 08:47:59	10985	94017j024_2	1438	45	23-JAN-1994 05:54:21
10196	93333j340_2	817	65	2-DEC-1993 09:53:33	10431	93347j354_2	1723	45	17-DEC-1993 20:34:43	10987	94017j024_2	388	65	23-JAN-1994 09:38:31
10202	93333j340_2	1224	45	2-DEC-1993 19:34:07	10433	93347j354_2	189	45	17-DEC-1993 23:43:55	10989	94017j024_2	1446	65	23-JAN-1994 12:12:17
10203	93333j340_2	1080	45	2-DEC-1993 21:28:31	10443	93347j354_2	1003	15	18-DEC-1993 15:53:45	10991	94017j024_2	1444	65	23-JAN-1994 15:21:21
10204	93333j340_2	1578	45	2-DEC-1993 22:31:23	10447	93347j354_2	1695	45	18-DEC-1993 21:48:39	10995	94017j024_2	1429	45	23-JAN-1994 21:39:57
10205	93333j340_2	1114	45	3-DEC-1993 00:36:41	10467	93347j354_2	1689	45	20-DEC-1993 05:20:39	11003	94024j031_2	1193	65	24-JAN-1994 10:16:31
10206	93333j340_2	1510	45	3-DEC-1993 01:42:13	10473	93354j361_2	1272	65	20-DEC-1993 14:49:53	11005	94024j031_2	1419	65	24-JAN-1994 13:25:33
10207	93333j340_2	1089	45	3-DEC-1993 03:45:53	10481	93354j361_2	1679	45	21-DEC-1993 03:25:17	11007	94024j031_2	405	65	24-JAN-1994 16:35:15
10208	93333j340_2	1580	45	3-DEC-1993 04:50:01	10483	93354j361_2	1670	45	21-DEC-1993 06:34:25	11011	94024j031_2	614	15	24-JAN-1994 22:55:49
10209	93333j340_2	899	45	3-DEC-1993 06:55:07	10485	93354j361_2	1672	65	21-DEC-1993 09:43:55	11021	94024j031_2	31	65	25-JAN-1994 15:06:05
10210	93333j340_2	1447	65	3-DEC-1993 08:04:11	10491	93354j361_2	359	45	21-DEC-1993 19:55:19	11029	94024j031_2	1538	45	26-JAN-1994 03:10:27
10219	93333j340_2	1005	45	3-DEC-1993 22:41:51	10493	93354j361_2	1675	45	21-DEC-1993 22:20:37	11031	94024j031_2	1535	45	26-JAN-1994 06:19:37
10220	93333j340_2	1160	45	3-DEC-1993 23:46:41	10495	93354j361_2	1674	45	22-DEC-1993 01:29:47	11037	94024j031_2	1532	65	26-JAN-1994 15:46:49
10225	93333j340_2	477	65	4-DEC-1993 08:29:37	10497	93354j361_2	1671	45	22-DEC-1993 04:38:59	11039	94024j031_2	1470	15	26-JAN-1994 18:57:57
10226	93333j340_2	1549	65	4-DEC-1993 09:13:57	10499	93354j361_2	383	45	22-DEC-1993 07:48:11	11041	94024j031_2	1525	15	26-JAN-1994 22:05:07
10227	93333j340_2	1088	65	4-DEC-1993 11:17:53	10503	93354j361_2	1399	45	22-DEC-1993 14:06:33	11049	94024j031_2	1529	65	27-JAN-1994 10:41:19
10228	93333j340_2	1544	65	4-DEC-1993 12:23:17	10507	93354j361_2	1669	45	22-DEC-1993 20:25:05	11051	94024j031_2	1530	65	27-JAN-1994 13:50:25
10229	93333j340_2	1086	45	4-DEC-1993 14:27:07	10517	93354j361_2	361	65	23-DEC-1993 12:54:35	11057	94024j031_2	834	15	27-JAN-1994 23:17:15
10230	93333j340_2	1067	15	4-DEC-1993 15:48:59	10519	93354j361_2	774	65	23-DEC-1993 15:20:27	11067	94024j031_2	873	65	28-JAN-1994 15:24:27
10231	93333j340_2	1096	15	4-DEC-1993 17:36:25	10523	93354j361_2	1647	45	23-DEC-1993 21:39:11	11069	94024j031_2	1519	15	28-JAN-1994 18:11:25
10232	93333j340_2	1169	15	4-DEC-1993 18:41:49	10525	93354j361_2	1671	45	24-DEC-1993 00:47:59	11071	94024j031_2	741	15	28-JAN-1994 21:20:25
10233	93333j340_2	1084	45	4-DEC-1993 20:45:55	10527	93354j361_2	1670	45	24-DEC-1993 03:57:11	11073	94024j031_2	1525	45	29-JAN-1994 00:29:57
10234	93333j340_2	1553	45	4-DEC-1993 21:51:19	10529	93354j361_2	1155	45	24-DEC-1993 07:06:35	11075	94024j031_2	1529	45	29-JAN-1994 03:38:51
10235	93333j340_2	1080	45	4-DEC-1993 23:54:53	10531	93354j361_2	1648	45	24-DEC-1993 10:15:43	11077	94024j031_2	1531	45	29-JAN-1994 06:47:51
10236	93333j340_2	1533	45	5-DEC-1993 01:00:35	10551	93354j361_2	373	15	25-DEC-1993 17:47:25	11079	94024j031_2	1363	65	29-JAN-1994 09:57:27
10237	93333j340_2	1073	45	5-DEC-1993 03:04:15	10553	93354j361_2	941	45	25-DEC-1993 20:56:51	11081	94024j031_2	1594	65	29-JAN-1994 13:02:37
10238	93333j340_2	1060	45	5-DEC-1993 04:09:53	10555	93354j361_2	1584	45	26-DEC-1993 00:08:31	11089	94024j031_2	1552	45	30-JAN-1994 01:40:15
10241	93333j340_2	1075	65	5-DEC-1993 09:22:35	10557	93354j361_2	1629	45	26-DEC-1993 03:15:31	11091	94024j031_2	1530	45	30-JAN-1994 04:48:59
10242	93333j340_2	1510	65	5-DEC-1993 10:29:01	10559	93354j361_2	1641	45	26-DEC-1993 06:24:25	11093	94024j031_2	200	45	30-JAN-1994 08:00:01
10243	93333j340_2	1075	45	5-DEC-1993 12:31:45	10561	93354j361_2	1615	65	26-DEC-1993 09:34:33	11095	94024j031_2	1622	65	30-JAN-1994 11:06:09
10244	93333j340_2	1063	65	5-DEC-1993 13:49:37	10563	93354j361_2	1643	45	26-DEC-1993 12:42:47	11097	94024j031_2	1624	65	30-JAN-1994 14:15:07
10248	93333j340_2	1532	45	5-DEC-1993 19:56:23	10569	93354j361_2	1634	45	26-DEC-1993 22:10:33	11109	94031j038_2	1202	65	31-JAN-1994 09:22:23
10249	93333j340_2	1063	45	5-DEC-1993 22:00:09	10571	93354j361_2	1637	15	27-DEC-1993 01:19:37	11111	94031j038_2	1573	65	31-JAN-1994 12:19:15
10256	93340j347_2	1532	65	6-DEC-1993 08:33:33	10573	93354j361_2	1637	45	27-DEC-1993 04:28:47	11115	94031j038_2	1546	65	31-JAN-1994 15:28:49
10264	93340j347_2	1500	45	6-DEC-1993 21:10:51	10575	93361j003_2	979	45	27-DEC-1993 07:37:55	11117	94031j038_2	1516	45	31-JAN-1994 21:48:17
10265	93340j347_2	912	45	6-DEC-1993 23:14:17	10577	93361j003_2	1636	65	27-DEC-1993 10:47:13	11125	94031j038_2	1605	65	1-FEB-1994 10:22:17
10271	93340j347_2	818	45	7-DEC-1993 08:49:43	10579	93361j003_2	1637	65	27-DEC-1993 13:56:23	11129	94031j038_2	1434	15	1-FEB-1994 16:45:21
10272	93340j347_2	1492	65	7-DEC-1993 09:48:43	10581	93361j003_2	1636	15	27-DEC-1993 17:05:25	11131	94031j038_2	1556	15	1-FEB-1994 19:50:31
10275	93340j347_2	216	15	7-DEC-1993 15:27:19	10583	93361j003_2	1641	45	27-DEC-1993 20:14:39	11133	94031j038_2	1543	15	1-FEB-1994 22:57:55
10276	93340j347_2	1475	15	7-DEC-1993 16:06:57	10585	93361j003_2	1633	45	27-DEC-1993 23:23:55	11135	94031j038_2	705	45	2-FEB-1994 02:24:49
10277	93340j347_2	1057	15	7-DEC-1993 18:09:07	10599	93361j003_2	1567	45	28-DEC-1993 21:30:17	11139	94031j038_2	1186	45	2-FEB-1994 05:30:37
10278	93340j347_2	877	15	7-DEC-1993 19:16:11	10601	93361j003_2	1646	45	29-DEC-1993 00:37:15	11141	94031j038_2	1608	65	2-FEB-1994 11:34:39
10279	93340j347_2	1035	45	7-DEC-1993 21:18:59	10607	93361j003_2	1627	65	29-DEC-1993 10:04:55	11143	94031j038_2	1607	65	2-FEB-1994 14:43:19
10280	93340j347_2	1469	45	7-DEC-1993 22:25:37	10609	93361j003_2	1626	65	29-DEC-1993 13:14:03	11149	94031j038_2	977	15	3-FEB-1994 00:11:29
10286	93340j347_2	691	65	8-DEC-1993 08:17:59	10611	93361j003_2	1636	15	29-DEC-1993 16:23:13	11157	94031j038_2	247	65	3-FEB-1994 13:32:23
10291	93340j347_2	1053	15	8-DEC-1993 16:13:13	10617	93361j003_2	1613	45	30-DEC-1993 01:51:07	11161	94031j038_2	1465	15	3-FEB-1994 15:59:59
10292	93340j347_2	1484	15	8-DEC-1993 17:20:47	10619	93361j003_2	1621	45	30-DEC-1993 05:00:01	11163	94031j038_2	831	45	3-FEB-1994 22:21:23
10293	93340j347_2	538	15	8-DEC-1993 19:22:23	10621	93361j003_2	502	45	30-DEC-1993 08:09:07	11165	94031j038_2	1037	45	4-FEB-1994 01:42:09
10294	93340j347_2	1217	45	8-DEC-1993 20:34:01	10625	93361j003_2	1617	65	30-DEC-1993 14:27:39	11167	94031j038_2	1557	45	4-FEB-1994 04:33:51
10295	93340j347_2	1033	45	8-DEC-1993 22:32:29	10629	93361j003_2	1602	45	30-DEC-1993 20:46:55	11169	94031j038_2	690	45	4-FEB-1994 07:42:45
10296	93340j347_2	1195	45	8-DEC-1993 23:40:09	10631	93361j003_2	1622	45	30-DEC-1993 23:54:59	11171	94031j038_2	1528	65	4-FEB-1994 10:51:53
10302	93340j347_2	1459	65	9-DEC-1993 09:07:25	10633	93361j003_2	1620	45	31-DEC-1993 03:04:09	11173	94031j038_2	1163	65	4-FEB-1994 14:11:53
10303	93340j347_2	1056	65	9-DEC-1993 11:08:27	10635	93361j003_2	1619	65	31-DEC-1993 09:22:35	11179	94031j038_2	1620	45	4-FEB-1994 23:25:59
10304	93340j347_2	1469	65	9-DEC-1993 12:16:57	10657	93361j003_2	91	15	1-JAN-1994 17:00:35	11181	94031j038_2	1633	45	5-FEB-1994 02:34:35
10305	93340j347_2	1050	65	9-DEC-1993 14:17:47	10659	93361j003_2	1419	45	1-JAN-1994 20:10:41	11183	94031j038_2	1635	45	5-FEB-1994 05:43:35
10306	93340j347_2	925	15	9-DEC-1993 15:44:21	10661	93361j003_2	1616	45	1-JAN-1994 23:12:35	11185	94031j038_2	191	65	5-FEB-1994 09:40:45
10307	93340j347_2	1000	15	9-DEC-1993 17:28:41	10663	93361j003_2	1614	45	2-JAN-1994 02:21:45	11187	94031j038_2	1607	65	5-FEB-1994 12:01:39
10308	93340j347_2	1465	15	9-DEC-1993 18:35:09	10665	93361j003_2	1614	45	2-JAN-1994 05:30:51	11189	94031j038_2	1605	65	5-FEB-1994 15:11:17
10313	93340j347_2	994	45	10-DEC-1993 02:56:09	10667	93361j003_2	659	65	2-JAN-1994 09:11:57	11193	94031j038_2	1625	45	5-FEB-1994 21:29:05
10314	93340j347_2	1458	45	10-DEC-1993 04:02:43	10669	93361j003_2	1609	65	2-JAN-1994 11:49:21	11195	94031j038_2	1618	45	6-FEB-1994 00:38:23
10315	93340j347_2	1011	45	10-DEC-1993 06:04:49	10671	93361j003_2	1610	65	2-JAN-1994 14:58:27	11197	94031j038_2	1636	45	6-FEB-1994 03:46:47
10316	93340j347_2	721	45	10-DEC-1993 07:11:55	10673	93361j003_2	1630	45	2-JAN-1994 00:26:05	11199	94031j038_2	1629	45	6-FEB-1994 06:56:05
10317	93340j347_2	425	65	10-DEC-1993 09:14:01	10683	94003j010_2	1596	65	3-JAN-1994 09:53:45	11209	94031j038_2	1635	45	6-FEB-1994 22:41:05
10324	93340j347_2	778	45	10-DEC-1993 20:11:37	10685	94003j010_2	1604	65	3-JAN-1994 13:02:45	11215	94038j045_2	810	45	7-FEB-1994 08:15:15
10325	93340j347_2	750	45	10-DEC-1993 21:51:43	10687	94003j010_2	1290	15	3-JAN-1994 16:22:23	11217	94038j045_2	909	65	7-FEB-1994 11:39:35
10326	93340j347_2	1451	45	10-DEC-1993 22:58:33	10691	94003j010_2	1398	45	3-JAN-1994 22:30:59	11219	94038j045_2	1549	65	7-FEB-1994 14:28:07
10327	93340j347_2	1026	45	11-DEC-1993 00:59:37	10693	94003j010_2	1618	45	4-JAN-1994 01:39:41	11221	94038j045_2	1618	45	7-FEB-1994 17:35:21
10328	93340j347_2	1454	45	11-DEC-1993 02:07:49	10695	94003j010_2	1617	45	4-JAN-1994 04:48:57	11223	94038j045_2	1641	15	7-FEB-1994 20:44:23
10329	93340j347_2	1013	45	11-DEC-1993 04:08:51	10697	94003j010_2	812	45	4-JAN-1994 07:58:07	11225	94038j045_2	1621	15	7-FEB-1994 23:53:21
10330	93340j347_2	1444	45	11-DEC-1993 05:16:59	10699	94003j010_2	578	65	4-JAN-1994 11:07:01	11231	94038j045_2	1592	45	8-FEB-1994 09:20:25
10332	93340j347_2	890	45	11-DEC-1993 08:45:31	10701	94003j010_2	1598	65	4-JAN-1994 14:15:45	11233	94038j045_2	1625	65	8-FEB-1994 12:29:27
10333	93340j347_2	1023	65	11-DEC-1993 10:27:01	10707	94003j010_2	501	45	5-JAN-1994 00:20:11	11235	94038j045_2	1604	65	8-FEB-1994 15:38:35
10334	93340j347_2	1439	65	11-DEC-1993 11:36:25	10709	94003j010_2	1601	45	5-JAN-1994 02:52:37	11237	94038j045_2	1617	15	8-FEB-1994 18:48:05
10335	93340j347_2	309	65	11-DEC-1993 13:36:13	10711	94003j010_2	1593	45	5-JAN-1994 06:01:27	11243	94038j045_2	1622	45	9-FEB-1994 04:14:43
10336	93340j347_2	1341	65	11-DEC-1993 14:45:35	10721	94003j010_2	1616	45	5-JAN-1994 21:47:49	11245	94038j045_2	902	45	9-FEB-1994 07:23:25
10339	93340j347_2	1033	45	11-DEC-1993 19:54:29	10723	94003j010_2	1627	45	6-JAN-1994 00:56:39	11247	94038j045_2	1604	65	9-FEB-1994 10:32:35
10352	93340j347_2	1454	15	12-DEC-1993 15:58:27	10725	94003j010_2	1616	45	6-JAN-1994 04:06:05	11255	94038j045_2	1622	15	9-FEB-1994 23:09:13
10353	93340j347_2	988	15	12-DEC-1993 17:59:53	10727	94003j010_2	1607	45	6-JAN-1994 07:15:31	11257	94038j045_2	559	45	10-FEB-1994 02:53:35
10354	93340j347_2	518	15	12-DEC-1993 19:08:05	10733	94003j010_2	864	15	6-JAN-1994 17:07:23	11259	94038j045_2	1634	45	10-FEB-1994 05:26:43
10354	93340j347_2	354	45	12-DEC-1993 19:44:05	10735	94003j010_2	308	45	6-JAN-1994 20:35:21	11261	94038j045_2	136	45	10-FEB-1994 08:35:45
10355	93340j347_2	854	15	12-DEC-1993 21:13:15	10737	94003j010_2	1606	45	7-JAN-1994 23:01:11	11271	94038j045_2	1637	45	11-FEB-1994 00:20:55
10356	93340j347_2	1435	45	12-DEC-1993 22:17:17	10739	94003j010_2	1535	45	7-JAN-1994 02:12:47	11273	94038j045_2	1635	45	11-FEB-1994 03:29:57
10357	93340j347_2	1012	45	13-DEC-1993 00:17:25	10741	94003j010_2	1607	45	7-JAN-1994 05:18:33	11275	94038j045_2	1625	45	11-FEB-1994 06:39:19
10358	93340j347_2	1455	45	13-DEC-1993 01:26:01	10747	94003j010_2	1586	65	7-JAN-1994 14:46:27	11277	94038j045_2	1639	65	11-FEB-1994 09:47:51
10359	93340j347_2	1005	45	13-DEC-1993 03:26:45	10751	94003j010_2	1587	45	7-JAN-1994 21:05:55	11279	94038j045_2	1639	65	11-FEB-1994 12:56:59
10360	93340j347_2	1459	45	13-DEC-1993 04:35:13	10753	94003j010_2	1579	45	8-JAN-1994 00:15:25	11281	94038j045_2	1644	15	11-FEB-1994 16:06:01
10361	93340j347_2	708	45	13-DEC-1993 06:36:25	10755	94003j010_2	1596	45	8-JAN-1994 03:24:01	11283	94038j045_2	1615	15	11-FEB-1994 19:15:55
10367	93347j354_2	1005	15	13-DEC-1993 16:03:37	10757	94003j010_2	1527	45	8-JAN-1994 06:33:19	11285	94038j045_2	1628	15	11-FEB-1994 22:24:27
10368	93347j354_2	265	15	13-DEC-1993 17:16:31	10759	94003j010_2	1177	65	8-JAN-1994 09:42:53	11289	94038j045_2	1606	45	12-FEB-1994 04:43:23
10369	93347j354_2	402	15	13-DEC-1993 19:12:57	10763	94003j010_2	209	15	8-JAN-1994 16:37:43	11291	94038j045_2	1491	45	12-FEB-1994 07:51:05
10375	93347j354_2	533	45	14-DEC-1993 04:55:37	10775	94003j010_2	233	65	9-JAN-1994 11:28:45	11293	94038j045_2	1639	45	12-FEB-1994 11:00:11
10376	93347j354_2	5	45	14-DEC-1993 05:52:47	10777	94003j010_2	1280	45	9-JAN-1994 14:04:51	11295	94038j045_2	1644	45	12-FEB-1994 14:09:09
10378	93347j354_2	1423	65	14-DEC-1993 08:58:19	10781	94003j010_2	277	45	9-JAN-1994 21:06:25	11297	94038j045_2	1635	15	12-FEB-1994 17:18:31
10379	93347j354_2	966	65	14-DEC-1993 10:59:33	10789	94010j017_2	452	65	10-JAN-1994 09:27:09	11299	94038j045_2	1636	15	12-FEB-1994 20:27:13
10385	93347j354_2	960	45	14-DEC-1993 20:27:05	10791	94010j017_2	310	45	10-JAN-1994 12:50:57	11311	94045j052_2	1652	15	13-FEB-1994 15:21:17
10386	93347j354_2	150	45	14-DEC-1993 21:37:01	10793	94010j017_2	1504	15	10-JAN-1994 15:17:59	11325	94045j052_2	1300	65	14-FEB-1994 13:37:05
10393	93347j354_2	771	65	15-DEC-1993 09:10:19	10821	94010j017_2	1431	65	12-JAN-1994 11:25:33	11327	94045j052_2	1655	65	14-FEB-1994 15:35:23
10394	93347j354_2	1310	65	15-DEC-1993 10:12:39	10823	94010j017_2	1546	65	12-JAN-1994 14:35:25	11331	94045j052_2	57	15	14-FEB-1994 23:33:29
10400	93347j354_2	231	45	15-DEC-1993 20:20:45	10829	94010j017_2	1549	45	13-JAN-1994 00:02:33	11339	94045j052_2	604	65	15-FEB-1994 11:27:33

11347	94045j052_2	1660	15	16-FEB-1994 00:03:31	11699	94066j073_2	1650	45	11-MAR-1994 02:30:51	11901	94080j087_2	1861	45	24-MAR-1994 07:08:27
11351	94045j052_2	679	45	16-FEB-1994 06:48:53	11701	94066j073_2	2116	45	11-MAR-1994 05:22:57	11908	94080j087_2	60	15	24-MAR-1994 19:40:09
11361	94045j052_2	1663	15	16-FEB-1994 22:06:35	11709	94066j073_2	2155	15	11-MAR-1994 17:56:49	11909	94080j087_2	2826	15	24-MAR-1994 19:42:09
11363	94045j052_2	143	15	17-FEB-1994 01:16:09	11715	94066j073_2	2184	45	12-MAR-1994 03:21:05	11910	94080j087_2	339	15	24-MAR-1994 21:16:21
11373	94045j052_2	1666	15	17-FEB-1994 17:00:41	11717	94066j073_2	2204	45	12-MAR-1994 06:29:21	11911	94080j087_2	2816	15	24-MAR-1994 22:50:35
11375	94045j052_2	1667	15	17-FEB-1994 20:09:49	11725	94066j073_2	1499	15	12-MAR-1994 16:06:41	11912	94080j087_2	375	15	25-MAR-1994 00:24:47
11377	94045j052_2	1669	45	17-FEB-1994 23:18:43	11727	94066j073_2	2249	15	12-MAR-1994 19:02:27	11913	94080j087_2	1103	15	25-MAR-1994 01:59:01
11379	94045j052_2	1634	45	18-FEB-1994 02:27:35	11729	94066j073_2	2232	15	12-MAR-1994 22:10:51	11915	94080j087_2	2634	45	25-MAR-1994 05:13:47
11381	94045j052_2	1674	15	18-FEB-1994 05:36:31	11731	94066j073_2	1542	15	13-MAR-1994 01:19:07	11916	94080j087_2	226	45	25-MAR-1994 06:41:39
11391	94045j052_2	866	15	18-FEB-1994 21:46:05	11733	94066j073_2	2277	45	13-MAR-1994 04:27:21	11922	94080j087_2	56	15	25-MAR-1994 17:38:31
11393	94045j052_2	1578	15	19-FEB-1994 00:31:27	11739	94066j073_2	1331	45	13-MAR-1994 07:35:35	11923	94080j087_2	2826	15	25-MAR-1994 17:41:09
11397	94045j052_2	1681	45	19-FEB-1994 06:48:29	11741	94066j073_2	2327	15	13-MAR-1994 17:00:19	11924	94080j087_2	414	15	25-MAR-1994 19:15:21
11403	94045j052_2	1348	15	19-FEB-1994 16:26:47	11743	94066j073_2	2323	15	13-MAR-1994 20:09:03	11925	94080j087_2	2825	15	25-MAR-1994 20:49:35
11405	94045j052_2	1670	15	19-FEB-1994 19:24:29	11745	94066j073_2	2354	45	13-MAR-1994 23:16:41	11926	94080j087_2	224	15	25-MAR-1994 22:23:47
11407	94045j052_2	1685	45	19-FEB-1994 22:33:31	11747	94066j073_2	2372	45	14-MAR-1994 02:24:49	11927	94080j087_2	2197	45	26-MAR-1994 00:18:35
11409	94045j052_2	1671	45	20-FEB-1994 01:42:31	11757	94073j080_2	1691	15	14-MAR-1994 21:39:23	11928	94080j087_2	399	45	26-MAR-1994 01:32:13
11411	94045j052_2	1690	45	20-FEB-1994 04:51:25	11758	94073j080_2	20	15	14-MAR-1994 22:35:45	11929	94080j087_2	2825	45	26-MAR-1994 03:06:25
11419	94045j052_2	1682	15	20-FEB-1994 17:27:47	11759	94073j080_2	2461	15	15-MAR-1994 00:21:13	11930	94080j087_2	300	45	26-MAR-1994 04:40:39
11421	94045j052_2	1683	15	20-FEB-1994 20:36:43	11760	94073j080_2	30	15	15-MAR-1994 01:44:13	11931	94080j087_2	2821	15	26-MAR-1994 06:14:51
11423	94045j052_2	1686	15	20-FEB-1994 23:45:39	11760	94073j080_2	234	15	15-MAR-1994 03:10:39	11932	94080j087_2	224	15	26-MAR-1994 07:49:05
11427	94045j052_2	1696	45	21-FEB-1994 06:03:21	11761	94073j080_2	2461	15	15-MAR-1994 03:18:25	11938	94080j087_2	177	15	26-MAR-1994 18:42:41
11435	94052j059_2	1687	15	21-FEB-1994 18:39:39	11762	94073j080_2	268	15	15-MAR-1994 04:52:39	11939	94080j087_2	2824	15	26-MAR-1994 18:48:35
11437	94052j059_2	1687	15	21-FEB-1994 21:48:13	11763	94073j080_2	2515	45	15-MAR-1994 06:26:53	11940	94080j087_2	415	15	26-MAR-1994 20:22:47
11439	94052j059_2	1112	15	22-FEB-1994 00:57:09	11764	94073j080_2	53	45	15-MAR-1994 08:01:05	11941	94080j087_2	2826	15	26-MAR-1994 21:56:59
11441	94052j059_2	1705	45	22-FEB-1994 04:06:11	11769	94073j080_2	2348	15	15-MAR-1994 16:07:47	11942	94080j087_2	412	15	26-MAR-1994 23:31:13
11443	94052j059_2	1694	45	22-FEB-1994 07:15:37	11770	94073j080_2	395	15	15-MAR-1994 17:26:25	11943	94080j087_2	2826	15	27-MAR-1994 01:05:25
11445	94052j059_2	1698	65	22-FEB-1994 10:24:21	11771	94073j080_2	920	15	15-MAR-1994 19:00:39	11944	94080j087_2	17	15	27-MAR-1994 02:39:39
11447	94052j059_2	1709	65	22-FEB-1994 13:33:09	11773	94073j080_2	1065	45	15-MAR-1994 23:07:45	11944	94080j087_2	188	45	27-MAR-1994 04:07:37
11449	94052j059_2	1709	65	22-FEB-1994 16:42:05	11774	94073j080_2	398	45	15-MAR-1994 23:43:19	11945	94080j087_2	2822	45	27-MAR-1994 04:13:51
11453	94052j059_2	960	15	22-FEB-1994 23:24:53	11775	94073j080_2	2826	45	16-MAR-1994 01:17:31	11946	94080j087_2	405	45	27-MAR-1994 05:48:03
11465	94052j059_2	1693	15	23-FEB-1994 17:54:49	11776	94073j080_2	401	45	16-MAR-1994 02:51:45	11947	94080j087_2	1335	45	27-MAR-1994 07:22:17
11467	94052j059_2	1701	15	23-FEB-1994 21:03:13	11777	94073j080_2	1481	45	16-MAR-1994 04:25:59	11952	94080j087_2	175	15	27-MAR-1994 16:41:31
11469	94052j059_2	1716	15	24-FEB-1994 00:11:59	11781	94073j080_2	2329	65	16-MAR-1994 10:59:23	11953	94080j087_2	2825	15	27-MAR-1994 16:47:33
11475	94052j059_2	1725	65	24-FEB-1994 09:38:53	11782	94073j080_2	148	65	16-MAR-1994 12:17:03	11955	94080j087_2	346	15	27-MAR-1994 18:21:47
11477	94052j059_2	1716	65	24-FEB-1994 12:47:53	11783	94073j080_2	1973	65	16-MAR-1994 14:19:45	11955	94080j087_2	2826	15	27-MAR-1994 19:55:59
11479	94052j059_2	1730	65	24-FEB-1994 15:56:47	11785	94073j080_2	397	65	16-MAR-1994 15:25:31	11956	94080j087_2	387	15	27-MAR-1994 21:30:13
11481	94052j059_2	1751	15	24-FEB-1994 19:03:33	11785	94073j080_2	2821	15	16-MAR-1994 16:59:43	11957	94080j087_2	2693	15	27-MAR-1994 23:04:25
11483	94052j059_2	1750	15	24-FEB-1994 22:12:17	11786	94073j080_2	145	65	16-MAR-1994 18:33:57	11958	94080j087_2	298	15	28-MAR-1994 00:41:49
11485	94052j059_2	405	15	25-FEB-1994 01:21:33	11794	94073j080_2	246	65	17-MAR-1994 08:33:43	11959	94080j087_2	814	15	28-MAR-1994 02:12:45
11491	94052j059_2	1755	65	25-FEB-1994 10:47:47	11795	94073j080_2	2827	65	17-MAR-1994 08:41:53	11984	94087j094_2	130	15	29-MAR-1994 18:57:55
11493	94052j059_2	1735	65	25-FEB-1994 13:57:21	11796	94073j080_2	396	65	17-MAR-1994 10:16:07	11985	94087j094_2	2825	15	29-MAR-1994 19:02:15
11495	94052j059_2	1395	65	25-FEB-1994 17:05:45	11797	94073j080_2	2826	65	17-MAR-1994 11:50:21	11986	94087j094_2	256	15	29-MAR-1994 20:36:27
11499	94052j059_2	1767	15	25-FEB-1994 23:23:39	11798	94073j080_2	306	65	17-MAR-1994 13:24:33	11990	94087j094_2	2483	15	30-MAR-1994 01:23:55
11507	94052j059_2	472	65	26-FEB-1994 12:43:17	11799	94073j080_2	2816	65	17-MAR-1994 14:58:47	11991	94087j094_2	626	45	30-MAR-1994 05:40:49
11509	94052j059_2	1791	65	26-FEB-1994 15:08:15	11800	94073j080_2	410	65	17-MAR-1994 16:32:59	11992	94087j094_2	392	45	30-MAR-1994 06:01:45
11511	94052j059_2	1793	65	26-FEB-1994 18:17:13	11801	94073j080_2	1193	65	17-MAR-1994 18:07:13	11993	94087j094_2	886	45	30-MAR-1994 07:35:57
11513	94052j059_2	1786	15	26-FEB-1994 21:26:05	11802	94073j080_2	167	15	17-MAR-1994 21:10:05	11998	94087j094_2	155	15	30-MAR-1994 16:55:41
11515	94052j059_2	1782	15	27-FEB-1994 00:35:01	11803	94073j080_2	2826	15	17-MAR-1994 21:15:39	11999	94087j094_2	2825	15	30-MAR-1994 17:01:15
11523	94052j059_2	1650	65	27-FEB-1994 13:15:03	11804	94073j080_2	394	15	17-MAR-1994 22:49:51	12000	94087j094_2	392	15	30-MAR-1994 18:35:27
11525	94052j059_2	404	15	27-FEB-1994 16:21:33	11805	94073j080_2	2825	15	18-MAR-1994 00:24:03	12001	94087j094_2	2826	15	30-MAR-1994 20:09:41
11527	94052j059_2	1774	15	27-FEB-1994 19:29:23	11806	94073j080_2	233	15	18-MAR-1994 01:58:17	12002	94087j094_2	394	15	30-MAR-1994 21:43:53
11529	94052j059_2	1804	15	27-FEB-1994 22:37:25	11809	94073j080_2	2424	15	18-MAR-1994 06:54:19	12003	94087j094_2	2825	15	30-MAR-1994 23:18:07
11537	94059j066_2	1793	65	28-FEB-1994 11:13:27	11810	94073j080_2	44	45	18-MAR-1994 08:15:09	12004	94087j094_2	397	15	31-MAR-1994 00:52:19
11539	94059j066_2	1800	65	28-FEB-1994 14:22:15	11811	94073j080_2	1459	15	18-MAR-1994 10:34:53	12005	94087j094_2	1009	15	31-MAR-1994 02:26:33
11541	94059j066_2	137	15	28-FEB-1994 18:27:01	11812	94073j080_2	338	65	18-MAR-1994 11:23:35	12006	94087j094_2	155	45	31-MAR-1994 05:29:27
11543	94059j066_2	1800	15	28-FEB-1994 20:40:19	11813	94073j080_2	2825	65	18-MAR-1994 12:57:47	12007	94087j094_2	2820	45	31-MAR-1994 05:34:59
11545	94059j066_2	1798	15	28-FEB-1994 23:49:31	11814	94073j080_2	384	45	18-MAR-1994 14:32:11	12008	94087j094_2	247	45	31-MAR-1994 09:09:11
11553	94059j066_2	1784	65	1-MAR-1994 12:25:27	11815	94073j080_2	2826	15	18-MAR-1994 16:06:13	12009	94087j094_2	234	65	31-MAR-1994 10:09:49
11555	94059j066_2	1810	65	1-MAR-1994 15:33:43	11816	94073j080_2	233	65	18-MAR-1994 17:40:27	12010	94087j094_2	264	65	31-MAR-1994 10:17:37
11557	94059j066_2	1822	15	1-MAR-1994 18:43:11	11819	94073j080_2	1933	45	18-MAR-1994 22:52:41	12013	94087j094_2	1247	15	31-MAR-1994 15:52:55
11563	94059j066_2	1810	45	2-MAR-1994 04:10:17	11820	94073j080_2	408	45	18-MAR-1994 23:57:17	12014	94087j094_2	1201	15	31-MAR-1994 16:34:29
11565	94059j066_2	1821	45	2-MAR-1994 07:19:11	11821	94073j080_2	2817	45	19-MAR-1994 01:31:31	12015	94087j094_2	2196	15	31-MAR-1994 18:29:43
11567	94059j066_2	550	65	2-MAR-1994 10:28:07	11822	94073j080_2	383	45	19-MAR-1994 03:05:43	12016	94087j094_2	1187	15	31-MAR-1994 19:42:55
11571	94059j066_2	1846	15	2-MAR-1994 16:45:17	11823	94073j080_2	2826	45	19-MAR-1994 04:39:57	12017	94087j094_2	2197	15	31-MAR-1994 21:38:07
11575	94059j066_2	1809	45	2-MAR-1994 23:04:31	11824	94073j080_2	396	45	19-MAR-1994 06:14:09	12018	94087j094_2	1201	15	31-MAR-1994 22:51:21
11577	94059j066_2	1831	45	3-MAR-1994 02:12:57	11825	94073j080_2	726	45	19-MAR-1994 07:48:23	12019	94087j094_2	2198	15	1-APR-1994 00:46:31
11579	94059j066_2	1824	45	3-MAR-1994 05:21:47	11826	94073j080_2	158	65	19-MAR-1994 10:51:33	12020	94087j094_2	1190	15	1-APR-1994 01:59:47
11583	94059j066_2	1331	65	3-MAR-1994 11:55:51	11827	94073j080_2	2826	15	19-MAR-1994 10:56:47	12022	94087j094_2	1042	45	1-APR-1994 05:13:01
11585	94059j066_2	1818	65	3-MAR-1994 14:48:33	11828	94073j080_2	412	65	19-MAR-1994 12:31:01	12023	94087j094_2	1869	45	1-APR-1994 07:03:23
11587	94059j066_2	1804	15	3-MAR-1994 17:58:55	11829	94073j080_2	2825	65	19-MAR-1994 14:05:13	12026	94087j094_2	619	65	1-APR-1994 11:44:27
11589	94059j066_2	1839	15	3-MAR-1994 21:06:23	11830	94073j080_2	375	65	19-MAR-1994 15:39:27	12027	94087j094_2	2198	65	1-APR-1994 13:20:15
11591	94059j066_2	1850	15	4-MAR-1994 00:15:21	11831	94073j080_2	2	65	19-MAR-1994 17:14:53	12028	94087j094_2	1187	65	1-APR-1994 14:33:31
11593	94059j066_2	979	45	4-MAR-1994 03:53:25	11834	94073j080_2	93	45	19-MAR-1994 23:27:25	12029	94087j094_2	211	15	1-APR-1994 17:34:55
11595	94059j066_2	1815	45	4-MAR-1994 06:32:53	11835	94073j080_2	2826	45	19-MAR-1994 23:30:31	12030	94087j094_2	1187	15	1-APR-1994 17:41:57
11597	94059j066_2	1843	65	4-MAR-1994 09:41:43	11836	94073j080_2	413	45	20-MAR-1994 01:04:43	12031	94087j094_2	2197	15	1-APR-1994 19:37:09
11599	94059j066_2	1846	65	4-MAR-1994 12:50:53	11837	94073j080_2	2825	45	20-MAR-1994 02:38:57	12032	94087j094_2	1182	15	1-APR-1994 20:50:23
11601	94059j066_2	1849	65	4-MAR-1994 15:59:49	11838	94073j080_2	226	45	20-MAR-1994 04:13:09	12033	94087j094_2	931	15	1-APR-1994 22:45:33
11603	94059j066_2	1102	15	4-MAR-1994 19:08:43	11848	94073j080_2	169	15	20-MAR-1994 21:23:53	12033	94087j094_2	1240	15	1-APR-1994 23:17:29
11615	94059j066_2	774	65	5-MAR-1994 14:03:51	11849	94073j080_2	2826	15	20-MAR-1994 21:29:29	12034	94087j094_2	1183	45	1-APR-1994 23:58:49
11617	94059j066_2	1907	15	5-MAR-1994 17:10:37	11850	94073j080_2	228	15	20-MAR-1994 23:03:41	12035	94087j094_2	2202	45	2-APR-1994 01:53:51
11619	94059j066_2	1917	15	5-MAR-1994 20:20:21	11850	94073j080_2	81	45	21-MAR-1994 00:35:15	12036	94087j094_2	1175	45	2-APR-1994 03:07:15
11621	94059j066_2	1937	15	5-MAR-1994 23:29:07	11851	94073j080_2	2825	45	21-MAR-1994 01:59:41	12037	94087j094_2	2203	45	2-APR-1994 05:02:15
11623	94059j066_2	1940	45	6-MAR-1994 02:37:59	11852	94073j080_2	231	45	21-MAR-1994 02:12:09	12038	94087j094_2	1176	45	2-APR-1994 06:15:41
11629	94059j066_2	1566	65	6-MAR-1994 12:17:43	11860	94080j087_2	99	15	21-MAR-1994 16:16:37	12040	94087j094_2	215	65	2-APR-1994 09:54:59
11631	94059j066_2	1919	65	6-MAR-1994 15:14:11	11861	94080j087_2	2825	15	21-MAR-1994 16:19:55	12041	94087j094_2	2166	65	2-APR-1994 11:20:21
11633	94059j066_2	1953	15	6-MAR-1994 18:23:43	11862	94080j087_2	391	15	21-MAR-1994 17:54:07	12042	94087j094_2	1185	65	2-APR-1994 12:32:33
11635	94059j066_2	1953	15	6-MAR-1994 21:32:31	11863	94080j087_2	532	15	21-MAR-1994 19:28:21	12043	94087j094_2	2203	65	2-APR-1994 14:27:33
11637	94059j066_2	1962	15	7-MAR-1994 00:41:25	11865	94080j087_2	1344	15	21-MAR-1994 23:25:59	12044	94087j094_2	1184	65	2-APR-1994 15:40:59
11643	94066j073_2	2000	65	7-MAR-1994 10:08:09	11866	94080j087_2	232	15	22-MAR-1994 00:10:39	12045	94087j094_2	2175	65	2-APR-1994 17:35:57
11645	94066j073_2	1975	65	7-MAR-1994 13:17:29	11869	94080j087_2	1963	45	22-MAR-1994 05:22:21	12046	94087j094_2	783	65	2-APR-1994 18:49:25
11647	94066j073_2	1950	15	7-MAR-1994 16:26:59	11870	94080j087_2	391	45	22-MAR-1994 07:27:49	12055	94087j094_2	1463	65	3-APR-1994 09:42:15
11649	94066j073_2	1974	15	7-MAR-1994 19:35:53	11871	94080j087_2	318	45	22-MAR-1994 08:02:03	12056	94087j094_2	1181	65	3-APR-1994 10:31:35
11651	94066j073_2	1488	15	7-MAR-1994 22:44:49	11876	94080j087_2	138	15	22-MAR-1994 17:22:45	12057	94087j094_2	2205	65	3-APR-1994 12:26:31
11653	94066j073_2	1039	45	8-MAR-1994 01:53:51	11877	94080j087_2	2825	15	22-MAR-1994 17:28:15	12058	94087j094_2	1172	65	3-APR-1994 13:40:01
11655	94066j073_2	1749	45	8-MAR-1994 05:09:07	11878	94080j087_2	365	15	22-MAR-1994 19:01:31	12059	94087j094_2	1816	15	3-APR-1994 15:47:55
11657	94066j073_2	462	45	8-MAR-1994 08:12:03	11879	94080j087_2	2825	45	22-MAR-1994 20:35:45	12059	94087j094_2	380	15	3-APR-1994 15:34:55
11663	94066j073_2	1993	15	8-MAR-1994 17:37:57	11880	94080j087_2	363	45	22-MAR-1994 22:09:57	12060	94087j094_2	1180	15	3-APR-1994 16:48:27
11669	94066j073_2	2000	15	9-MAR-1994 03:05:11	11881	94080j087_2	2821	15	22-MAR-1994 23:44:11	12061	94087j094_2	2205	15	3-APR-1994 18:43:23
11671	94066j073_2	2017	45	9-MAR-1994 06:13:25	11882	94080j087_2	230	15	23-MAR-1994 01:18:23	12062	94087j094_2	1168	15	3-APR-1994 19:56:53
11677	94066j073_2	1089	15	9-MAR-1994 16:11:51	11882	94080j087_2	185	45	23-MAR-1994 02:46:27	12063	94087j094_2	2113	15	3-APR-1994 21:54:53
11679	94066j073_2	2049	15	9-MAR-1994 18:48:45	11883	94080j087_2	2826	45	23-MAR-1994 02:52:35	12064	94087j094_2	1180	15	3-APR-1994 23:05:17
11681	94066j073_2	2085	15	9-MAR-1994 21:56:47	11884	94080j087_2	416	45	23-MAR-1994 04:26:49	12065	94087j094_2	184	45	4-APR-1994 02:07:37
11683	94066j073_2	2083	45	10-MAR-1994 01:05:35	11885	94080j087_2	2816	45	23-MAR-1994 06:01:01	12066	94087j094_2	1171	45	4-APR-1994 03:12:43
11685	94066j073_2	2100	45	10-MAR-1994 04:14:23	11886	94080j087_2	227	45	23-MAR-1994 07:35:15	12067	94087j094_2	2183	45	4-APR-1994 04:08:57
11687	94066j073_2	2097	45	10-MAR-1994 07:23:27	11897	94080j087_2	1363	45	24-MAR-1994 01:40:19	12068	94087j094_2	1181	45	4-APR-1994 05:22:01
11693	94066j073_2	1707	15	10-MAR-1994 16:52:41	11898	94080j087_2	388	45	24-MAR-1994 02:25:47	12069	94094j101_2	1339	45	4-APR-1994 07:16:59
11695	94066j073_2	2102	15	10-MAR-1994 19:57:59	11899	94080j087_2	2826	45	24-MAR-1994 04:00:01	12071	94094j101_2	2171	65	4-APR-1994 10:26:31
11697	94066j073_2	2132	15	10-MAR-1994 23:06:15	11900	94080j087_2	400	45	24-MAR-1994 05:34:13	12072	94094j101_2	1181	65	4-APR-1994 11:38:53

12073	94094j101_2	2166	65	4-APR-1994 13:35:07	12197	94101j108_2	109	15	12-APR-1994 15:56:55	12344	94108j115_2	857	45	22-APR-1994 03:20:33
12074	94094j101_2	1179	65	4-APR-1994 14:47:19	12198	94101j108_2	1720	15	12-APR-1994 16:42:51	12345	94108j115_2	1555	45	22-APR-1994 04:30:41
12075	94094j101_2	1543	15	4-APR-1994 16:42:17	12199	94101j108_2	98	15	12-APR-1994 19:04:15	12346	94108j115_2	1801	45	22-APR-1994 05:54:57
12076	94094j101_2	315	15	4-APR-1994 18:14:21	12200	94101j108_2	1711	15	12-APR-1994 19:49:05	12347	94108j115_2	832	15	22-APR-1994 07:27:21
12077	94094j101_2	1517	15	4-APR-1994 19:51:43	12201	94101j108_2	104	15	12-APR-1994 22:10:53	12348	94108j115_2	1609	65	22-APR-1994 09:07:15
12078	94094j101_2	73	15	4-APR-1994 21:22:39	12202	94101j108_2	1685	15	12-APR-1994 22:56:03	12349	94108j115_2	1827	65	22-APR-1994 10:32:55
12079	94094j101_2	2153	15	4-APR-1994 23:02:07	12203	94101j108_2	231	15	13-APR-1994 01:17:29	12350	94108j115_2	1821	65	22-APR-1994 12:05:59
12080	94094j101_2	435	15	5-APR-1994 00:13:53	12204	94101j108_2	2193	15	13-APR-1994 01:45:01	12351	94108j115_2	1823	65	22-APR-1994 13:38:55
12081	94094j101_2	1607	15	5-APR-1994 02:10:15	12205	94101j108_2	225	45	13-APR-1994 04:23:33	12352	94108j115_2	1670	65	22-APR-1994 15:11:55
12085	94094j101_2	123	65	5-APR-1994 09:37:37	12206	94101j108_2	2163	45	13-APR-1994 04:51:43	12353	94108j115_2	1817	65	22-APR-1994 16:44:55
12086	94094j101_2	1157	65	5-APR-1994 09:41:43	12209	94101j108_2	358	65	13-APR-1994 10:35:19	12354	94108j115_2	1836	65	22-APR-1994 18:17:45
12087	94094j101_2	2247	65	5-APR-1994 11:36:07	12210	94101j108_2	2154	65	13-APR-1994 11:03:45	12355	94108j115_2	907	15	22-APR-1994 20:21:11
12088	94094j101_2	1154	65	5-APR-1994 12:50:59	12211	94101j108_2	224	65	13-APR-1994 13:41:11	12356	94108j115_2	476	65	22-APR-1994 19:50:45
12089	94094j101_2	850	65	5-APR-1994 15:31:51	12212	94101j108_2	1772	65	13-APR-1994 14:22:25	12357	94108j115_2	1721	15	22-APR-1994 21:27:33
12090	94094j101_2	1166	65	5-APR-1994 16:00:09	12213	94101j108_2	222	65	13-APR-1994 16:47:03	12358	94108j115_2	1834	15	22-APR-1994 22:56:45
12091	94094j101_2	728	65	5-APR-1994 18:44:39	12214	94101j108_2	2128	65	13-APR-1994 17:16:29	12359	94108j115_2	1833	15	23-APR-1994 00:29:43
12092	94094j101_2	478	65	5-APR-1994 19:08:55	12216	94101j108_2	2106	15	13-APR-1994 20:22:47	12360	94108j115_2	1829	15	23-APR-1994 02:02:39
12093	94094j101_2	897	15	5-APR-1994 21:05:59	12217	94101j108_2	249	15	13-APR-1994 22:58:41	12361	94108j115_2	1532	15	23-APR-1994 03:45:35
12094	94094j101_2	1218	15	5-APR-1994 22:17:15	12218	94101j108_2	2008	15	13-APR-1994 23:31:13	12362	94108j115_2	1830	45	23-APR-1994 05:08:37
12095	94094j101_2	2106	15	6-APR-1994 00:15:05	12219	94101j108_2	241	15	14-APR-1994 02:04:45	12363	94108j115_2	1828	45	23-APR-1994 06:41:31
12096	94094j101_2	1213	15	6-APR-1994 01:25:15	12220	94101j108_2	2112	15	14-APR-1994 02:35:23	12364	94108j115_2	1593	65	23-APR-1994 08:21:53
12097	94094j101_2	2119	45	6-APR-1994 03:22:39	12221	94101j108_2	243	45	14-APR-1994 05:10:39	12365	94108j115_2	1808	65	23-APR-1994 09:47:39
12098	94094j101_2	1213	45	6-APR-1994 04:33:15	12222	94101j108_2	2105	45	14-APR-1994 05:41:37	12366	94108j115_2	1808	65	23-APR-1994 11:20:37
12099	94094j101_2	2164	45	6-APR-1994 06:29:09	12224	94101j108_2	1600	65	14-APR-1994 09:03:55	12367	94108j115_2	1807	65	23-APR-1994 12:53:33
12100	94094j101_2	414	45	6-APR-1994 07:41:17	12225	94101j108_2	241	65	14-APR-1994 11:22:15	12368	94108j115_2	1803	65	23-APR-1994 14:26:33
12101	94094j101_2	2125	65	6-APR-1994 09:38:29	12226	94101j108_2	2088	65	14-APR-1994 11:53:37	12369	94108j115_2	1820	15	23-APR-1994 15:59:33
12102	94094j101_2	1213	65	6-APR-1994 10:49:17	12227	94101j108_2	259	65	14-APR-1994 14:28:07	12370	94108j115_2	1811	15	23-APR-1994 17:32:29
12103	94094j101_2	2166	65	6-APR-1994 12:45:07	12228	94101j108_2	2096	65	14-APR-1994 14:59:47	12371	94108j115_2	1814	15	23-APR-1994 19:05:35
12104	94094j101_2	1202	65	6-APR-1994 13:57:17	12229	94101j108_2	262	65	14-APR-1994 17:33:59	12372	94108j115_2	1815	15	23-APR-1994 20:38:33
12105	94094j101_2	2165	65	6-APR-1994 15:53:09	12230	94101j108_2	2073	65	14-APR-1994 18:06:01	12373	94108j115_2	1812	15	23-APR-1994 22:11:35
12106	94094j101_2	1211	65	6-APR-1994 17:05:19	12232	94101j108_2	1424	15	14-APR-1994 21:34:09	12374	94108j115_2	1470	15	23-APR-1994 23:44:19
12107	94094j101_2	854	65	6-APR-1994 19:02:23	12233	94101j108_2	106	15	14-APR-1994 23:51:07	12375	94108j115_2	1814	45	24-APR-1994 01:17:23
12110	94094j101_2	1047	45	6-APR-1994 23:26:47	12234	94101j108_2	2082	15	15-APR-1994 00:18:11	12376	94108j115_2	1815	45	24-APR-1994 02:50:17
12111	94094j101_2	2166	45	7-APR-1994 01:17:09	12235	94101j108_2	267	15	15-APR-1994 02:51:47	12377	94108j115_2	1814	45	24-APR-1994 04:23:17
12112	94094j101_2	1204	45	7-APR-1994 02:29:21	12244	94101j108_2	2050	15	15-APR-1994 15:49:11	12378	94108j115_2	1813	45	24-APR-1994 05:56:17
12113	94094j101_2	2167	45	7-APR-1994 04:25:09	12245	94101j108_2	283	15	15-APR-1994 18:21:05	12378	94108j115_2	642	45	24-APR-1994 07:29:19
12114	94094j101_2	1210	45	7-APR-1994 05:37:21	12246	94101j108_2	2042	15	15-APR-1994 18:55:27	12378	94108j115_2	396	65	24-APR-1994 08:16:31
12115	94094j101_2	662	45	7-APR-1994 07:33:05	12247	94101j108_2	287	15	15-APR-1994 21:26:55	12379	94108j115_2	1794	65	24-APR-1994 09:02:15
12115	94094j101_2	756	65	7-APR-1994 08:20:11	12248	94101j108_2	2042	15	15-APR-1994 22:01:35	12380	94108j115_2	1781	65	24-APR-1994 10:35:15
12116	94094j101_2	1199	65	7-APR-1994 08:45:21	12249	94101j108_2	280	15	16-APR-1994 00:32:49	12381	94108j115_2	1792	65	24-APR-1994 12:08:11
12117	94094j101_2	2152	65	7-APR-1994 10:41:39	12250	94101j108_2	1991	15	16-APR-1994 01:08:33	12382	94108j115_2	1787	65	24-APR-1994 13:41:11
12118	94094j101_2	1208	65	7-APR-1994 11:53:23	12252	94101j108_2	1997	45	16-APR-1994 04:14:17	12383	94108j115_2	1786	65	24-APR-1994 15:14:09
12119	94094j101_2	2145	15	7-APR-1994 13:49:53	12253	94101j108_2	281	45	16-APR-1994 06:44:37	12384	94108j115_2	1807	15	24-APR-1994 16:47:07
12120	94094j101_2	1209	65	7-APR-1994 15:01:23	12254	94101j108_2	1079	45	16-APR-1994 07:20:13	12385	94108j115_2	1803	15	24-APR-1994 18:20:07
12121	94094j101_2	2140	65	7-APR-1994 16:58:05	12259	94101j108_2	290	15	16-APR-1994 16:02:19	12386	94108j115_2	1802	15	24-APR-1994 19:53:07
12122	94094j101_2	1101	65	7-APR-1994 18:09:23	12260	94101j108_2	1998	15	16-APR-1994 16:38:43	12387	94108j115_2	1803	15	24-APR-1994 21:26:03
12125	94094j101_2	1760	45	7-APR-1994 23:26:45	12261	94101j108_2	292	15	16-APR-1994 19:08:09	12388	94108j115_2	1803	15	24-APR-1994 22:58:59
12126	94094j101_2	1211	45	8-APR-1994 00:25:25	12262	94101j108_2	1976	15	16-APR-1994 19:44:05	12389	94108j115_2	1809	15	25-APR-1994 00:31:47
12127	94094j101_2	2171	45	8-APR-1994 02:21:03	12263	94101j108_2	297	15	16-APR-1994 22:14:01	12390	94108j115_2	1803	15	25-APR-1994 02:04:53
12128	94094j101_2	1213	45	8-APR-1994 03:33:25	12264	94101j108_2	1963	15	16-APR-1994 22:51:09	12394	94108j115_2	1776	65	25-APR-1994 08:16:53
12129	94094j101_2	2171	45	8-APR-1994 05:29:03	12265	94101j108_2	296	45	17-APR-1994 01:19:59	12395	94115j122_2	1777	65	25-APR-1994 09:49:51
12130	94094j101_2	566	45	8-APR-1994 06:41:25	12266	94101j108_2	1950	45	17-APR-1994 01:57:19	12396	94115j122_2	1775	65	25-APR-1994 11:22:49
12135	94094j101_2	1138	65	8-APR-1994 15:27:31	12267	94101j108_2	282	45	17-APR-1994 04:25:55	12397	94115j122_2	1774	65	25-APR-1994 12:55:49
12136	94094j101_2	1196	65	8-APR-1994 16:05:25	12268	94101j108_2	1947	45	17-APR-1994 05:03:25	12398	94115j122_2	1774	65	25-APR-1994 14:28:43
12137	94094j101_2	2169	65	8-APR-1994 18:01:09	12269	94101j108_2	281	45	17-APR-1994 07:31:45	12399	94115j122_2	1646	65	25-APR-1994 16:01:43
12138	94094j101_2	704	65	8-APR-1994 19:13:25	12275	94101j108_2	310	15	17-APR-1994 16:46:31	12400	94115j122_2	1771	65	25-APR-1994 17:34:41
12141	94094j101_2	2167	65	9-APR-1994 00:17:13	12276	94101j108_2	1922	15	17-APR-1994 17:25:01	12401	94115j122_2	1758	65	25-APR-1994 19:07:41
12142	94094j101_2	1211	45	9-APR-1994 01:29:27	12277	94101j108_2	325	15	17-APR-1994 19:52:19	12406	94115j122_2	331	45	26-APR-1994 03:40:23
12143	94094j101_2	749	45	9-APR-1994 03:25:15	12278	94101j108_2	1938	15	17-APR-1994 20:30:45	12407	94115j122_2	1662	45	26-APR-1994 04:28:53
12144	94094j101_2	2616	45	9-APR-1994 03:50:19	12279	94101j108_2	321	15	17-APR-1994 22:58:17	12408	94115j122_2	1752	45	26-APR-1994 05:58:27
12145	94094j101_2	754	45	9-APR-1994 06:33:13	12280	94101j108_2	1953	15	17-APR-1994 23:36:49	12409	94115j122_2	529	45	26-APR-1994 07:32:21
12146	94094j101_2	1699	45	9-APR-1994 06:58:21	12281	94101j108_2	326	15	18-APR-1994 02:04:03	12409	94115j122_2	848	65	26-APR-1994 08:02:33
12147	94094j101_2	469	65	9-APR-1994 09:41:23	12282	94101j108_2	1269	15	18-APR-1994 02:42:47	12410	94115j122_2	1776	65	26-APR-1994 09:04:31
12148	94094j101_2	2627	65	9-APR-1994 10:06:21	12283	94101j108_2	254	45	18-APR-1994 05:11:43	12411	94115j122_2	1776	65	26-APR-1994 10:37:27
12149	94094j101_2	750	65	9-APR-1994 12:49:21	12284	94101j108_2	1844	45	18-APR-1994 05:48:49	12412	94115j122_2	1776	65	26-APR-1994 12:10:25
12150	94094j101_2	2633	65	9-APR-1994 13:14:19	12287	94108j115_2	300	15	18-APR-1994 11:21:55	12413	94115j122_2	1776	65	26-APR-1994 13:43:23
12151	94094j101_2	743	15	9-APR-1994 15:57:35	12288	94108j115_2	1892	65	18-APR-1994 12:00:51	12414	94115j122_2	1773	65	26-APR-1994 15:16:23
12152	94094j101_2	2633	15	9-APR-1994 16:22:19	12289	94108j115_2	305	65	18-APR-1994 14:27:47	12415	94115j122_2	1756	65	26-APR-1994 16:49:19
12153	94094j101_2	636	15	9-APR-1994 19:05:33	12290	94108j115_2	1899	65	18-APR-1994 15:07:21	12416	94115j122_2	1774	65	26-APR-1994 18:22:19
12154	94094j101_2	2546	15	9-APR-1994 19:33:11	12291	94108j115_2	318	15	18-APR-1994 17:33:21	12417	94115j122_2	669	65	26-APR-1994 19:55:17
12155	94094j101_2	599	15	9-APR-1994 22:13:33	12292	94108j115_2	470	15	18-APR-1994 18:15:41	12418	94115j122_2	1764	15	26-APR-1994 21:28:11
12156	94094j101_2	2571	15	9-APR-1994 22:40:17	12293	94108j115_2	318	15	18-APR-1994 20:39:13	12419	94115j122_2	1774	15	26-APR-1994 23:01:09
12157	94094j101_2	502	45	10-APR-1994 01:24:01	12294	94108j115_2	1915	15	18-APR-1994 21:18:45	12420	94115j122_2	1772	15	27-APR-1994 00:34:07
12158	94094j101_2	2600	45	10-APR-1994 01:47:29	12295	94108j115_2	312	15	18-APR-1994 23:45:07	12421	94115j122_2	1763	15	27-APR-1994 02:07:05
12159	94094j101_2	534	45	10-APR-1994 04:30:33	12296	94108j115_2	1904	15	19-APR-1994 00:25:17	12422	94115j122_2	1720	45	27-APR-1994 03:41:45
12160	94094j101_2	2575	45	10-APR-1994 04:56:19	12297	94108j115_2	316	15	19-APR-1994 02:50:55	12423	94115j122_2	1755	45	27-APR-1994 05:13:09
12161	94094j101_2	544	45	10-APR-1994 07:37:45	12298	94108j115_2	528	45	19-APR-1994 04:16:43	12424	94115j122_2	1747	45	27-APR-1994 06:46:07
12163	94094j101_2	456	65	10-APR-1994 10:48:19	12299	94108j115_2	298	45	19-APR-1994 05:57:23	12430	94115j122_2	1742	65	27-APR-1994 16:03:55
12164	94094j101_2	2522	65	10-APR-1994 11:13:25	12300	94108j115_2	1908	45	19-APR-1994 06:36:41	12431	94115j122_2	1230	65	27-APR-1994 17:36:57
12165	94094j101_2	512	65	10-APR-1994 13:54:19	12302	94108j115_2	1894	15	19-APR-1994 09:42:43	12433	94115j122_2	1722	15	27-APR-1994 20:43:09
12166	94094j101_2	2492	65	10-APR-1994 14:21:51	12303	94108j115_2	304	65	19-APR-1994 12:08:41	12434	94115j122_2	1720	15	27-APR-1994 22:16:07
12167	94094j101_2	508	15	10-APR-1994 17:01:49	12304	94108j115_2	1891	65	19-APR-1994 12:48:39	12435	94115j122_2	1737	15	27-APR-1994 23:48:47
12168	94094j101_2	2466	15	10-APR-1994 17:30:23	12305	94108j115_2	304	65	19-APR-1994 15:14:33	12436	94115j122_2	1744	15	28-APR-1994 01:21:51
12169	94094j101_2	504	15	10-APR-1994 20:09:49	12306	94108j115_2	1887	65	19-APR-1994 15:54:37	12437	94115j122_2	1342	15	28-APR-1994 02:54:45
12170	94094j101_2	2434	15	10-APR-1994 20:39:11	12307	94108j115_2	303	65	19-APR-1994 18:20:23	12438	94115j122_2	1755	45	28-APR-1994 04:27:49
12171	94094j101_2	490	15	10-APR-1994 23:17:51	12308	94108j115_2	1883	65	19-APR-1994 19:00:37	12439	94115j122_2	1726	45	28-APR-1994 06:00:43
12172	94094j101_2	2420	15	10-APR-1994 23:47:41	12311	94108j115_2	309	45	20-APR-1994 00:32:13	12440	94115j122_2	667	45	28-APR-1994 07:33:37
12173	94094j101_2	487	15	11-APR-1994 02:25:53	12312	94108j115_2	1885	45	20-APR-1994 01:12:35	12445	94115j122_2	854	15	28-APR-1994 15:47:41
12174	94094j101_2	446	15	11-APR-1994 02:56:01	12313	94108j115_2	305	45	20-APR-1994 03:38:01	12446	94115j122_2	1682	15	28-APR-1994 16:53:43
12178	94101j108_2	2353	15	11-APR-1994 09:12:39	12314	94108j115_2	1869	45	20-APR-1994 04:18:41	12447	94115j122_2	1745	15	28-APR-1994 18:24:31
12179	94101j108_2	451	65	11-APR-1994 11:49:57	12321	94108j115_2	308	15	20-APR-1994 16:01:43	12448	94115j122_2	1738	15	28-APR-1994 19:57:27
12180	94101j108_2	2365	65	11-APR-1994 12:20:03	12322	94108j115_2	1865	15	20-APR-1994 16:42:43	12449	94115j122_2	1744	15	28-APR-1994 21:30:27
12181	94101j108_2	97	15	11-APR-1994 14:58:19	12323	94108j115_2	313	15	20-APR-1994 19:07:29	12450	94115j122_2	1757	15	28-APR-1994 23:03:21
12182	94101j108_2	1875	15	11-APR-1994 15:43:51	12324	94108j115_2	1872	15	20-APR-1994 19:48:21	12451	94115j122_2	1717	15	29-APR-1994 00:37:03
12183	94101j108_2	125	15	11-APR-1994 18:06:11	12325	94108j115_2	313	15	20-APR-1994 22:13:21	12452	94115j122_2	1752	15	29-APR-1994 02:09:03
12184	94101j108_2	1885	15	11-APR-1994 18:50:59	12326	94108j115_2	1870	15	20-APR-1994 22:54:21	12453	94115j122_2	1729	45	29-APR-1994 03:42:21
12185	94101j108_2	128	15	11-APR-1994 21:14:11	12327	94108j115_2	312	15	21-APR-1994 01:19:15	12454	94115j122_2	1736	45	29-APR-1994 05:15:23
12186	94101j108_2	629	15	11-APR-1994 22:39:39	12328	94108j115_2	1517	15	21-APR-1994 02:12:01	12455	94115j122_2	1720	45	29-APR-1994 06:48:11
12187	94101j108_2	348	15	12-APR-1994 00:22:15	12329	94108j115_2	310	45	21-APR-1994 04:25:07	12456	94115j122_2	1740	65	29-APR-1994 08:21:13
12188	94101j108_2	2279	45	12-APR-1994 00:50:59	12330	94108j115_2	1864	45	21-APR-1994 05:06:17	12457	94115j122_2	1745	65	29-APR-1994 09:54:09
12189	94101j108_2	343	45	12-APR-1994 03:29:07	12331	94108j115_2	299	45	21-APR-1994 07:30:59	12458	94115j122_2	1740	65	29-APR-1994 11:27:09
12190	94101j108_2	2277	45	12-APR-1994 03:57:59	12332	94108j115_2	1429	65	21-APR-1994 08:22:03	12459	94115j122_2	1739	15	29-APR-1994 13:00:07
12191	94101j108_2	340	45	12-APR-1994 06:35:53	12333	94108j115_2	297	65	21-APR-1994 10:36:55	12460	94115j122_2	1746	65	29-APR-1994 14:33:05
12192	94101j108_2	1522	45	12-APR-1994 07:04:43	12334	94108j115_2	1851	65	21-APR-1994 11:18:13	12461	94115j122_2	1725	15	29-APR-1994 16:06:01
12193	94101j108_2	329	45	12-APR-1994 09:42:55	12335	94108j115_2	1826	15	21-APR-1994 12:51:49	12462	94115j122_2	1737	15	29-APR-1994 17:39:01
12194	94101j108_2	2243	15	12-APR-1994 10:12:07	12337	94108j115_2	1852	65	21-APR-1994 14:24:11	12463	94115j122_2	1737	15	29-APR-1994 19:11:55
12195	94101j108_2	343	15	12-APR-1994 12:49:53	12338	94108j115_2	1841	15	21-APR-1994 15:57:09	12464	94115j122_2	1736	15	29-APR-1994 20:44:55
12196	94101j108_2	2242	15	12-APR-1994 13:19:23	12338	94108j115_2	541	65	21-APR-1994 17:30:09	12465	94115j122_2	1739	15	29-APR-1994 22:17:49

ID	File	Val1	Val2	Date	Time
12466	94115j122_2	1745	15	29-APR-1994	23:50:35
12467	94115j122_2	1742	15	30-APR-1994	01:23:35
12468	94115j122_2	1647	15	30-APR-1994	02:56:31
12469	94115j122_2	1718	45	30-APR-1994	04:29:31
12470	94115j122_2	1722	45	30-APR-1994	06:02:43
12471	94115j122_2	534	45	30-APR-1994	07:35:41
12471	94115j122_2	64	65	30-APR-1994	08:03:15
12472	94115j122_2	1735	65	30-APR-1994	09:08:43
12473	94115j122_2	1730	65	30-APR-1994	10:41:41
12474	94115j122_2	1735	65	30-APR-1994	12:14:41
12475	94115j122_2	1738	65	30-APR-1994	13:47:35
12476	94115j122_2	1735	65	30-APR-1994	15:20:35
12477	94115j122_2	1491	15	30-APR-1994	16:53:03
12478	94115j122_2	1701	15	30-APR-1994	18:26:23
12479	94115j122_2	853	15	30-APR-1994	19:59:29
12484	94115j122_2	1727	45	1-MAY-1994	03:44:15
12485	94115j122_2	1738	45	1-MAY-1994	05:17:01
12486	94115j122_2	1733	45	1-MAY-1994	06:50:11
12497	94115j122_2	1467	15	2-MAY-1994	00:01:07
12498	94115j122_2	1706	15	2-MAY-1994	01:25:45
12499	94115j122_2	1417	15	2-MAY-1994	02:58:47
12502	94122j129_2	132	65	2-MAY-1994	08:30:43
12503	94122j129_2	1686	65	2-MAY-1994	09:11:51
12504	94122j129_2	1674	65	2-MAY-1994	10:45:07
12505	94122j129_2	1674	65	2-MAY-1994	12:17:47
12506	94122j129_2	1679	65	2-MAY-1994	13:50:51
12507	94122j129_2	1679	65	2-MAY-1994	15:23:45
12508	94122j129_2	427	65	2-MAY-1994	16:56:45
12509	94122j129_2	1679	65	2-MAY-1994	18:29:41
12510	94122j129_2	1620	15	2-MAY-1994	20:04:15
12511	94122j129_2	1678	15	2-MAY-1994	21:35:37
12512	94122j129_2	1676	15	2-MAY-1994	23:08:37
12513	94122j129_2	1676	15	3-MAY-1994	00:41:35
12514	94122j129_2	1675	15	3-MAY-1994	02:14:29
12515	94122j129_2	265	15	3-MAY-1994	03:47:33
12515	94122j129_2	126	45	3-MAY-1994	04:39:11
12516	94122j129_2	1668	45	3-MAY-1994	05:20:35
12517	94122j129_2	1669	45	3-MAY-1994	06:53:33
12518	94122j129_2	1566	65	3-MAY-1994	08:30:07
12519	94122j129_2	1670	65	3-MAY-1994	09:59:27
12520	94122j129_2	1671	65	3-MAY-1994	11:32:21
12522	94122j129_2	1668	65	3-MAY-1994	14:38:19
12523	94122j129_2	1638	15	3-MAY-1994	16:11:19
12524	94122j129_2	1670	15	3-MAY-1994	17:44:11
12525	94122j129_2	1667	15	3-MAY-1994	19:17:11
12526	94122j129_2	219	15	3-MAY-1994	20:50:11
12527	94122j129_2	1581	15	3-MAY-1994	22:25:37
12528	94122j129_2	1666	15	3-MAY-1994	23:56:07
12529	94122j129_2	1652	15	4-MAY-1994	01:29:09
12530	94122j129_2	1653	15	4-MAY-1994	03:02:05
12531	94122j129_2	1611	45	4-MAY-1994	04:36:43
12532	94122j129_2	1663	45	4-MAY-1994	06:08:01
12533	94122j129_2	487	45	4-MAY-1994	07:40:57
12537	94122j129_2	461	65	4-MAY-1994	14:32:51
12538	94122j129_2	1656	65	4-MAY-1994	15:25:51
12539	94122j129_2	1657	65	4-MAY-1994	16:58:49
12540	94122j129_2	1656	65	4-MAY-1994	18:31:47
12541	94122j129_2	211	15	4-MAY-1994	20:52:57
12541	94122j129_2	931	65	4-MAY-1994	20:04:45
12542	94122j129_2	1640	15	4-MAY-1994	21:37:41
12543	94122j129_2	1652	15	4-MAY-1994	23:10:43
12544	94122j129_2	1655	45	5-MAY-1994	00:43:37
12545	94122j129_2	1647	45	5-MAY-1994	02:16:35
12546	94122j129_2	1653	45	5-MAY-1994	03:49:35
12547	94122j129_2	1652	45	5-MAY-1994	05:22:33
12548	94122j129_2	1602	45	5-MAY-1994	06:55:29
12549	94122j129_2	1654	65	5-MAY-1994	08:27:27
12550	94122j129_2	1649	65	5-MAY-1994	10:00:25
12551	94122j129_2	1651	65	5-MAY-1994	11:33:25
12552	94122j129_2	1653	65	5-MAY-1994	13:06:21
12553	94122j129_2	1649	65	5-MAY-1994	14:39:19
12554	94122j129_2	1658	15	5-MAY-1994	16:12:19
12555	94122j129_2	1659	15	5-MAY-1994	17:45:15
12556	94122j129_2	1583	15	5-MAY-1994	19:18:17
12557	94122j129_2	1657	15	5-MAY-1994	20:51:15
12558	94122j129_2	1657	15	5-MAY-1994	22:24:11
12559	94122j129_2	1658	15	5-MAY-1994	23:57:11
12560	94122j129_2	1655	15	6-MAY-1994	01:30:11
12561	94122j129_2	1513	45	6-MAY-1994	03:03:11
12562	94122j129_2	1643	45	6-MAY-1994	04:36:05
12563	94122j129_2	1633	45	6-MAY-1994	06:09:03
12564	94122j129_2	419	45	6-MAY-1994	07:41:59
12565	94122j129_2	1643	45	6-MAY-1994	09:14:59
12566	94122j129_2	1644	65	6-MAY-1994	10:47:55
12567	94122j129_2	1640	65	6-MAY-1994	12:20:55
12568	94122j129_2	1586	65	6-MAY-1994	13:53:53
12569	94122j129_2	1617	65	6-MAY-1994	15:26:49
12570	94122j129_2	1587	65	6-MAY-1994	16:59:47
12571	94122j129_2	1648	65	6-MAY-1994	18:32:47
12572	94122j129_2	804	65	6-MAY-1994	20:05:45
12575	94122j129_2	1638	15	7-MAY-1994	00:44:39
12576	94122j129_2	1645	15	7-MAY-1994	02:17:37
12577	94122j129_2	1645	45	7-MAY-1994	03:50:41
12578	94122j129_2	1624	45	7-MAY-1994	05:23:37
12579	94122j129_2	1465	45	7-MAY-1994	06:56:35
12580	94122j129_2	1607	65	7-MAY-1994	08:29:35
12581	94122j129_2	1628	65	7-MAY-1994	10:02:33
12582	94122j129_2	1639	65	7-MAY-1994	11:35:31
12583	94122j129_2	1583	65	7-MAY-1994	13:08:29
12584	94122j129_2	1625	65	7-MAY-1994	14:41:27
12585	94122j129_2	1640	65	7-MAY-1994	16:14:25
12586	94122j129_2	1654	65	7-MAY-1994	17:47:21
12587	94122j129_2	1627	65	7-MAY-1994	19:20:19
12595	94122j129_2	794	65	8-MAY-1994	08:11:47
12596	94122j129_2	1607	65	8-MAY-1994	09:17:39
12597	94122j129_2	1624	65	8-MAY-1994	10:49:59
12598	94122j129_2	1625	65	8-MAY-1994	12:22:57
12599	94122j129_2	1620	65	8-MAY-1994	13:55:57
12600	94122j129_2	1598	65	8-MAY-1994	15:29:31
12601	94122j129_2	1620	65	8-MAY-1994	17:01:53
12602	94122j129_2	1599	65	8-MAY-1994	18:34:51
12605	94122j129_2	1628	15	8-MAY-1994	23:13:45
12606	94122j129_2	1629	15	9-MAY-1994	00:46:41
12607	94122j129_2	1627	15	9-MAY-1994	02:19:39
12608	94122j129_2	392	15	9-MAY-1994	03:52:35
12611	94129j136_2	1616	65	9-MAY-1994	08:31:37
12612	94129j136_2	1665	65	9-MAY-1994	10:03:37
12613	94129j136_2	1655	65	9-MAY-1994	11:36:33
12614	94129j136_2	1656	65	9-MAY-1994	13:09:31
12621	94129j136_2	1072	15	10-MAY-1994	00:20:01
12622	94129j136_2	1628	15	10-MAY-1994	01:33:55
12623	94129j136_2	1630	15	10-MAY-1994	03:06:49
12624	94129j136_2	382	15	10-MAY-1994	05:21:29
12625	94129j136_2	1653	45	10-MAY-1994	06:12:01
12631	94129j136_2	606	15	10-MAY-1994	16:04:17
12632	94129j136_2	1565	15	10-MAY-1994	17:05:29
12633	94129j136_2	1646	15	10-MAY-1994	18:35:47
12634	94129j136_2	1628	15	10-MAY-1994	20:08:45
12635	94129j136_2	1649	15	10-MAY-1994	21:41:41
12636	94129j136_2	1645	15	10-MAY-1994	23:14:39
12637	94129j136_2	1580	15	11-MAY-1994	00:47:55
12638	94129j136_2	1618	15	11-MAY-1994	02:20:55
12639	94129j136_2	1614	45	11-MAY-1994	03:54:39
12640	94129j136_2	1632	45	11-MAY-1994	05:26:35
12641	94129j136_2	1644	45	11-MAY-1994	06:59:29
12645	94129j136_2	1637	65	11-MAY-1994	13:11:35
12646	94129j136_2	1653	65	11-MAY-1994	14:44:19
12647	94129j136_2	1617	15	11-MAY-1994	16:17:15
12648	94129j136_2	1638	15	11-MAY-1994	17:50:11
12649	94129j136_2	1609	15	11-MAY-1994	19:23:09
12650	94129j136_2	1637	15	11-MAY-1994	20:56:07
12651	94129j136_2	1595	15	11-MAY-1994	22:30:27
12652	94129j136_2	1297	15	12-MAY-1994	00:02:49
12653	94129j136_2	1613	15	12-MAY-1994	01:35:21
12654	94129j136_2	1629	45	12-MAY-1994	03:07:59
12655	94129j136_2	1572	45	12-MAY-1994	04:42:27
12656	94129j136_2	1627	45	12-MAY-1994	06:13:23
12657	94129j136_2	914	45	12-MAY-1994	08:10:41
12658	94129j136_2	1628	65	12-MAY-1994	09:19:49
12659	94129j136_2	1628	65	12-MAY-1994	10:52:45
12660	94129j136_2	1626	65	12-MAY-1994	12:25:45
12661	94129j136_2	1626	65	12-MAY-1994	13:58:43
12662	94129j136_2	1656	65	12-MAY-1994	15:31:39
12663	94129j136_2	1616	65	12-MAY-1994	17:04:31
12664	94129j136_2	1626	65	12-MAY-1994	18:37:09
12665	94129j136_2	1027	65	12-MAY-1994	20:10:07
12668	94129j136_2	1571	45	13-MAY-1994	00:51:15
12669	94129j136_2	1640	45	13-MAY-1994	02:21:55
12670	94129j136_2	1626	45	13-MAY-1994	03:54:55
12671	94129j136_2	1628	45	13-MAY-1994	05:27:49
12672	94129j136_2	1637	45	13-MAY-1994	07:00:47
12673	94129j136_2	1651	65	13-MAY-1994	08:33:51
12674	94129j136_2	1648	65	13-MAY-1994	10:06:47
12675	94129j136_2	1650	65	13-MAY-1994	11:39:45
12676	94129j136_2	1656	65	13-MAY-1994	13:12:41
12677	94129j136_2	1655	65	13-MAY-1994	14:45:39
12678	94129j136_2	1631	15	13-MAY-1994	16:18:37
12679	94129j136_2	1627	15	13-MAY-1994	17:51:31
12680	94129j136_2	1632	15	13-MAY-1994	19:24:31
12681	94129j136_2	1633	15	13-MAY-1994	20:57:27
12682	94129j136_2	1629	15	13-MAY-1994	22:30:27
12683	94129j136_2	1630	15	14-MAY-1994	00:03:23
12684	94129j136_2	1554	15	14-MAY-1994	01:36:17
12685	94129j136_2	1612	45	14-MAY-1994	03:09:17
12686	94129j136_2	1633	45	14-MAY-1994	04:42:15
12687	94129j136_2	1632	45	14-MAY-1994	06:15:11
12688	94129j136_2	407	45	14-MAY-1994	07:48:07
12688	94129j136_2	889	65	14-MAY-1994	08:12:31
12689	94129j136_2	1618	65	14-MAY-1994	09:21:09
12690	94129j136_2	1617	65	14-MAY-1994	10:54:07
12691	94129j136_2	1609	65	14-MAY-1994	12:27:03
12692	94129j136_2	1616	65	14-MAY-1994	14:00:03
12693	94129j136_2	1611	65	14-MAY-1994	15:33:01
12694	94129j136_2	1628	15	14-MAY-1994	17:05:55
12695	94129j136_2	1628	15	14-MAY-1994	18:38:55
12696	94129j136_2	1629	15	14-MAY-1994	20:11:49
12697	94129j136_2	1628	15	14-MAY-1994	21:44:49
12698	94129j136_2	1628	15	14-MAY-1994	23:17:45
12699	94129j136_2	1627	15	15-MAY-1994	00:50:43
12700	94129j136_2	1626	15	15-MAY-1994	02:23:41
12701	94129j136_2	569	15	15-MAY-1994	03:56:37
12702	94129j136_2	1622	45	15-MAY-1994	05:29:49
12703	94129j136_2	1616	45	15-MAY-1994	07:02:31
12704	94129j136_2	1651	65	15-MAY-1994	08:35:33
12705	94129j136_2	1649	65	15-MAY-1994	10:08:33
12706	94129j136_2	1647	65	15-MAY-1994	11:41:31
12707	94129j136_2	1627	65	15-MAY-1994	13:15:09
12708	94129j136_2	1637	65	15-MAY-1994	14:47:23
12709	94129j136_2	1623	15	15-MAY-1994	16:20:21
12710	94129j136_2	1624	15	15-MAY-1994	17:53:17
12711	94129j136_2	1616	15	15-MAY-1994	19:26:13
12712	94129j136_2	1622	15	15-MAY-1994	20:59:13
12713	94129j136_2	1626	15	15-MAY-1994	22:32:07
12714	94129j136_2	1205	15	16-MAY-1994	00:05:31
12715	94129j136_2	1610	15	16-MAY-1994	01:38:29
12716	94129j136_2	1607	15	16-MAY-1994	03:11:27
12717	94129j136_2	831	45	16-MAY-1994	05:10:07
12718	94129j136_2	1612	45	16-MAY-1994	06:16:55
12719	94136j143_2	461	15	16-MAY-1994	07:49:53
12719	94136j143_2	761	65	16-MAY-1994	08:19:07
12720	94136j143_2	1619	65	16-MAY-1994	09:52:03
12721	94136j143_2	1636	65	16-MAY-1994	10:55:51
12722	94136j143_2	1635	65	16-MAY-1994	12:28:53
12723	94136j143_2	1635	65	16-MAY-1994	14:01:49
12724	94136j143_2	1637	65	16-MAY-1994	15:34:43
12725	94136j143_2	1634	65	16-MAY-1994	17:07:43
12726	94136j143_2	1639	65	16-MAY-1994	18:40:37
12727	94136j143_2	933	65	16-MAY-1994	20:13:35
12730	94136j143_2	1623	45	17-MAY-1994	00:52:27
12731	94136j143_2	1606	45	17-MAY-1994	02:25:27
12732	94136j143_2	1617	45	17-MAY-1994	03:58:25
12733	94136j143_2	1606	45	17-MAY-1994	05:31:19
12734	94136j143_2	1605	45	17-MAY-1994	07:04:19
12735	94136j143_2	1609	65	17-MAY-1994	08:37:15
12736	94136j143_2	1605	65	17-MAY-1994	10:10:17
12737	94136j143_2	1598	65	17-MAY-1994	11:43:27
12738	94136j143_2	1605	65	17-MAY-1994	13:16:11
12739	94136j143_2	1605	65	17-MAY-1994	14:49:07
12740	94136j143_2	1620	15	17-MAY-1994	16:22:03
12741	94136j143_2	1472	15	17-MAY-1994	17:55:03
12742	94136j143_2	1617	15	17-MAY-1994	19:28:01
12743	94136j143_2	1614	15	17-MAY-1994	21:00:57
12744	94136j143_2	1613	15	17-MAY-1994	22:33:55
12745	94136j143_2	1508	15	18-MAY-1994	00:10:27
12746	94136j143_2	1013	15	18-MAY-1994	01:39:49
12746	94136j143_2	587	45	18-MAY-1994	02:13:47
12747	94136j143_2	1616	45	18-MAY-1994	03:12:49
12748	94136j143_2	1604	45	18-MAY-1994	04:45:43
12749	94136j143_2	1602	45	18-MAY-1994	06:18:43
12750	94136j143_2	392	45	18-MAY-1994	07:51:57
12754	94136j143_2	8	65	18-MAY-1994	14:18:53
12755	94136j143_2	1606	65	18-MAY-1994	15:36:31
12756	94136j143_2	1611	15	18-MAY-1994	17:09:27
12757	94136j143_2	1604	15	18-MAY-1994	18:42:21
12758	94136j143_2	1526	15	18-MAY-1994	20:15:21
12759	94136j143_2	1573	15	18-MAY-1994	21:48:19
12760	94136j143_2	1612	15	18-MAY-1994	23:21:13
12761	94136j143_2	1611	15	19-MAY-1994	00:54:13
12762	94136j143_2	1598	45	19-MAY-1994	02:27:17
12763	94136j143_2	1584	45	19-MAY-1994	04:00:17
12764	94136j143_2	1593	45	19-MAY-1994	05:33:09
12765	94136j143_2	1601	45	19-MAY-1994	07:05:59
12766	94136j143_2	1578	65	19-MAY-1994	08:39:05
12767	94136j143_2	1600	65	19-MAY-1994	10:11:59
12768	94136j143_2	1597	65	19-MAY-1994	11:44:59
12769	94136j143_2	1599	65	19-MAY-1994	13:17:55
12770	94136j143_2	1600	65	19-MAY-1994	14:50:53
12771	94136j143_2	1500	65	19-MAY-1994	16:23:53
12772	94136j143_2	1599	65	19-MAY-1994	17:56:49
12773	94136j143_2	1596	65	19-MAY-1994	19:29:45
12774	94136j143_2	971	15	19-MAY-1994	21:23:55
12775	94136j143_2	1603	15	19-MAY-1994	22:35:47
12776	94136j143_2	1599	15	20-MAY-1994	00:08:53
12777	94136j143_2	1595	15	20-MAY-1994	01:41:59
12778	94136j143_2	1591	15	20-MAY-1994	03:14:51
12781	94136j143_2	886	45	20-MAY-1994	08:16:59
12782	94136j143_2	1587	65	20-MAY-1994	09:26:21
12783	94136j143_2	1597	65	20-MAY-1994	10:59:21
12784	94136j143_2	1598	65	20-MAY-1994	12:32:15
12785	94136j143_2	1596	65	20-MAY-1994	14:05:13
12786	94136j143_2	1600	65	20-MAY-1994	15:38:09
12787	94136j143_2	1605	15	20-MAY-1994	17:11:11
12788	94136j143_2	1590	15	20-MAY-1994	18:44:45
12789	94136j143_2	1576	15	20-MAY-1994	20:17:13
12790	94136j143_2	1585	15	20-MAY-1994	21:50:05
12791	94136j143_2	1603	15	20-MAY-1994	23:23:03
12792	94136j143_2	1604	15	21-MAY-1994	00:55:59
12793	94136j143_2	1614	45	21-MAY-1994	02:28:57
12794	94136j143_2	1611	45	21-MAY-1994	04:01:51
12795	94136j143_2	1606	45	21-MAY-1994	05:34:53
12796	94136j143_2	1585	45	21-MAY-1994	07:08:29
12797	94136j143_2	1586	65	21-MAY-1994	08:40:47
12798	94136j143_2	1593	65	21-MAY-1994	10:13:47
12799	94136j143_2	1595	65	21-MAY-1994	11:46:41
12800	94136j143_2	1591	65	21-MAY-1994	13:19:41
12801	94136j143_2	1395	65	21-MAY-1994	14:52:39
12802	94136j143_2	194	15	21-MAY-1994	16:25:35
12803	94136j143_2	1624	15	21-MAY-1994	17:57:55
12804	94136j143_2	1603	15	21-MAY-1994	19:31:29
12805	94136j143_2	1583	15	21-MAY-1994	21:04:49
12806	94136j143_2	1602	15	22-MAY-1994	22:37:27
12807	94136j143_2	1601	15	22-MAY-1994	00:10:23
12808	94136j143_2	1587	15	22-MAY-1994	01:43:21
12811	94136j143_2	1483	45	22-MAY-1994	06:23:47
12812	94136j143_2	944	65	22-MAY-1994	08:16:45
12813	94136j143_2	1591	65	22-MAY-1994	09:28:09
12814	94136j143_2	1592	65	22-MAY-1994	11:01:03
12815	94136j143_2	1592	65	22-MAY-1994	12:34:03
12816	94136j143_2	1594	65	22-MAY-1994	14:06:57
12817	94136j143_2	1586	65	22-MAY-1994	15:40:05
12818	94136j143_2	1599	15	22-MAY-1994	17:12:59
12819	94136j143_2	1602	15	22-MAY-1994	18:45:55
12820	94136j143_2	1595	15	22-MAY-1994	20:18:53
12821	94136j143_2	1600	15	22-MAY-1994	21:51:47
12822	94136j143_2	1600	15	22-MAY-1994	23:24:47
12823	94136j143_2	1598	15	23-MAY-1994	00:57:49
12824	94136j143_2	1597	15	23-MAY-1994	02:30:43
12825	94136j143_2	789	15	23-MAY-1994	04:03:43
12826	94136j143_2	1523	45	23-MAY-1994	05:38:45
12827	94143j150_2	1587	45	23-MAY-1994	07:09:33
12828	94143j150_2	1587	65	23-MAY-1994	08:42:37
12829	94143j150_2	1587	65	23-MAY-1994	10:15:31
12830	94143j150_2	1586	65	23-MAY-1994	11:48:31
12831	94143j150_2	1586	65	23-MAY-1994	13:21:25
12832	94143j150_2	1586	65	23-MAY-1994	14:54:25
12833	94143j150_2	1584	65	23-MAY-1994	16:27:25
12834	94143j150_2	1576	65	23-MAY-1994	18:00:19
12835	94143j150_2	1584	65	23-MAY-1994	19:33:17
12836	94143j150_2	1220	15	23-MAY-1994	21:18:47
12837	94143j150_2	1592	15	23-MAY-1994	22:39:17
12838	94143j150_2	1585	15	24-MAY-1994	00:12:09
12839	94143j150_2	1589	45	24-MAY-1994	01:45:11
12840	94143j150_2	1585	45	24-MAY-1994	03:18:05
12841	94143j150_2	1597	45	24-MAY-1994	04:51:01

12842	94143j150_2	1598	45	24-MAY-1994	06:23:59	12963	94150j157_2	1549	45	1-JUN-1994	01:52:01	13113	94157j164_2	1572 15 10-JUN-1994 18:15:07
12843	94143j150_2	385	45	24-MAY-1994	07:57:03	12964	94150j157_2	1560	45	1-JUN-1994	03:25:43	13114	94157j164_2	841 15 10-JUN-1994 20:12:29
12843	94143j150_2	771	65	24-MAY-1994	08:24:05	12965	94150j157_2	1435	45	1-JUN-1994	04:58:07	13115	94157j164_2	1570 15 10-JUN-1994 21:21:03
12844	94143j150_2	1584	65	24-MAY-1994	09:29:57	12971	94150j157_2	1558	65	1-JUN-1994	14:15:59	13116	94157j164_2	832 15 10-JUN-1994 23:18:35
12845	94143j150_2	1585	65	24-MAY-1994	11:02:55	12972	94150j157_2	1568	65	1-JUN-1994	15:48:37	13117	94157j164_2	1571 15 11-JUN-1994 00:26:57
12846	94143j150_2	1584	65	24-MAY-1994	12:35:51	12973	94150j157_2	1569	65	1-JUN-1994	17:21:35	13118	94157j164_2	822 45 11-JUN-1994 02:24:21
12847	94143j150_2	1583	65	24-MAY-1994	14:08:49	12974	94150j157_2	1551	65	1-JUN-1994	18:54:43	13119	94157j164_2	1560 45 11-JUN-1994 03:32:51
12848	94143j150_2	1585	45	24-MAY-1994	15:41:45	12975	94150j157_2	1484	65	1-JUN-1994	20:27:29	13120	94157j164_2	838 45 11-JUN-1994 05:30:03
12849	94143j150_2	1591	15	24-MAY-1994	17:14:49	12979	94150j157_2	1567	45	2-JUN-1994	02:39:19	13121	94157j164_2	1563 45 11-JUN-1994 06:38:45
12850	94143j150_2	1592	15	24-MAY-1994	18:47:45	12980	94150j157_2	1142	45	2-JUN-1994	04:12:17	13122	94157j164_2	433 65 11-JUN-1994 08:50:21
12851	94143j150_2	1591	15	24-MAY-1994	20:20:43	12982	94150j157_2	1571	45	2-JUN-1994	07:18:07	13123	94157j164_2	1594 65 11-JUN-1994 09:44:39
12852	94143j150_2	1589	15	24-MAY-1994	21:53:41	12983	94150j157_2	1566	65	2-JUN-1994	08:51:05	13124	94157j164_2	856 65 11-JUN-1994 11:42:07
12853	94143j150_2	1593	15	24-MAY-1994	23:26:37	12984	94150j157_2	1567	65	2-JUN-1994	10:24:07	13125	94157j164_2	1586 65 11-JUN-1994 12:50:35
12854	94143j150_2	1438	15	25-MAY-1994	00:59:39	12985	94150j157_2	1564	65	2-JUN-1994	11:57:11	13126	94157j164_2	840 65 11-JUN-1994 14:48:55
12855	94143j150_2	1592	45	25-MAY-1994	02:32:33	12986	94150j157_2	1567	65	2-JUN-1994	13:30:01	13127	94157j164_2	1558 65 11-JUN-1994 15:56:31
12856	94143j150_2	1595	45	25-MAY-1994	04:05:29	12987	94150j157_2	1565	45	2-JUN-1994	15:02:59	13128	94157j164_2	838 15 11-JUN-1994 17:53:53
12857	94143j150_2	1586	45	25-MAY-1994	05:38:45	12988	94150j157_2	1506	65	2-JUN-1994	16:35:53	13129	94157j164_2	1571 15 11-JUN-1994 19:02:23
12858	94143j150_2	1578	45	25-MAY-1994	07:11:25	12989	94150j157_2	1556	65	2-JUN-1994	18:09:11	13130	94157j164_2	842 15 11-JUN-1994 20:59:35
12864	94143j150_2	1590	15	25-MAY-1994	16:29:09	12990	94150j157_2	207	65	2-JUN-1994	20:06:15	13131	94157j164_2	1568 15 11-JUN-1994 22:08:17
12865	94143j150_2	1593	15	25-MAY-1994	18:02:07	12991	94150j157_2	1575	15	2-JUN-1994	21:14:49	13132	94157j164_2	843 15 12-JUN-1994 00:05:31
12866	94143j150_2	1592	15	25-MAY-1994	19:35:03	12992	94150j157_2	839	15	2-JUN-1994	23:12:53	13133	94157j164_2	1561 45 12-JUN-1994 01:14:09
12867	94143j150_2	1590	15	25-MAY-1994	21:08:01	12993	94150j157_2	1577	15	3-JUN-1994	00:20:45	13134	94157j164_2	837 45 12-JUN-1994 03:11:27
12868	94143j150_2	1593	15	25-MAY-1994	22:40:53	12994	94150j157_2	817	45	3-JUN-1994	02:18:23	13135	94157j164_2	1559 45 12-JUN-1994 04:20:05
12869	94143j150_2	1575	15	26-MAY-1994	00:14:05	12995	94150j157_2	1577	45	3-JUN-1994	03:26:37	13136	94157j164_2	841 15 12-JUN-1994 06:17:23
12870	94143j150_2	1565	45	26-MAY-1994	01:47:29	12998	94150j157_2	705	65	3-JUN-1994	08:33:49	13137	94157j164_2	1560 45 12-JUN-1994 07:25:59
12871	94143j150_2	1582	45	26-MAY-1994	03:19:47	12999	94150j157_2	1566	65	3-JUN-1994	09:38:25	13138	94157j164_2	866 65 12-JUN-1994 09:23:29
12872	94143j150_2	1555	45	26-MAY-1994	04:53:43	13000	94150j157_2	815	65	3-JUN-1994	11:36:03	13139	94157j164_2	1556 65 12-JUN-1994 10:31:57
12873	94143j150_2	1581	45	26-MAY-1994	06:25:43	13001	94150j157_2	1566	65	3-JUN-1994	12:44:19	13140	94157j164_2	856 65 12-JUN-1994 12:29:21
12874	94143j150_2	515	45	26-MAY-1994	07:58:41	13002	94150j157_2	831	65	3-JUN-1994	14:41:45	13141	94157j164_2	1560 65 12-JUN-1994 13:37:49
12874	94143j150_2	878	65	26-MAY-1994	08:22:07	13003	94150j157_2	1488	65	3-JUN-1994	15:50:11	13142	94157j164_2	831 65 12-JUN-1994 15:35:01
12875	94143j150_2	1581	65	26-MAY-1994	09:31:39	13004	94150j157_2	831	65	3-JUN-1994	17:47:41	13143	94157j164_2	1572 15 12-JUN-1994 16:43:41
12876	94143j150_2	1579	65	26-MAY-1994	11:04:39	13005	94150j157_2	1562	65	3-JUN-1994	18:56:07	13144	94157j164_2	843 15 12-JUN-1994 18:40:59
12877	94143j150_2	1573	65	26-MAY-1994	12:37:37	13006	94150j157_2	702	65	3-JUN-1994	20:53:37	13145	94157j164_2	1570 15 12-JUN-1994 19:49:35
12878	94143j150_2	1579	65	26-MAY-1994	14:10:33	13007	94150j157_2	1575	15	3-JUN-1994	22:02:05	13146	94157j164_2	842 15 12-JUN-1994 21:46:53
12879	94143j150_2	1579	65	26-MAY-1994	15:43:31	13008	94150j157_2	839	15	3-JUN-1994	23:59:35	13147	94157j164_2	1571 15 12-JUN-1994 22:55:31
12880	94143j150_2	1579	65	26-MAY-1994	17:16:27	13009	94150j157_2	1490	45	4-JUN-1994	01:10:55	13148	94157j164_2	839 15 13-JUN-1994 00:52:51
12881	94143j150_2	1578	65	26-MAY-1994	18:49:25	13010	94150j157_2	829	45	4-JUN-1994	03:05:29	13149	94157j164_2	1567 45 13-JUN-1994 02:01:25
12882	94143j150_2	1577	65	26-MAY-1994	20:22:25	13011	94150j157_2	1567	45	4-JUN-1994	04:13:49	13150	94157j164_2	828 45 13-JUN-1994 03:58:49
12883	94143j150_2	1589	15	26-MAY-1994	21:55:25	13012	94150j157_2	833	45	4-JUN-1994	06:11:13	13151	94157j164_2	1566 45 13-JUN-1994 05:07:19
12884	94143j150_2	1568	15	26-MAY-1994	23:28:59	13013	94150j157_2	1556	45	4-JUN-1994	07:19:29	13152	94164j171_2	848 15 13-JUN-1994 07:04:29
12885	94143j150_2	1549	15	27-MAY-1994	01:01:13	13014	94150j157_2	831	65	4-JUN-1994	09:17:15	13153	94164j171_2	958 45 13-JUN-1994 08:13:11
12886	94143j150_2	1558	45	27-MAY-1994	02:34:53	13015	94150j157_2	1565	45	4-JUN-1994	10:25:43	13164	94164j171_2	834 15 14-JUN-1994 01:39:59
12887	94143j150_2	1593	45	27-MAY-1994	04:07:11	13016	94150j157_2	828	65	4-JUN-1994	12:23:11	13165	94164j171_2	1546 45 14-JUN-1994 02:48:33
12888	94143j150_2	632	45	27-MAY-1994	05:40:11	13017	94150j157_2	1564	45	4-JUN-1994	13:31:39	13166	94164j171_2	823 45 14-JUN-1994 04:46:07
12890	94143j150_2	1578	65	27-MAY-1994	08:46:05	13018	94150j157_2	828	65	4-JUN-1994	15:29:07	13167	94164j171_2	1561 45 14-JUN-1994 05:54:25
12891	94143j150_2	1568	65	27-MAY-1994	10:19:03	13019	94150j157_2	1562	65	4-JUN-1994	16:37:33	13168	94164j171_2	557 45 14-JUN-1994 07:51:53
12892	94143j150_2	1578	65	27-MAY-1994	11:52:01	13020	94150j157_2	825	45	4-JUN-1994	18:35:03	13169	94164j171_2	1543 45 14-JUN-1994 09:01:01
12893	94143j150_2	1577	65	27-MAY-1994	13:24:57	13021	94150j157_2	1563	45	4-JUN-1994	19:43:27	13170	94164j171_2	809 65 14-JUN-1994 10:58:03
12894	94143j150_2	1577	65	27-MAY-1994	14:57:55	13024	94150j157_2	843	15	5-JUN-1994	00:46:41	13171	94164j171_2	1397 65 14-JUN-1994 12:11:45
12895	94143j150_2	1577	65	27-MAY-1994	16:30:51	13025	94150j157_2	1488	45	5-JUN-1994	01:58:19	13172	94164j171_2	819 65 14-JUN-1994 14:03:53
12896	94143j150_2	1575	65	27-MAY-1994	18:03:51	13026	94150j157_2	842	45	5-JUN-1994	03:52:37	13173	94164j171_2	1506 65 14-JUN-1994 15:13:57
12897	94143j150_2	1575	65	27-MAY-1994	19:36:49	13027	94150j157_2	1577	45	5-JUN-1994	05:01:11	13174	94164j171_2	837 15 14-JUN-1994 17:09:33
12901	94143j150_2	882	15	28-MAY-1994	01:48:33	13029	94150j157_2	664	65	5-JUN-1994	08:37:05	13175	94164j171_2	1572 15 14-JUN-1994 18:18:05
12901	94143j150_2	640	45	28-MAY-1994	02:19:33	13030	94150j157_2	818	65	5-JUN-1994	10:04:33	13176	94164j171_2	836 15 14-JUN-1994 20:15:29
12903	94143j150_2	1529	45	28-MAY-1994	04:56:19	13031	94150j157_2	1546	65	5-JUN-1994	11:13:01	13177	94164j171_2	1572 15 14-JUN-1994 21:23:57
12904	94143j150_2	1583	45	28-MAY-1994	06:27:33	13032	94150j157_2	549	65	5-JUN-1994	12:45:57	13178	94164j171_2	1376 15 15-JUN-1994 00:36:17
12905	94143j150_2	459	45	28-MAY-1994	08:00:29	13040	94150j157_2	617	15	6-JUN-1994	01:09:31	13180	94164j171_2	825 15 15-JUN-1994 02:27:41
12905	94143j150_2	829	65	28-MAY-1994	08:25:23	13041	94150j157_2	1574	15	6-JUN-1994	02:42:31	13181	94164j171_2	1567 15 15-JUN-1994 03:35:49
12906	94143j150_2	1576	65	28-MAY-1994	09:33:27	13042	94150j157_2	805	45	6-JUN-1994	04:40:41	13182	94164j171_2	822 45 15-JUN-1994 05:33:19
12907	94143j150_2	1578	65	28-MAY-1994	11:06:23	13043	94150j157_2	1532	45	6-JUN-1994	05:49:21	13183	94164j171_2	1559 45 15-JUN-1994 06:41:41
12908	94143j150_2	1575	65	28-MAY-1994	12:39:23	13051	94157j164_2	847	45	6-JUN-1994	07:45:41	13188	94164j171_2	462 15 15-JUN-1994 16:36:19
12909	94143j150_2	1574	65	28-MAY-1994	14:12:19	13052	94157j164_2	147	15	6-JUN-1994	20:32:33	13190	94164j171_2	828 15 15-JUN-1994 17:57:01
12910	94143j150_2	1560	65	28-MAY-1994	15:45:23	13053	94157j164_2	1568	15	6-JUN-1994	21:17:57	13191	94164j171_2	1549 15 15-JUN-1994 19:05:41
12911	94143j150_2	1586	15	28-MAY-1994	17:18:15	13054	94157j164_2	846	15	6-JUN-1994	23:15:11	13192	94164j171_2	825 15 15-JUN-1994 21:03:01
12912	94143j150_2	1585	15	28-MAY-1994	18:51:13	13055	94157j164_2	1573	15	7-JUN-1994	00:23:47	13193	94164j171_2	1557 15 15-JUN-1994 22:11:43
12913	94143j150_2	1586	15	28-MAY-1994	20:24:09	13056	94157j164_2	843	15	7-JUN-1994	02:21:07	13194	94164j171_2	824 15 16-JUN-1994 00:08:57
12914	94143j150_2	1586	15	28-MAY-1994	21:57:05	13057	94157j164_2	1543	45	7-JUN-1994	03:29:45	13195	94164j171_2	1553 15 16-JUN-1994 01:17:39
12915	94143j150_2	1579	15	28-MAY-1994	23:30:15	13058	94157j164_2	844	15	7-JUN-1994	05:27:07	13196	94164j171_2	815 15 16-JUN-1994 03:15:07
12916	94143j150_2	896	15	29-MAY-1994	01:03:01	13059	94157j164_2	1553	45	7-JUN-1994	06:35:43	13197	94164j171_2	801 15 16-JUN-1994 04:23:17
12916	94143j150_2	678	45	29-MAY-1994	01:32:57	13060	94157j164_2	632	65	7-JUN-1994	08:40:43	13199	94164j171_2	773 45 16-JUN-1994 07:31:19
12917	94143j150_2	1564	45	29-MAY-1994	02:36:45	13061	94157j164_2	1564	65	7-JUN-1994	09:41:37	13207	94164j171_2	873 15 16-JUN-1994 20:15:43
12918	94143j150_2	1575	45	29-MAY-1994	04:08:59	13062	94157j164_2	805	65	7-JUN-1994	11:39:47	13208	94164j171_2	810 15 16-JUN-1994 21:50:47
12919	94143j150_2	1560	45	29-MAY-1994	05:42:11	13063	94157j164_2	1561	65	7-JUN-1994	12:47:33	13209	94164j171_2	1571 15 16-JUN-1994 22:58:19
12921	94143j150_2	1574	65	29-MAY-1994	08:47:51	13064	94157j164_2	858	65	7-JUN-1994	14:44:59	13210	94164j171_2	831 15 17-JUN-1994 00:55:57
12922	94143j150_2	1576	65	29-MAY-1994	10:20:45	13065	94157j164_2	1531	65	7-JUN-1994	15:53:23	13211	94164j171_2	1572 15 17-JUN-1994 02:04:11
12923	94143j150_2	1572	65	29-MAY-1994	11:53:47	13071	94157j164_2	1389	45	8-JUN-1994	01:16:55	13212	94164j171_2	822 15 17-JUN-1994 04:02:01
12924	94143j150_2	1577	65	29-MAY-1994	13:26:41	13072	94157j164_2	812	45	8-JUN-1994	03:09:31	13214	94164j171_2	771 45 17-JUN-1994 07:09:21
12925	94143j150_2	1572	65	29-MAY-1994	14:59:39	13073	94157j164_2	1561	45	8-JUN-1994	04:17:03	13215	94164j171_2	1153 45 17-JUN-1994 08:17:03
12926	94143j150_2	1567	15	29-MAY-1994	16:33:15	13074	94157j164_2	840	45	8-JUN-1994	06:14:27	13216	94164j171_2	814 65 17-JUN-1994 10:13:53
12927	94143j150_2	1585	15	29-MAY-1994	18:05:33	13075	94157j164_2	1549	45	8-JUN-1994	07:23:21	13217	94164j171_2	1560 65 17-JUN-1994 11:21:59
12928	94143j150_2	1560	15	29-MAY-1994	19:39:19	13081	94157j164_2	1565	15	8-JUN-1994	16:40:59	13218	94164j171_2	818 65 17-JUN-1994 13:19:41
12929	94143j150_2	1396	15	29-MAY-1994	21:11:27	13082	94157j164_2	814	15	8-JUN-1994	18:38:17	13219	94164j171_2	1558 65 17-JUN-1994 14:27:53
12930	94143j150_2	1571	15	29-MAY-1994	22:44:27	13083	94157j164_2	1539	15	8-JUN-1994	19:47:45	13220	94164j171_2	817 65 17-JUN-1994 16:25:37
12931	94143j150_2	1584	15	30-MAY-1994	00:17:23	13084	94157j164_2	835	15	8-JUN-1994	21:44:03	13221	94164j171_2	1561 65 17-JUN-1994 17:33:41
12932	94143j150_2	1580	15	30-MAY-1994	01:50:21	13085	94157j164_2	1571	15	8-JUN-1994	22:52:29	13222	94164j171_2	759 65 17-JUN-1994 19:31:37
12933	94143j150_2	1579	15	30-MAY-1994	03:23:19	13086	94157j164_2	837	15	9-JUN-1994	00:49:53	13223	94164j171_2	1284 65 17-JUN-1994 20:39:43
12934	94143j150_2	1124	45	30-MAY-1994	05:09:09	13087	94157j164_2	1572	15	9-JUN-1994	01:58:19	13224	94164j171_2	713 15 17-JUN-1994 22:37:33
12935	94143j150_2	920	45	30-MAY-1994	06:29:19	13088	94157j164_2	812	45	9-JUN-1994	03:56:47	13225	94164j171_2	1570 15 17-JUN-1994 23:45:35
12939	94150j157_2	1570	65	30-MAY-1994	12:41:09	13089	94157j164_2	1575	45	9-JUN-1994	05:04:15	13226	94164j171_2	813 15 18-JUN-1994 01:43:25
12940	94150j157_2	1572	65	30-MAY-1994	14:14:01	13090	94157j164_2	837	45	9-JUN-1994	07:01:47	13227	94164j171_2	1561 45 18-JUN-1994 02:51:25
12941	94150j157_2	1561	65	30-MAY-1994	15:47:01	13091	94157j164_2	628	15	9-JUN-1994	08:10:07	13231	94164j171_2	1557 65 18-JUN-1994 09:03:21
12942	94150j157_2	1582	15	30-MAY-1994	17:19:59	13092	94157j164_2	861	65	9-JUN-1994	10:07:33	13232	94164j171_2	803 65 18-JUN-1994 11:01:09
12943	94150j157_2	1580	15	30-MAY-1994	18:52:57	13093	94157j164_2	1591	65	9-JUN-1994	11:16:11	13233	94164j171_2	1520 65 18-JUN-1994 12:10:29
12944	94150j157_2	1582	15	30-MAY-1994	20:25:51	13094	94157j164_2	826	65	9-JUN-1994	13:13:31	13234	94164j171_2	808 65 18-JUN-1994 14:07:09
12945	94150j157_2	1580	15	30-MAY-1994	21:58:51	13095	94157j164_2	1490	65	9-JUN-1994	14:22:03	13235	94164j171_2	1560 65 18-JUN-1994 15:15:03
12946	94150j157_2	1579	15	30-MAY-1994	23:31:47	13097	94157j164_2	1596	65	9-JUN-1994	17:27:57	13236	94164j171_2	796 65 18-JUN-1994 17:13:03
12947	94150j157_2	1580	15	31-MAY-1994	01:04:43	13098	94157j164_2	724	65	9-JUN-1994	19:25:09	13237	94164j171_2	1518 65 18-JUN-1994 18:21:57
12948	94150j157_2	1447	45	31-MAY-1994	02:38:37	13099	94157j164_2	1571	15	9-JUN-1994	20:33:49	13238	94164j171_2	778 65 18-JUN-1994 20:18:59
12949	94150j157_2	1560	45	31-MAY-1994	04:10:59	13100	94157j164_2	842	15	9-JUN-1994	22:31:07	13247	94164j171_2	1559 45 19-JUN-1994 09:50:29
12950	94150j157_2	1574	45	31-MAY-1994	05:43:31	13101	94157j164_2	1556	15	9-JUN-1994	23:40:11	13248	94164j171_2	809 65 19-JUN-1994 11:48:25
12951	94150j157_2	1560	45	31-MAY-1994	07:16:37	13102	94157j164_2	757	45	10-JUN-1994	01:39:49	13249	94164j171_2	1545 15 19-JUN-1994 12:56:21
12952	94150j157_2	1543	65	31-MAY-1994	08:50:29	13103	94157j164_2	1557	45	10-JUN-1994	02:45:37	13250	94164j171_2	813 65 19-JUN-1994 14:54:09
12953	94150j157_2	1563	65	31-MAY-1994	10:22:37	13107	94157j164_2	1559	65	10-JUN-1994	08:57:29	13251	94164j171_2	1557 65 19-JUN-1994 16:02:17
12958	94150j157_2	1063	15	31-MAY-1994	18:07:17	13108	94157j164_2	867	65	10-JUN-1994	10:55:01	13252	94164j171_2	803 65 19-JUN-1994 18:00:05
12959	94150j157_2	1577	15	31-MAY-1994	19:40:11	13109	94157j164_2	1561	65	10-JUN-1994	12:03:21	13253	94164j171_2	1539 65 19-JUN-1994 19:08:09
12960	94150j157_2	1579	15	31-MAY-1994	21:13:09	13110	94157j164_2	728	65	10-JUN-1994	14:04:05	13254	94164j171_2	439 65 19-JUN-1994 21:06:01
12961	94150j157_2	1579	15	31-MAY-1994	22:46:03	13111	94157j164_2	1559	65	10-JUN-1994	15:09:15	13256	94164j171_2	668 15 20-JUN-1994 00:16:45
12962	94150j157_2	1579	15	1-JUN-1994	00:19:03	13112	94157j164_2	842	15	10-JUN-1994	17:06:31	13257	94164j171_2	1549 15 20-JUN-1994 01:19:49

13258	94164j171_2	802	15	20-JUN-1994	03:17:57	13412	94178j185_2	350	15	30-JUN-1994	02:08:07	13539	94185j192_2	440	45	8-JUL-1994	06:11:13
13259	94164j171_2	724	15	20-JUN-1994	04:25:55	13413	94178j185_2	1250	15	30-JUN-1994	03:00:01	13541	94185j192_2	604	65	8-JUL-1994	09:37:47
13262	94171j178_2	832	65	20-JUN-1994	09:29:43	13423	94178j185_2	1259	65	30-JUN-1994	18:29:33	13542	94185j192_2	495	65	8-JUL-1994	11:05:03
13263	94171j178_2	1575	65	20-JUN-1994	10:37:39	13424	94178j185_2	484	15	30-JUN-1994	20:17:59	13543	94185j192_2	1220	65	8-JUL-1994	12:23:07
13264	94171j178_2	830	65	20-JUN-1994	12:35:39	13425	94178j185_2	1246	15	30-JUN-1994	21:35:29	13544	94185j192_2	90	65	8-JUL-1994	14:10:59
13265	94171j178_2	1559	65	20-JUN-1994	13:43:33	13426	94178j185_2	481	15	30-JUN-1994	23:23:53	13545	94185j192_2	1218	65	8-JUL-1994	15:28:57
13266	94171j178_2	806	65	20-JUN-1994	15:41:35	13427	94178j185_2	1241	15	1-JUL-1994	00:41:37	13546	94185j192_2	363	15	8-JUL-1994	17:43:07
13267	94171j178_2	1556	65	20-JUN-1994	16:49:29	13428	94178j185_2	482	15	1-JUL-1994	02:29:45	13548	94185j192_2	513	15	8-JUL-1994	20:22:37
13268	94171j178_2	831	65	20-JUN-1994	18:47:31	13429	94178j185_2	1237	15	1-JUL-1994	03:47:19	13549	94185j192_2	1228	15	8-JUL-1994	21:40:49
13269	94171j178_2	1558	65	20-JUN-1994	19:55:21	13433	94178j185_2	1207	15	1-JUL-1994	09:59:59	13550	94185j192_2	526	15	8-JUL-1994	23:28:31
13272	94171j178_2	811	15	21-JUN-1994	00:59:19	13434	94178j185_2	70	65	1-JUL-1994	11:48:07	13551	94185j192_2	1065	15	9-JUL-1994	00:46:33
13273	94171j178_2	1563	15	21-JUN-1994	02:07:15	13435	94178j185_2	1235	65	1-JUL-1994	13:04:59	13552	94185j192_2	471	45	9-JUL-1994	02:34:25
13274	94171j178_2	804	15	21-JUN-1994	04:05:11	13436	94178j185_2	365	65	1-JUL-1994	15:18:09	13554	94185j192_2	511	45	9-JUL-1994	05:40:19
13277	94171j178_2	112	65	21-JUN-1994	09:07:13	13437	94178j185_2	1235	65	1-JUL-1994	16:10:49	13557	94185j192_2	1219	65	9-JUL-1994	10:04:15
13278	94171j178_2	802	65	21-JUN-1994	10:17:07	13438	94178j185_2	484	15	1-JUL-1994	17:59:13	13558	94185j192_2	499	65	9-JUL-1994	11:52:09
13279	94171j178_2	1559	65	21-JUN-1994	11:24:49	13439	94178j185_2	1248	15	1-JUL-1994	19:16:43	13559	94185j192_2	1220	65	9-JUL-1994	13:10:07
13280	94171j178_2	710	65	21-JUN-1994	13:23:03	13440	94178j185_2	477	15	1-JUL-1994	21:05:09	13560	94185j192_2	427	45	9-JUL-1994	14:57:55
13293	94171j178_2	1426	65	22-JUN-1994	09:11:37	13441	94178j185_2	1244	15	1-JUL-1994	22:22:39	13561	94185j192_2	1221	65	9-JUL-1994	16:15:59
13294	94171j178_2	829	65	22-JUN-1994	11:04:25	13442	94178j185_2	475	15	2-JUL-1994	00:10:59	13562	94185j192_2	516	15	9-JUL-1994	18:03:49
13295	94171j178_2	1592	65	22-JUN-1994	12:11:57	13443	94178j185_2	1249	15	2-JUL-1994	01:28:27	13563	94185j192_2	1229	65	9-JUL-1994	19:21:59
13297	94171j178_2	446	65	22-JUN-1994	15:34:49	13444	94178j185_2	479	15	2-JUL-1994	03:16:51	13564	94185j192_2	518	15	9-JUL-1994	20:09:43
13298	94171j178_2	1255	15	22-JUN-1994	16:51:39	13445	94178j185_2	1233	45	2-JUL-1994	04:34:25	13565	94185j192_2	1230	15	9-JUL-1994	22:27:47
13299	94171j178_2	409	15	22-JUN-1994	18:40:29	13446	94178j185_2	376	45	2-JUL-1994	06:26:11	13566	94185j192_2	517	15	10-JUL-1994	00:15:39
13300	94171j178_2	1267	15	22-JUN-1994	19:57:01	13447	94178j185_2	1247	15	2-JUL-1994	07:40:15	13567	94185j192_2	1218	45	10-JUL-1994	01:33:41
13301	94171j178_2	407	15	22-JUN-1994	21:46:31	13448	94178j185_2	467	65	2-JUL-1994	09:28:37	13568	94185j192_2	518	45	10-JUL-1994	03:21:31
13302	94171j178_2	1255	15	22-JUN-1994	23:03:21	13449	94178j185_2	1235	65	2-JUL-1994	10:46:11	13569	94185j192_2	1226	45	10-JUL-1994	04:39:31
13303	94171j178_2	346	15	23-JUN-1994	00:52:41	13450	94178j185_2	471	65	2-JUL-1994	12:34:31	13572	94185j192_2	409	65	10-JUL-1994	09:57:53
13304	94171j178_2	1267	15	23-JUN-1994	02:08:49	13451	94178j185_2	1193	65	2-JUL-1994	13:51:27	13573	94185j192_2	1221	65	10-JUL-1994	10:51:19
13305	94171j178_2	329	15	23-JUN-1994	04:23:05	13452	94178j185_2	484	65	2-JUL-1994	15:40:23	13574	94185j192_2	503	65	10-JUL-1994	12:39:07
13307	94171j178_2	350	45	23-JUN-1994	07:28:57	13453	94178j185_2	904	65	2-JUL-1994	16:57:59	13575	94185j192_2	1224	65	10-JUL-1994	13:57:13
13308	94171j178_2	1197	45	23-JUN-1994	08:20:37	13454	94178j185_2	479	15	2-JUL-1994	18:46:19	13576	94185j192_2	501	65	10-JUL-1994	15:45:01
13311	94171j178_2	388	65	23-JUN-1994	13:16:43	13455	94178j185_2	902	15	2-JUL-1994	20:03:53	13577	94185j192_2	831	15	10-JUL-1994	17:16:17
13312	94171j178_2	1254	15	23-JUN-1994	14:32:23	13456	94178j185_2	478	15	2-JUL-1994	21:52:13	13577	94185j192_2	348	65	10-JUL-1994	17:03:07
13313	94171j178_2	404	65	23-JUN-1994	16:21:59	13457	94178j185_2	1007	15	2-JUL-1994	23:09:47	13578	94185j192_2	519	15	10-JUL-1994	18:50:55
13314	94171j178_2	1244	65	23-JUN-1994	17:38:17	13458	94178j185_2	103	15	3-JUL-1994	00:57:59	13579	94185j192_2	1230	15	10-JUL-1994	20:08:59
13315	94171j178_2	273	15	23-JUN-1994	19:52:37	13458	94178j185_2	380	45	3-JUL-1994	01:22:51	13580	94185j192_2	538	15	10-JUL-1994	21:56:41
13316	94171j178_2	910	15	23-JUN-1994	20:56:31	13459	94178j185_2	1236	45	3-JUL-1994	02:15:45	13581	94185j192_2	1232	15	10-JUL-1994	23:14:49
13317	94171j178_2	375	15	23-JUN-1994	22:34:51	13460	94178j185_2	480	45	3-JUL-1994	04:03:59	13582	94185j192_2	532	15	11-JUL-1994	01:02:33
13318	94171j178_2	1260	15	23-JUN-1994	23:50:07	13461	94178j185_2	1239	45	3-JUL-1994	05:21:37	13583	94185j192_2	1231	15	11-JUL-1994	02:20:39
13319	94171j178_2	367	15	24-JUN-1994	01:40:11	13464	94178j185_2	393	65	3-JUL-1994	10:40:21	13584	94185j192_2	209	15	11-JUL-1994	04:08:21
13320	94171j178_2	1264	15	24-JUN-1994	02:55:53	13465	94178j185_2	1230	65	3-JUL-1994	11:33:23	13588	94192j199_2	511	65	11-JUL-1994	10:20:19
13324	94171j178_2	759	65	24-JUN-1994	09:13:27	13467	94178j185_2	1232	65	3-JUL-1994	14:39:15	13589	94192j199_2	1221	65	11-JUL-1994	18:18:23
13325	94171j178_2	430	65	24-JUN-1994	10:57:19	13469	94178j185_2	1232	15	3-JUL-1994	17:45:09	13590	94192j199_2	504	65	11-JUL-1994	13:26:13
13326	94171j178_2	1225	65	24-JUN-1994	12:13:39	13470	94178j185_2	495	15	3-JUL-1994	19:33:19	13591	94192j199_2	1217	65	11-JUL-1994	14:44:17
13327	94171j178_2	433	65	24-JUN-1994	14:03:13	13471	94178j185_2	1243	15	3-JUL-1994	20:50:59	13592	94192j199_2	509	65	11-JUL-1994	16:31:59
13328	94171j178_2	1260	65	24-JUN-1994	15:20:11	13472	94178j185_2	499	15	3-JUL-1994	22:39:07	13593	94192j199_2	1218	65	11-JUL-1994	17:50:09
13334	94171j178_2	113	15	25-JUN-1994	00:53:51	13473	94178j185_2	1242	15	3-JUL-1994	23:56:51	13594	94192j199_2	505	65	11-JUL-1994	19:37:57
13334	94171j178_2	337	45	25-JUN-1994	01:18:11	13474	94178j185_2	83	45	4-JUL-1994	01:45:27	13595	94192j199_2	288	65	11-JUL-1994	20:56:01
13335	94171j178_2	1248	15	25-JUN-1994	02:11:01	13475	94178j185_2	1242	45	4-JUL-1994	03:02:51	13597	94192j199_2	868	15	12-JUL-1994	00:13:57
13336	94171j178_2	427	45	25-JUN-1994	03:59:51	13476	94178j185_2	476	45	4-JUL-1994	04:51:03	13598	94192j199_2	527	15	12-JUL-1994	01:49:41
13337	94171j178_2	1252	45	25-JUN-1994	05:16:43	13477	94178j185_2	1122	45	4-JUL-1994	06:08:45	13599	94192j199_2	1218	15	12-JUL-1994	03:07:45
13338	94171j178_2	403	45	25-JUN-1994	07:05:41	13479	94185j192_2	730	65	4-JUL-1994	09:31:11	13603	94192j199_2	466	65	12-JUL-1994	09:44:43
13339	94171j178_2	1261	45	25-JUN-1994	08:22:03	13480	94185j192_2	480	65	4-JUL-1994	11:02:41	13604	94192j199_2	501	65	12-JUL-1994	11:07:29
13340	94171j178_2	428	15	25-JUN-1994	10:11:09	13481	94185j192_2	1228	65	4-JUL-1994	12:20:29	13605	94192j199_2	1219	65	12-JUL-1994	12:22:29
13341	94171j178_2	1274	65	25-JUN-1994	11:27:53	13482	94185j192_2	478	65	4-JUL-1994	14:08:33	13606	94192j199_2	501	65	12-JUL-1994	14:13:25
13342	94171j178_2	467	65	25-JUN-1994	13:16:53	13483	94185j192_2	1228	65	4-JUL-1994	15:26:21	13607	94192j199_2	1220	65	12-JUL-1994	15:31:19
13343	94171j178_2	1280	65	25-JUN-1994	14:33:51	13484	94185j192_2	461	65	4-JUL-1994	17:14:27	13608	94192j199_2	517	15	12-JUL-1994	17:19:17
13344	94171j178_2	452	65	25-JUN-1994	16:23:27	13485	94185j192_2	1231	15	4-JUL-1994	18:32:31	13609	94192j199_2	1231	15	12-JUL-1994	18:37:13
13347	94171j178_2	1259	15	25-JUN-1994	20:45:45	13486	94185j192_2	491	15	4-JUL-1994	20:20:21	13610	94192j199_2	521	15	12-JUL-1994	20:25:07
13348	94171j178_2	395	15	25-JUN-1994	22:35:53	13487	94185j192_2	1238	15	4-JUL-1994	21:38:09	13611	94192j199_2	1233	15	12-JUL-1994	21:43:05
13349	94171j178_2	1265	15	25-JUN-1994	23:51:31	13488	94185j192_2	480	15	4-JUL-1994	23:26:15	13612	94192j199_2	522	15	12-JUL-1994	23:31:01
13350	94171j178_2	403	45	26-JUN-1994	01:41:23	13489	94185j192_2	1236	15	5-JUL-1994	00:44:01	13613	94192j199_2	1224	15	13-JUL-1994	00:49:01
13352	94171j178_2	82	45	26-JUN-1994	04:46:27	13490	94185j192_2	95	45	5-JUL-1994	02:32:13	13614	94192j199_2	526	15	13-JUL-1994	02:36:47
13353	94171j178_2	1226	45	26-JUN-1994	06:04:25	13491	94185j192_2	1237	15	5-JUL-1994	03:49:57	13615	94192j199_2	1229	15	13-JUL-1994	03:54:47
13355	94171j178_2	1026	65	26-JUN-1994	09:17:39	13492	94185j192_2	493	45	5-JUL-1994	05:37:57	13616	94192j199_2	424	45	13-JUL-1994	06:07:29
13356	94171j178_2	439	65	26-JUN-1994	10:58:55	13493	94185j192_2	1226	45	5-JUL-1994	06:55:49	13617	94192j199_2	1225	45	13-JUL-1994	07:00:57
13357	94171j178_2	1184	65	26-JUN-1994	12:14:53	13494	94185j192_2	485	45	5-JUL-1994	08:43:51	13618	94192j199_2	519	45	13-JUL-1994	08:48:39
13359	94171j178_2	1256	65	26-JUN-1994	15:20:43	13495	94185j192_2	1214	65	5-JUL-1994	10:02:09	13622	94192j199_2	514	15	13-JUL-1994	15:00:31
13360	94171j178_2	336	15	26-JUN-1994	17:35:15	13496	94185j192_2	475	65	5-JUL-1994	11:49:55	13623	94192j199_2	1217	65	13-JUL-1994	16:18:23
13361	94171j178_2	1246	15	26-JUN-1994	18:27:07	13497	94185j192_2	1227	65	5-JUL-1994	13:07:21	13624	94192j199_2	512	15	13-JUL-1994	18:06:39
13362	94171j178_2	325	15	26-JUN-1994	20:41:25	13498	94185j192_2	491	65	5-JUL-1994	14:55:49	13625	94192j199_2	1214	15	13-JUL-1994	19:24:19
13363	94171j178_2	1258	15	26-JUN-1994	21:32:57	13499	94185j192_2	1240	65	5-JUL-1994	16:13:31	13626	94192j199_2	532	15	13-JUL-1994	21:12:15
13364	94171j178_2	91	65	26-JUN-1994	23:46:49	13500	94185j192_2	482	15	5-JUL-1994	18:01:41	13627	94192j199_2	1231	15	13-JUL-1994	22:30:05
13365	94171j178_2	1257	15	27-JUN-1994	00:38:43	13501	94185j192_2	1236	15	5-JUL-1994	19:19:23	13628	94192j199_2	531	15	14-JUL-1994	00:18:09
13366	94171j178_2	403	15	27-JUN-1994	02:28:15	13502	94185j192_2	499	15	5-JUL-1994	21:07:21	13629	94192j199_2	1231	15	14-JUL-1994	01:35:57
13367	94171j178_2	1256	15	27-JUN-1994	03:44:43	13503	94185j192_2	1238	15	5-JUL-1994	22:25:15	13630	94192j199_2	527	15	14-JUL-1994	03:23:55
13369	94178j185_2	239	45	27-JUN-1994	07:24:27	13504	94185j192_2	522	15	6-JUL-1994	00:12:27	13632	94192j199_2	524	45	14-JUL-1994	06:29:49
13370	94178j185_2	406	45	27-JUN-1994	08:39:59	13505	94185j192_2	1253	15	6-JUL-1994	01:30:49	13633	94192j199_2	1222	45	14-JUL-1994	07:47:45
13371	94178j185_2	1256	65	27-JUN-1994	09:56:25	13506	94185j192_2	483	15	6-JUL-1994	03:19:01	13634	94192j199_2	97	45	14-JUL-1994	09:35:45
13372	94178j185_2	224	65	27-JUN-1994	11:45:49	13507	94185j192_2	289	15	6-JUL-1994	04:36:41	13634	94192j199_2	421	65	14-JUL-1994	10:00:15
13373	94178j185_2	1260	65	27-JUN-1994	13:02:07	13507	94185j192_2	130	45	6-JUL-1994	05:13:47	13635	94192j199_2	1220	65	14-JUL-1994	10:53:39
13374	94178j185_2	443	65	27-JUN-1994	14:51:41	13508	94185j192_2	497	45	6-JUL-1994	06:25:03	13636	94192j199_2	509	65	14-JUL-1994	12:41:33
13375	94178j185_2	1258	65	27-JUN-1994	16:08:01	13509	94185j192_2	1201	45	6-JUL-1994	07:44:13	13637	94192j199_2	1542	65	14-JUL-1994	13:59:31
13376	94178j185_2	420	65	27-JUN-1994	17:57:39	13513	94185j192_2	622	65	6-JUL-1994	14:14:45	13638	94192j199_2	458	65	14-JUL-1994	15:48:15
13377	94178j185_2	1259	65	27-JUN-1994	19:13:57	13514	94185j192_2	487	65	6-JUL-1994	15:42:39	13639	94192j199_2	1564	65	14-JUL-1994	17:04:35
13378	94178j185_2	72	65	27-JUN-1994	21:03:37	13515	94185j192_2	1196	65	6-JUL-1994	18:48:35	13640	94192j199_2	477	15	14-JUL-1994	18:54:07
13379	94178j185_2	843	15	27-JUN-1994	22:33:45	13516	94185j192_2	498	15	6-JUL-1994	18:48:35	13641	94192j199_2	1575	15	14-JUL-1994	20:10:33
13380	94178j185_2	410	15	28-JUN-1994	00:09:19	13517	94185j192_2	1233	15	6-JUL-1994	20:06:31	13642	94192j199_2	483	15	14-JUL-1994	22:00:01
13381	94178j185_2	1259	15	28-JUN-1994	01:25:45	13518	94185j192_2	497	15	6-JUL-1994	21:54:31	13643	94192j199_2	1574	15	14-JUL-1994	23:16:29
13382	94178j185_2	80	15	28-JUN-1994	03:39:55	13519	94185j192_2	1230	15	6-JUL-1994	23:12:27	13644	94192j199_2	491	45	15-JUL-1994	01:05:55
13383	94178j185_2	615	15	28-JUN-1994	04:31:49	13520	94185j192_2	395	15	7-JUL-1994	00:01:41	13645	94192j199_2	1582	45	15-JUL-1994	02:22:15
13385	94178j185_2	1168	45	28-JUN-1994	07:40:09	13521	94185j192_2	1252	15	7-JUL-1994	02:17:37	13646	94192j199_2	499	45	15-JUL-1994	04:11:41
13389	94178j185_2	517	15	28-JUN-1994	14:13:37	13522	94185j192_2	498	15	7-JUL-1994	03:35:19	13647	94192j199_2	1574	15	15-JUL-1994	05:28:11
13390	94178j185_2	349	65	28-JUN-1994	16:02:57	13523	94185j192_2	1240	45	7-JUL-1994	05:24:09	13648	94192j199_2	490	45	15-JUL-1994	07:17:35
13391	94178j185_2	1235	65	28-JUN-1994	16:55:31	13524	94185j192_2	503	45	7-JUL-1994	07:12:09	13649	94192j199_2	1578	45	15-JUL-1994	08:33:57
13394	94178j185_2	463	15	28-JUN-1994	21:49:49	13525	94185j192_2	1221	45	7-JUL-1994	08:30:03	13650	94192j199_2	466	65	15-JUL-1994	10:23:29
13395	94178j185_2	1257	15	28-JUN-1994	23:06:59	13526	94185j192_2	497	65	7-JUL-1994	10:17:57	13651	94192j199_2	1562	65	15-JUL-1994	11:40:01
13396	94178j185_2	428	15	29-JUN-1994	00:55:41	13527	94185j192_2	1223	65	7-JUL-1994	11:35:57	13652	94192j199_2	464	65	15-JUL-1994	13:29:25
13397	94178j185_2	1258	15	29-JUN-1994	02:12:51	13528	94185j192_2	490	65	7-JUL-1994	13:23:53	13653	94192j199_2	1429	65	15-JUL-1994	14:45:41
13398	94178j185_2	457	15	29-JUN-1994	04:01:35	13529	94185j192_2	1226	65	7-JUL-1994	14:41:47	13654	94192j199_2	462	65	15-JUL-1994	16:35:19
13400	94178j185_2	223	45	29-JUN-1994	07:36:45	13530	94185j192_2	489	65	7-JUL-1994	16:29:45	13655	94192j199_2	1563	65	15-JUL-1994	17:51:43
13401	94178j185_2	1241	45	29-JUN-1994	08:24:41	13531	94185j192_2	1218	65	7-JUL-1994	17:48:15	13656	94192j199_2	472	65	15-JUL-1994	19:41:03
13403	94178j185_2	1186	65	29-JUN-1994	11:32:31	13532	94185j192_2	489	65	7-JUL-1994	19:35:39	13657	94192j199_2	1574	15	15-JUL-1994	20:57:37
13407	94178j185_2	1252	15	29-JUN-1994	17:42:25	13533	94185j192_2	531	65	7-JUL-1994	20:53:39	13658	94192j199_2	490	15	15-JUL-1994	22:46:59
13408	94178j185_2	370	15	29-JUN-1994	19:55:43	13535	94185j192_2	776	15	8-JUL-1994	00:14:15	13659	94192j199_2	1575	15	16-JUL-1994	00:03:27
13409	94178j185_2	1248	15	29-JUN-1994	20:48:29	13536	94185j192_2	508	45	8-JUL-1994	01:47:27	13660	94192j199_2	104	15	16-JUL-1994	01:52:53
13410	94178j185_2	472	15	29-JUN-1994	22:36:53	13537	94185j192_2	1224	45	8-JUL-1994	03:05:19	13665	94192j199_2	588	45	16-JUL-1994	09:21:03
13411	94178j185_2	1218	15	29-JUN-1994	23:55:15	13538	94185j192_2	490	45	8-JUL-1994	04:53:15	13665	94192j199_2	691	65	16-JUL-1994	09:50:15

13666	94192j199_2	464	65	16-JUL-1994	11:10:43	13823	94206j213_2	1592	65	26-JUL-1994	14:05:51	13992	94213j220_2	1134	65	6-AUG-1994	11:52:41
13667	94192j199_2	1564	65	16-JUL-1994	12:27:01	13824	94206j213_2	411	65	26-JUL-1994	16:20:43	13993	94213j220_2	1778	65	6-AUG-1994	13:25:35
13668	94192j199_2	468	65	16-JUL-1994	14:16:31	13825	94206j213_2	1595	65	26-JUL-1994	17:11:37	13994	94213j220_2	1129	65	6-AUG-1994	14:58:39
13669	94192j199_2	260	15	16-JUL-1994	15:39:23	13826	94206j213_2	426	15	26-JUL-1994	19:26:31	13995	94213j220_2	1758	65	6-AUG-1994	16:31:25
13670	94192j199_2	391	15	16-JUL-1994	17:47:01	13827	94206j213_2	1593	15	26-JUL-1994	20:17:33	13996	94213j220_2	742	65	6-AUG-1994	18:17:31
13670	94192j199_2	92	65	16-JUL-1994	17:22:25	13832	94206j213_2	379	45	27-JUL-1994	04:45:43	13997	94213j220_2	1695	65	6-AUG-1994	19:37:17
13671	94192j199_2	1577	15	16-JUL-1994	18:38:43	13833	94206j213_2	1596	45	27-JUL-1994	05:35:07	14001	94213j220_2	1788	45	7-AUG-1994	01:49:01
13672	94192j199_2	496	15	16-JUL-1994	20:28:17	13834	94206j213_2	412	45	27-JUL-1994	07:50:11	14002	94213j220_2	1149	45	7-AUG-1994	03:21:55
13673	94192j199_2	1580	15	16-JUL-1994	21:44:35	13839	94206j213_2	1598	15	27-JUL-1994	14:52:41	14003	94213j220_2	1802	45	7-AUG-1994	04:54:57
13674	94192j199_2	492	15	16-JUL-1994	23:34:11	13840	94206j213_2	476	45	27-JUL-1994	17:07:41	14004	94213j220_2	1188	45	7-AUG-1994	06:27:43
13675	94192j199_2	1576	15	17-JUL-1994	00:50:27	13841	94206j213_2	1600	65	27-JUL-1994	17:58:31	14007	94213j220_2	1785	65	7-AUG-1994	11:06:31
13676	94192j199_2	486	15	17-JUL-1994	02:40:03	13842	94206j213_2	497	15	27-JUL-1994	20:13:35	14008	94213j220_2	1161	65	7-AUG-1994	12:39:31
13677	94192j199_2	1020	15	17-JUL-1994	03:56:21	13843	94206j213_2	1580	15	27-JUL-1994	21:04:45	14009	94213j220_2	228	15	7-AUG-1994	14:12:25
13680	94192j199_2	485	45	17-JUL-1994	08:51:49	13844	94206j213_2	471	15	27-JUL-1994	23:19:27	14011	94213j220_2	802	15	7-AUG-1994	17:51:21
13681	94192j199_2	1568	65	17-JUL-1994	10:08:07	13845	94206j213_2	1536	15	28-JUL-1994	00:10:57	14012	94213j220_2	865	15	7-AUG-1994	19:01:13
13682	94192j199_2	470	65	17-JUL-1994	11:57:41	13846	94206j213_2	98	45	28-JUL-1994	02:00:43	14013	94213j220_2	1823	15	7-AUG-1994	20:24:07
13683	94192j199_2	1566	65	17-JUL-1994	13:13:59	13851	94206j213_2	318	65	28-JUL-1994	10:10:13	14014	94213j220_2	1185	15	7-AUG-1994	21:57:05
13684	94192j199_2	467	15	17-JUL-1994	15:03:37	13852	94206j213_2	426	65	28-JUL-1994	11:43:23	14015	94213j220_2	1818	15	7-AUG-1994	23:29:57
13685	94192j199_2	1197	65	17-JUL-1994	16:19:51	13856	94206j213_2	479	15	28-JUL-1994	17:54:37	14016	94213j220_2	1193	15	8-AUG-1994	01:02:51
13686	94192j199_2	481	15	17-JUL-1994	18:09:31	13857	94206j213_2	1594	15	28-JUL-1994	18:45:43	14017	94213j220_2	1821	15	8-AUG-1994	02:35:45
13687	94192j199_2	1579	15	17-JUL-1994	19:25:43	13858	94206j213_2	495	15	28-JUL-1994	21:00:33	14018	94213j220_2	578	45	8-AUG-1994	04:49:41
13688	94192j199_2	509	15	17-JUL-1994	21:14:57	13859	94206j213_2	1605	15	28-JUL-1994	21:51:15	14019	94213j220_2	1824	45	8-AUG-1994	05:41:35
13689	94192j199_2	1581	15	17-JUL-1994	22:31:37	13860	94206j213_2	484	15	29-JUL-1994	00:06:25	14020	94220j227_2	1157	45	8-AUG-1994	07:14:33
13690	94192j199_2	502	15	18-JUL-1994	00:21:07	13861	94206j213_2	1607	15	29-JUL-1994	00:57:03	14022	94220j227_2	953	65	8-AUG-1994	10:27:05
13691	94192j199_2	1582	15	18-JUL-1994	01:37:27	13862	94206j213_2	401	45	29-JUL-1994	03:13:31	14023	94220j227_2	1806	65	8-AUG-1994	11:53:21
13692	94192j199_2	505	15	18-JUL-1994	03:27:03	13863	94206j213_2	1573	45	29-JUL-1994	04:04:03	14024	94220j227_2	1133	65	8-AUG-1994	13:27:11
13696	94199j206_2	355	65	18-JUL-1994	10:04:25	13864	94206j213_2	431	45	29-JUL-1994	06:18:15	14025	94220j227_2	1800	65	8-AUG-1994	14:59:11
13697	94199j206_2	1560	15	18-JUL-1994	10:55:31	13865	94206j213_2	1608	45	29-JUL-1994	07:08:47	14026	94220j227_2	1162	65	8-AUG-1994	16:32:09
13698	94199j206_2	368	15	18-JUL-1994	12:45:05	13871	94206j213_2	1609	65	29-JUL-1994	16:26:25	14027	94220j227_2	1811	65	8-AUG-1994	18:04:59
13699	94199j206_2	1511	15	18-JUL-1994	14:03:19	13872	94206j213_2	422	65	29-JUL-1994	18:41:45	14028	94220j227_2	574	15	8-AUG-1994	20:19:13
13700	94199j206_2	468	65	18-JUL-1994	15:50:57	13873	94206j213_2	1592	65	29-JUL-1994	19:32:17	14028	94220j227_2	579	65	8-AUG-1994	19:37:57
13701	94199j206_2	892	15	18-JUL-1994	17:28:21	13874	94206j213_2	445	15	29-JUL-1994	21:47:43	14029	94220j227_2	1817	15	8-AUG-1994	21:10:49
13702	94199j206_2	480	15	18-JUL-1994	18:56:37	13875	94206j213_2	1611	15	29-JUL-1994	22:38:07	14037	94220j227_2	237	65	9-AUG-1994	10:27:07
13703	94199j206_2	1578	15	18-JUL-1994	20:12:45	13876	94206j213_2	438	45	30-JUL-1994	00:53:33	14038	94220j227_2	1180	65	9-AUG-1994	11:07:09
13712	94199j206_2	488	65	19-JUL-1994	10:25:59	13877	94206j213_2	1608	45	30-JUL-1994	01:44:05	14044	94220j227_2	594	15	9-AUG-1994	21:06:07
13713	94199j206_2	1581	65	19-JUL-1994	11:42:09	13878	94206j213_2	423	45	30-JUL-1994	03:59:25	14045	94220j227_2	1159	15	9-AUG-1994	21:57:35
13714	94199j206_2	483	15	19-JUL-1994	13:31:55	13879	94206j213_2	1610	45	30-JUL-1994	04:49:55	14046	94220j227_2	1811	15	9-AUG-1994	23:30:27
13715	94199j206_2	1584	65	19-JUL-1994	14:48:01	13880	94206j213_2	436	45	30-JUL-1994	07:05:11	14047	94220j227_2	453	45	10-AUG-1994	01:03:27
13716	94199j206_2	478	65	19-JUL-1994	16:37:47	13881	94206j213_2	441	45	30-JUL-1994	07:55:49	14048	94220j227_2	1719	45	10-AUG-1994	02:40:31
13718	94199j206_2	446	15	19-JUL-1994	19:43:47	13882	94206j213_2	346	65	30-JUL-1994	10:13:49	14049	94220j227_2	1162	45	10-AUG-1994	04:09:17
13719	94199j206_2	1573	15	19-JUL-1994	20:59:43	13883	94206j213_2	1615	65	30-JUL-1994	11:01:33	14050	94220j227_2	1844	45	10-AUG-1994	05:42:09
13720	94199j206_2	419	15	19-JUL-1994	22:49:35	13884	94206j213_2	419	65	30-JUL-1994	13:17:15	14055	94220j227_2	565	65	10-AUG-1994	14:09:25
13721	94199j206_2	1585	15	20-JUL-1994	00:05:37	13885	94206j213_2	1615	65	30-JUL-1994	14:07:25	14056	94220j227_2	1410	65	10-AUG-1994	14:59:49
13722	94199j206_2	484	15	20-JUL-1994	01:55:27	13886	94206j213_2	449	45	30-JUL-1994	16:22:59	14062	94220j227_2	1862	45	11-AUG-1994	00:17:09
13723	94199j206_2	1582	15	20-JUL-1994	03:11:29	13887	94206j213_2	1617	65	30-JUL-1994	17:13:13	14063	94220j227_2	1169	45	11-AUG-1994	01:50:09
13725	94199j206_2	1474	45	20-JUL-1994	06:20:33	13888	94206j213_2	408	45	30-JUL-1994	19:29:05	14064	94220j227_2	1704	45	11-AUG-1994	03:23:03
13726	94199j206_2	395	45	20-JUL-1994	08:31:57	13889	94206j213_2	921	65	30-JUL-1994	20:19:11	14065	94220j227_2	1180	45	11-AUG-1994	04:55:57
13729	94199j206_2	121	65	20-JUL-1994	13:18:11	13892	94206j213_2	447	15	31-JUL-1994	01:40:37	14066	94220j227_2	1091	45	11-AUG-1994	06:28:53
13730	94199j206_2	464	15	20-JUL-1994	14:18:41	13893	94206j213_2	1617	15	31-JUL-1994	02:30:51	14070	94220j227_2	904	65	11-AUG-1994	13:12:31
13731	94199j206_2	1590	65	20-JUL-1994	15:35:09	13898	94206j213_2	563	65	31-JUL-1994	10:31:49	14071	94220j227_2	577	65	11-AUG-1994	14:13:29
13732	94199j206_2	517	15	20-JUL-1994	17:24:27	13899	94206j213_2	515	15	31-JUL-1994	12:07:11	14075	94220j227_2	28	15	11-AUG-1994	21:17:47
13733	94199j206_2	1584	65	20-JUL-1994	18:41:01	13900	94206j213_2	574	65	31-JUL-1994	13:37:39	14076	94220j227_2	1853	15	11-AUG-1994	21:58:03
13734	94199j206_2	105	65	20-JUL-1994	20:30:31	13901	94206j213_2	1611	15	31-JUL-1994	14:54:25	14077	94220j227_2	460	15	11-AUG-1994	23:00:31
13736	94199j206_2	486	15	20-JUL-1994	23:36:33	13902	94206j213_2	571	65	31-JUL-1994	16:43:33	14077	94220j227_2	256	45	12-AUG-1994	00:12:47
13737	94199j206_2	1579	15	21-JUL-1994	00:52:27	13903	94206j213_2	829	65	31-JUL-1994	18:26:11	14078	94220j227_2	1873	45	12-AUG-1994	01:03:53
13738	94199j206_2	477	15	21-JUL-1994	02:42:25	13904	94206j213_2	572	65	31-JUL-1994	19:49:23	14079	94220j227_2	1209	45	12-AUG-1994	02:06:23
13739	94199j206_2	802	15	21-JUL-1994	03:58:27	13914	94213j220_2	551	15	1-AUG-1994	11:18:51	14080	94220j227_2	12	45	12-AUG-1994	03:39:19
13740	94199j206_2	395	45	21-JUL-1994	06:13:03	13915	94213j220_2	785	65	1-AUG-1994	13:03:21	14086	94220j227_2	1654	65	12-AUG-1994	13:34:29
13741	94199j206_2	1572	65	21-JUL-1994	07:04:19	13916	94213j220_2	549	65	1-AUG-1994	14:24:39	14087	94220j227_2	1208	65	12-AUG-1994	14:29:49
13743	94199j206_2	1595	65	21-JUL-1994	10:10:13	13917	94213j220_2	1620	65	1-AUG-1994	15:41:45	14088	94220j227_2	1876	65	12-AUG-1994	16:02:45
13744	94199j206_2	487	65	21-JUL-1994	12:00:11	13922	94213j220_2	575	15	2-AUG-1994	23:42:23	14089	94220j227_2	1193	15	12-AUG-1994	18:05:57
13745	94199j206_2	1574	65	21-JUL-1994	13:16:07	13923	94213j220_2	1615	15	2-AUG-1994	00:59:15	14089	94220j227_2	9	65	12-AUG-1994	17:35:41
13746	94199j206_2	486	65	21-JUL-1994	15:05:57	13924	94213j220_2	615	15	2-AUG-1994	02:48:11	14090	94220j227_2	1898	15	12-AUG-1994	19:08:37
13747	94199j206_2	1596	65	21-JUL-1994	16:21:59	13925	94213j220_2	964	45	2-AUG-1994	04:08:27	14091	94220j227_2	1214	15	12-AUG-1994	20:41:33
13748	94199j206_2	442	15	21-JUL-1994	18:11:51	13926	94213j220_2	460	45	2-AUG-1994	06:20:09	14092	94220j227_2	1904	15	12-AUG-1994	22:14:29
13750	94199j206_2	475	15	21-JUL-1994	21:17:35	13927	94213j220_2	1057	45	2-AUG-1994	07:11:19	14093	94220j227_2	21	15	12-AUG-1994	23:47:27
13751	94199j206_2	1587	15	21-JUL-1994	22:33:37	13931	94213j220_2	77	65	2-AUG-1994	13:54:21	14093	94220j227_2	1191	45	13-AUG-1994	00:17:37
13752	94199j206_2	496	15	22-JUL-1994	00:23:45	13932	94213j220_2	429	65	2-AUG-1994	15:37:49	14094	94220j227_2	1883	45	13-AUG-1994	01:20:23
13753	94199j206_2	1587	15	22-JUL-1994	01:39:31	13933	94213j220_2	1609	65	2-AUG-1994	16:28:45	14095	94220j227_2	1219	45	13-AUG-1994	02:53:19
13754	94199j206_2	39	15	22-JUL-1994	03:29:21	13934	94213j220_2	381	15	2-AUG-1994	18:43:37	14096	94220j227_2	1864	45	13-AUG-1994	04:26:15
13758	94199j206_2	348	65	22-JUL-1994	10:05:39	13942	94213j220_2	426	45	3-AUG-1994	07:07:11	14097	94220j227_2	616	45	13-AUG-1994	05:59:11
13759	94199j206_2	1569	65	22-JUL-1994	10:57:31	13944	94213j220_2	232	65	3-AUG-1994	10:18:53	14101	94220j227_2	737	65	13-AUG-1994	12:56:51
13760	94199j206_2	498	65	22-JUL-1994	12:46:45	13945	94213j220_2	1635	65	3-AUG-1994	11:03:47	14102	94220j227_2	1897	65	13-AUG-1994	13:43:53
13761	94199j206_2	1503	65	22-JUL-1994	14:02:59	13946	94213j220_2	426	65	3-AUG-1994	13:18:55	14103	94220j227_2	1231	65	13-AUG-1994	15:16:49
13762	94199j206_2	409	65	22-JUL-1994	15:55:41	13947	94213j220_2	1634	65	3-AUG-1994	14:09:43	14104	94220j227_2	1912	65	13-AUG-1994	16:49:45
13763	94199j206_2	1528	65	22-JUL-1994	17:08:49	13948	94213j220_2	441	65	3-AUG-1994	16:24:51	14105	94220j227_2	1213	15	13-AUG-1994	18:52:35
13764	94199j206_2	472	65	22-JUL-1994	18:59:01	13949	94213j220_2	1370	65	3-AUG-1994	17:15:33	14105	94220j227_2	21	65	13-AUG-1994	18:22:41
13765	94199j206_2	696	15	22-JUL-1994	20:16:43	13950	94213j220_2	444	65	3-AUG-1994	19:30:43	14106	94220j227_2	1928	15	13-AUG-1994	19:55:37
13766	94199j206_2	498	15	22-JUL-1994	22:04:55	13954	94213j220_2	433	45	4-AUG-1994	01:42:29	14107	94220j227_2	39	15	13-AUG-1994	21:28:33
13767	94199j206_2	1096	15	22-JUL-1994	23:20:45	13955	94213j220_2	1603	45	4-AUG-1994	02:33:19	14109	94220j227_2	1221	45	14-AUG-1994	01:04:15
13768	94199j206_2	472	45	23-JUL-1994	01:10:47	13956	94213j220_2	449	45	4-AUG-1994	04:48:47	14110	94220j227_2	1928	45	14-AUG-1994	02:07:23
13769	94199j206_2	1528	45	23-JUL-1994	02:26:43	13957	94213j220_2	1575	45	4-AUG-1994	03:40:19	14111	94220j227_2	1259	45	14-AUG-1994	03:40:19
13770	94199j206_2	489	45	23-JUL-1994	04:16:31	13960	94213j220_2	554	65	4-AUG-1994	11:00:45	14112	94220j227_2	1939	45	14-AUG-1994	05:13:17
13771	94199j206_2	683	45	23-JUL-1994	06:01:37	13961	94213j220_2	1620	65	4-AUG-1994	11:50:47	14113	94220j227_2	43	45	14-AUG-1994	06:46:13
13772	94199j206_2	555	45	23-JUL-1994	07:21:47	13962	94213j220_2	437	65	4-AUG-1994	14:06:01	14117	94220j227_2	1015	45	14-AUG-1994	13:34:35
13773	94199j206_2	1002	45	23-JUL-1994	08:38:17	13963	94213j220_2	1787	65	4-AUG-1994	14:56:35	14118	94220j227_2	1934	65	14-AUG-1994	14:30:55
13774	94199j206_2	521	65	23-JUL-1994	10:27:43	13964	94213j220_2	1100	65	4-AUG-1994	16:30:51	14119	94220j227_2	1254	65	14-AUG-1994	16:03:53
13775	94199j206_2	1587	65	23-JUL-1994	11:44:29	13965	94213j220_2	1763	15	4-AUG-1994	18:03:43	14120	94220j227_2	31	65	14-AUG-1994	17:36:49
13776	94199j206_2	541	65	23-JUL-1994	13:33:37	13966	94213j220_2	1061	15	4-AUG-1994	19:36:29	14131	94227j234_2	1082	15	15-AUG-1994	11:14:41
13777	94199j206_2	1578	15	23-JUL-1994	14:50:31	13967	94213j220_2	1714	15	4-AUG-1994	21:09:27	14132	94227j234_2	1958	65	15-AUG-1994	12:41:09
13778	94199j206_2	500	65	23-JUL-1994	16:40:15	13968	94213j220_2	1138	15	4-AUG-1994	22:42:17	14133	94227j234_2	1280	15	15-AUG-1994	14:14:05
13779	94199j206_2	757	65	23-JUL-1994	18:23:57	13969	94213j220_2	1796	15	5-AUG-1994	00:14:41	14134	94227j234_2	1959	65	15-AUG-1994	15:47:01
13783	94199j206_2	380	45	24-JUL-1994	00:46:49	13970	94213j220_2	1106	15	5-AUG-1994	01:48:47	14138	94227j234_2	1457	15	15-AUG-1994	22:15:45
13784	94199j206_2	499	45	24-JUL-1994	01:57:51	13971	94213j220_2	1762	15	5-AUG-1994	03:21:05	14139	94227j234_2	603	15	15-AUG-1994	23:31:33
13785	94199j206_2	1587	45	24-JUL-1994	03:13:51	13972	94213j220_2	1112	45	5-AUG-1994	04:54:39	14139	94227j234_2	634	45	16-AUG-1994	00:16:05
13786	94199j206_2	529	45	24-JUL-1994	05:03:01	13973	94213j220_2	1821	15	5-AUG-1994	06:26:57	14140	94227j234_2	359	65	16-AUG-1994	01:13:21
13798	94199j206_2	339	15	25-JUL-1994	00:03:13	13975	94213j220_2	303	65	5-AUG-1994	10:21:39	14141	94227j234_2	1283	45	16-AUG-1994	02:37:27
13800	94199j206_2	491	15	25-JUL-1994	02:46:17	13976	94213j220_2	1138	65	5-AUG-1994	11:05:49	14142	94227j234_2	1949	65	16-AUG-1994	04:10:49
13805	94206j213_2	933	65	25-JUL-1994	10:15:19	13977	94213j220_2	1773	65	5-AUG-1994	12:38:49	14147	94227j234_2	1141	65	16-AUG-1994	12:00:07
13806	94206j213_2	397	65	25-JUL-1994	12:28:09	13978	94213j220_2	1141	65	5-AUG-1994	14:11:41	14148	94227j234_2	1987	65	16-AUG-1994	13:27:41
13807	94206j213_2	1578	15	25-JUL-1994	13:19:07	13979	94213j220_2	1777	65	5-AUG-1994	15:44:37	14149	94227j234_2	1292	65	16-AUG-1994	15:00:33
13808	94206j213_2	366	65	25-JUL-1994	15:33:35	13984	94213j220_2	396	45	6-AUG-1994	00:15:09	14150	94227j234_2	1991	65	16-AUG-1994	16:33:31
13809	94206j213_2	638	65	25-JUL-1994	16:25:01	13985	94213j220_2	1669	45	6-AUG-1994	01:05:13	14151	94227j234_2	496	15	16-AUG-1994	18:06:25
13814	94206j213_2	404	15	26-JUL-1994	00:51:45	13986	94213j220_2	1116	45	6-AUG-1994	02:35:03	14155	94227j234_2	1036	15	17-AUG-1994	00:26:55
13815	94206j213_2	1578	15	26-JUL-1994	01:42:45	13987	94213j220_2	1774	45	6-AUG-1994	04:08:03	14156	94227j234_2	1179	15	17-AUG-1994	01:50:55
13816	94206j213_2	381	15	26-JUL-1994	03:57:27	13988	94213j220_2	1151	45	6-AUG-1994	05:40:57	14163	94227j234_2	499	15	17-AUG-1994	13:31:55
13821	94206j213_2	1587	65	26-JUL-1994	11:00:05	13989	94213j220_2	355	45	6-AUG-1994	07:13:53	14164	94227j234_2	2022	65	17-AUG-1994	14:14:07
13822	94206j213_2	402	65	26-JUL-1994	13:15:09	13991	94213j220_2	1664	65	6-AUG-1994	10:23:31	14165	94227j234_2	1330	65	17-AUG-1994	15:47:03

14166	94227j234_2	2027	65	17-AUG-1994 17:19:53		14650	94255j262_2	2020	15	17-SEP-1994 22:07:39
14169	94227j234_2	1171	15	17-AUG-1994 22:04:27		14651	94255j262_2	696	15	17-SEP-1994 23:15:15
14170	94227j234_2	2014	15	17-AUG-1994 23:31:43		14660	94255j262_2	1971	65	18-SEP-1994 13:34:51
14171	94227j234_2	611	15	18-AUG-1994 01:05:01		14661	94255j262_2	681	15	18-SEP-1994 14:41:23
14177	94227j234_2	264	65	18-AUG-1994 11:21:29		14666	94255j262_2	1791	15	18-SEP-1994 22:57:23
14178	94227j234_2	2044	65	18-AUG-1994 11:54:49		14667	94255j262_2	696	15	18-SEP-1994 23:57:03
14179	94227j234_2	1344	65	18-AUG-1994 13:28:53		14674	94262j269_2	1770	65	19-SEP-1994 11:18:33
14180	94227j234_2	2059	65	18-AUG-1994 15:00:27		14675	94262j269_2	759	65	19-SEP-1994 12:17:31
14181	94227j234_2	1385	65	18-AUG-1994 16:33:31		14680	94262j269_2	1960	15	19-SEP-1994 20:27:51
14186	94227j234_2	1914	15	19-AUG-1994 00:23:39		14681	94262j269_2	692	15	19-SEP-1994 21:33:11
14187	94227j234_2	632	15	19-AUG-1994 01:51:11		14690	94262j269_2	1995	65	20-SEP-1994 11:52:47
14193	94227j234_2	756	65	19-AUG-1994 11:52:45		14691	94262j269_2	688	65	20-SEP-1994 12:59:15
14197	94227j234_2	1432	15	19-AUG-1994 17:19:39		14696	94262j269_2	1738	15	20-SEP-1994 21:16:59
14198	94227j234_2	2116	15	19-AUG-1994 18:52:33		14697	94262j269_2	696	15	20-SEP-1994 22:14:53
14202	94227j234_2	2129	15	20-AUG-1994 01:04:03		14706	94262j269_2	1975	65	21-SEP-1994 12:35:09
14208	94227j234_2	1519	65	20-AUG-1994 10:42:03		14707	94262j269_2	693	65	21-SEP-1994 13:40:57
14212	94227j234_2	2153	65	20-AUG-1994 16:32:53		14712	94262j269_2	1974	15	21-SEP-1994 21:49:51
14224	94227j234_2	1978	65	21-AUG-1994 11:14:31		14713	94262j269_2	702	15	21-SEP-1994 22:56:35
14228	94227j234_2	1759	65	21-AUG-1994 17:33:57		14722	94262j269_2	1924	65	22-SEP-1994 13:18:31
14244	94234j241_2	2118	65	22-AUG-1994 18:10:09		14723	94262j269_2	705	65	22-SEP-1994 14:22:39
14246	94234j241_2	2031	15	22-AUG-1994 21:18:49		14728	94262j269_2	1844	15	22-SEP-1994 22:36:49
14256	94234j241_2	2156	65	23-AUG-1994 12:45:19		14729	94262j269_2	675	15	22-SEP-1994 23:38:15
14260	94234j241_2	1771	15	23-AUG-1994 19:11:53		14738	94262j269_2	2039	65	23-SEP-1994 13:56:21
14260	94234j241_2	193	65	23-AUG-1994 19:04:53		14739	94262j269_2	676	65	23-SEP-1994 15:04:17
14261	94234j241_2	463	15	23-AUG-1994 20:10:53		14742	94262j269_2	2023	15	23-SEP-1994 20:07:17
14272	94234j241_2	1871	65	24-AUG-1994 13:34:21		14743	94262j269_2	691	15	23-SEP-1994 21:14:43
14276	94234j241_2	1578	15	24-AUG-1994 20:07:57		14758	94262j269_2	2050	15	24-SEP-1994 20:47:45
14277	94234j241_2	296	15	24-AUG-1994 21:00:31		14759	94262j269_2	668	15	24-SEP-1994 21:56:21
14302	94234j241_2	1379	65	26-AUG-1994 13:03:45		14768	94262j269_2	2051	65	25-SEP-1994 12:13:59
14303	94234j241_2	1095	65	26-AUG-1994 13:49:51		14769	94262j269_2	668	65	25-SEP-1994 13:22:21
14309	94234j241_2	1754	15	26-AUG-1994 23:36:35		14774	94262j269_2	2043	15	25-SEP-1994 21:29:51
14316	94234j241_2	195	65	27-AUG-1994 11:21:21		14784	94269j276_2	1858	65	26-SEP-1994 12:56:15
14317	94234j241_2	2604	65	27-AUG-1994 11:27:51		14785	94269j276_2	417	65	26-SEP-1994 14:12:33
14332	94234j241_2	269	15	28-AUG-1994 12:02:17		14790	94269j276_2	1860	15	26-SEP-1994 22:10:59
14333	94234j241_2	1121	65	28-AUG-1994 12:11:15		14791	94269j276_2	344	15	26-SEP-1994 23:28:33
14337	94234j241_2	1186	15	28-AUG-1994 19:07:29		14800	94269j276_2	1856	65	27-SEP-1994 13:36:49
14356	94241j248_2	787	45	30-AUG-1994 00:49:37		14801	94269j276_2	365	65	27-SEP-1994 14:55:11
14357	94241j248_2	1081	45	30-AUG-1994 01:15:51		14804	94269j276_2	1757	15	27-SEP-1994 19:50:17
14363	94241j248_2	1430	65	30-AUG-1994 11:06:09		14805	94269j276_2	375	15	27-SEP-1994 21:05:47
14370	94241j248_2	539	15	30-AUG-1994 22:35:21		14814	94269j276_2	1766	65	28-SEP-1994 11:15:03
14371	94241j248_2	1194	15	30-AUG-1994 22:53:39		14815	94269j276_2	80	65	28-SEP-1994 12:32:33
14381	94241j248_2	1531	65	31-AUG-1994 14:21:43		14820	94269j276_2	1895	15	28-SEP-1994 20:25:47
14386	94241j248_2	280	15	31-AUG-1994 23:27:27		14821	94269j276_2	197	15	28-SEP-1994 21:47:41
14387	94241j248_2	1373	15	31-AUG-1994 23:36:47		14832	94269j276_2	1867	65	29-SEP-1994 14:54:59
14400	94241j248_2	250	15	1-SEP-1994 21:06:07		14833	94269j276_2	189	65	29-SEP-1994 16:16:47
14401	94241j248_2	1429	15	1-SEP-1994 21:14:25		14836	94269j276_2	1719	15	29-SEP-1994 21:08:59
14411	94241j248_2	1413	65	2-SEP-1994 13:23:13		14837	94269j276_2	168	15	29-SEP-1994 22:26:27
14412	94241j248_2	370	65	2-SEP-1994 15:34:19		14846	94269j276_2	1948	65	30-SEP-1994 12:25:13
14413	94241j248_2	1344	65	2-SEP-1994 15:46:39		14852	94269j276_2	1936	15	30-SEP-1994 21:39:55
14418	94241j248_2	336	15	3-SEP-1994 00:51:33		14853	94269j276_2	126	15	30-SEP-1994 23:05:45
14419	94241j248_2	370	15	3-SEP-1994 01:02:43		14862	94269j276_2	2018	65	1-OCT-1994 13:05:07
14426	94241j248_2	39	65	3-SEP-1994 13:22:53		14863	94269j276_2	41	65	1-OCT-1994 14:28:07
14427	94241j248_2	1703	65	3-SEP-1994 13:24:09		14866	94269j276_2	1885	15	1-OCT-1994 19:13:57
14434	94241j248_2	306	45	4-SEP-1994 01:35:23		14867	94269j276_2	172	15	1-OCT-1994 20:38:59
14435	94241j248_2	1462	45	4-SEP-1994 01:45:33		14878	94269j276_2	1885	15	2-OCT-1994 13:42:19
14438	94241j248_2	355	45	4-SEP-1994 07:44:27		14882	94269j276_2	1975	15	2-OCT-1994 19:53:41
14439	94241j248_2	1440	45	4-SEP-1994 07:56:15		14892	94276j283_2	1918	65	3-OCT-1994 11:14:09
14452	94241j248_2	489	45	5-SEP-1994 05:17:21		14898	94276j283_2	1848	15	3-OCT-1994 20:30:29
14453	94241j248_2	1245	45	5-SEP-1994 05:33:39		14908	94276j283_2	1889	65	4-OCT-1994 11:52:25
14460	94248j255_2	320	65	5-SEP-1994 17:43:53		14914	94276j283_2	1772	15	4-OCT-1994 21:10:39
14461	94248j255_2	793	15	5-SEP-1994 18:16:37		14915	94276j283_2	112	15	4-OCT-1994 22:33:45
14461	94248j255_2	638	65	5-SEP-1994 17:54:33		14926	94276j283_2	1774	15	5-OCT-1994 15:38:19
14466	94248j255_2	280	45	6-SEP-1994 03:01:13		14927	94276j283_2	89	65	5-OCT-1994 17:02:29
14467	94248j255_2	1456	45	6-SEP-1994 03:10:31		14930	94276j283_2	1656	15	5-OCT-1994 21:51:57
14476	94248j255_2	298	15	6-SEP-1994 18:27:15		14940	94276j283_2	1442	65	6-OCT-1994 13:22:15
14477	94248j255_2	1420	15	6-SEP-1994 18:37:09		14941	94276j283_2	108	65	6-OCT-1994 14:35:39
14478	94248j255_2	398	15	6-SEP-1994 21:29:05		14944	94276j283_2	1978	15	6-OCT-1994 19:19:45
14479	94248j255_2	1382	15	6-SEP-1994 21:42:29		14956	94276j283_2	1400	65	7-OCT-1994 13:48:45
14492	94248j255_2	369	15	7-SEP-1994 19:07:11		14957	94276j283_2	85	65	7-OCT-1994 15:13:27
14493	94248j255_2	1352	15	7-SEP-1994 19:19:43		14960	94276j283_2	1860	15	7-OCT-1994 19:55:39
14494	94248j255_2	239	15	7-SEP-1994 22:17:05		14961	94276j283_2	87	15	7-OCT-1994 21:23:29
14495	94248j255_2	1367	15	7-SEP-1994 22:25:03		14970	94276j283_2	1694	65	8-OCT-1994 11:24:07
14508	94248j255_2	78	15	8-SEP-1994 19:59:39		14971	94276j283_2	93	65	8-OCT-1994 12:46:33
14509	94248j255_2	1393	15	8-SEP-1994 20:02:13		14976	94276j283_2	1997	15	8-OCT-1994 20:33:51
14510	94248j255_2	272	15	8-SEP-1994 22:58:29		14986	94276j283_2	1712	65	9-OCT-1994 11:58:15
14511	94248j255_2	1384	15	8-SEP-1994 23:07:31		14992	94276j283_2	1814	15	9-OCT-1994 21:11:53
14522	94248j255_2	399	65	9-SEP-1994 17:26:05						
14523	94248j255_2	1381	65	9-SEP-1994 17:39:21						
14526	94248j255_2	286	45	9-SEP-1994 23:40:07						
14527	94248j255_2	1366	45	9-SEP-1994 23:49:57						
14534	94248j255_2	235	65	10-SEP-1994 12:03:21						
14535	94248j255_2	1389	65	10-SEP-1994 12:11:09						
14538	94248j255_2	391	65	10-SEP-1994 18:08:45						
14539	94248j255_2	1280	65	10-SEP-1994 18:21:45						
14550	94248j255_2	410	65	11-SEP-1994 12:39:53						
14551	94248j255_2	1374	65	11-SEP-1994 12:53:31						
14554	94248j255_2	394	15	11-SEP-1994 18:50:59						
14555	94248j255_2	1399	15	11-SEP-1994 19:04:07						
14570	94255j262_2	276	15	12-SEP-1994 19:32:41						
14571	94255j262_2	1398	15	12-SEP-1994 19:45:33						
14572	94255j262_2	54	15	12-SEP-1994 22:40:47						
14573	94255j262_2	2377	15	12-SEP-1994 22:50:49						
14586	94255j262_2	259	15	13-SEP-1994 20:18:57						
14587	94255j262_2	1423	15	13-SEP-1994 20:27:35						
14588	94255j262_2	301	15	13-SEP-1994 23:22:49						
14589	94255j262_2	1415	15	13-SEP-1994 23:32:49						
14602	94255j262_2	419	15	14-SEP-1994 20:55:37						
14603	94255j262_2	1433	15	14-SEP-1994 21:09:33						
14612	94255j262_2	396	65	15-SEP-1994 12:22:19						
14613	94255j262_2	1411	65	15-SEP-1994 12:35:47						
14618	94255j262_2	747	15	15-SEP-1994 21:26:37						
14619	94255j262_2	1026	15	15-SEP-1994 21:51:31						
14628	94255j262_2	715	65	16-SEP-1994 12:53:45						
14629	94255j262_2	1065	65	16-SEP-1994 13:17:41						
14634	94255j262_2	798	15	16-SEP-1994 22:06:49						
14635	94255j262_2	681	15	16-SEP-1994 22:33:25						
14644	94255j262_2	1977	65	17-SEP-1994 12:53:41						
14645	94255j262_2	689	65	17-SEP-1994 13:59:33						

Appendix F
PVO and Magellan Data Arcs

The following information is included in this appendix:

1. Start and stop times for each data arc for PVO low altitude.
2. Start and stop times for each data arc for PVO high altitude.
3. PVO maneuvers.
3. Start and stop times for each data arc for Magellan cycle4.
4. Start and stop times for each data arc for Magellan cycle5.
5. Hide information for Magellan cycle 4.
6. Hide information for Magellan cycles 5 and 6.

PVO low altitude data arcs:

09-DEC-78 10:00	11-DEC-78 04:00
11-DEC-78 06:00	13-DEC-78 04:00
16-DEC-78 01:00	20-DEC-78 00:00
20-DEC-78 07:00	22-DEC-78 22:00
23-DEC-78 01:00	27-DEC-78 23:00
28-DEC-78 04:00	03-JAN-79 05:00
11-JAN-79 10:00	16-JAN-79 09:59
17-JAN-79 12:00	24-JAN-79 07:00
24-JAN-79 13:00	31-JAN-79 07:00
31-JAN-79 11:00	02-FEB-79 00:00
02-FEB-79 03:00	07-FEB-79 07:00
07-FEB-79 11:00	13-FEB-79 20:00
14-FEB-79 11:00	17-FEB-79 00:00
17-FEB-79 05:00	21-FEB-79 04:59
22-FEB-79 23:00	27-FEB-79 06:00
27-FEB-79 11:00	03-MAR-79 00:00
03-MAR-79 04:00	08-MAR-79 04:00
08-MAR-79 04:00	13-MAR-79 12:00
14-MAR-79 01:00	18-MAR-79 00:59
18-MAR-79 01:00	22-MAR-79 23:00
23-MAR-79 00:00	28-MAR-79 23:59
29-MAR-79 00:00	03-APR-79 23:30
04-APR-79 02:00	10-APR-79 01:59
10-APR-79 02:00	15-APR-79 01:59
15-APR-79 02:00	19-APR-79 01:59
19-APR-79 02:00	24-APR-79 01:59
25-APR-79 12:00	30-APR-79 02:00
30-APR-79 13:00	08-MAY-79 06:00
11-MAY-79 03:00	16-MAY-79 07:00
16-MAY-79 10:00	24-MAY-79 07:00
24-MAY-79 12:00	28-MAY-79 07:00
29-MAY-79 10:00	31-MAY-79 17:00
01-JUN-79 02:00	06-JUN-79 06:00
06-JUN-79 10:00	14-JUN-79 07:00
14-JUN-79 12:00	22-JUN-79 07:00
22-JUN-79 12:00	26-JUN-79 00:00
26-JUN-79 04:00	01-JUL-79 03:59
01-JUL-79 04:00	06-JUL-79 17:00
06-JUL-79 17:30	11-JUL-79 17:00
19-DEC-79 04:00	21-DEC-79 22:00
21-DEC-79 22:00	27-DEC-79 12:00
27-DEC-79 22:00	03-JAN-80 10:00
03-JAN-80 22:00	08-JAN-80 10:00
09-JAN-80 09:00	15-JAN-80 04:00
15-JAN-80 22:00	22-JAN-80 12:00
23-JAN-80 00:00	29-JAN-80 12:00
29-JAN-80 23:00	05-FEB-80 18:45
06-FEB-80 09:00	07-FEB-80 19:00
08-FEB-80 08:00	13-FEB-80 07:59
13-FEB-80 08:00	18-FEB-80 17:59
18-FEB-80 18:00	23-FEB-80 17:59
23-FEB-80 18:00	28-FEB-80 12:00
28-FEB-80 20:00	07-MAR-80 19:59
07-MAR-80 20:00	14-MAR-80 14:00
14-MAR-80 21:30	19-MAR-80 21:29
19-MAR-80 21:30	23-MAR-80 21:29
23-MAR-80 21:30	28-MAR-80 21:29
28-MAR-80 21:30	01-APR-80 16:00
01-APR-80 22:50	08-APR-80 00:00
09-APR-80 15:00	15-APR-80 22:00
19-APR-80 17:00	22-APR-80 15:00
22-APR-80 22:45	29-APR-80 03:00
30-APR-80 00:00	05-MAY-80 23:59
07-MAY-80 00:00	13-MAY-80 14:00
14-MAY-80 10:00	20-MAY-80 02:00
21-MAY-80 11:00	25-MAY-80 21:00
25-MAY-80 21:00	30-MAY-80 20:59
30-MAY-80 21:00	03-JUN-80 20:59
03-JUN-80 22:00	08-JUN-80 21:59
08-JUN-80 22:00	12-JUN-80 20:00
12-JUN-80 20:25	19-JUN-80 08:00
19-JUN-80 21:55	26-JUN-80 07:59
26-JUN-80 20:10	03-JUL-80 20:09
03-JUL-80 21:00	10-JUL-80 02:00
10-JUL-80 21:00	13-JUL-80 09:00
13-JUL-80 22:00	21-JUL-80 21:59
21-JUL-80 22:00	28-JUL-80 10:00
31-JUL-80 00:00	09-AUG-80 10:00
09-AUG-80 10:00	17-AUG-80 00:00
17-AUG-80 00:00	26-AUG-80 23:59
27-AUG-80 00:00	05-SEP-80 23:59
06-SEP-80 00:00	15-SEP-80 23:59
16-SEP-80 00:00	23-SEP-80 10:00
25-SEP-80 01:00	05-OCT-80 00:59
05-OCT-80 01:00	15-OCT-80 00:59
15-OCT-80 01:00	25-OCT-80 00:59
25-OCT-80 01:00	04-NOV-80 00:59
04-NOV-80 01:00	14-NOV-80 00:59
14-NOV-80 01:00	23-NOV-80 00:00
23-NOV-80 01:00	04-DEC-80 00:00

PVO high altitude data arcs:

```
06-NOV-81 13:49    09-NOV-81 06:00
12-NOV-81 05:00    16-NOV-81 06:00
19-NOV-81 23:00    23-NOV-81 06:00
26-NOV-81 23:00    30-NOV-81 06:00
02-DEC-81 13:51    09-DEC-81 10:00
09-DEC-81 10:01    15-DEC-81 10:00
15-DEC-81 10:01    22-DEC-81 10:00
22-DEC-81 10:00    29-DEC-81 02:00
29-DEC-81 10:01    05-JAN-82 10:00
05-JAN-82 10:01    12-JAN-82 10:00
12-JAN-82 10:00    20-JAN-82 05:00
20-JAN-82 10:00    23-JAN-82 06:00
24-JAN-82 23:00    29-JAN-82 06:00
30-JAN-82 23:00    03-FEB-82 05:00
03-FEB-82 10:00    05-FEB-82 06:00
06-FEB-82 23:00    10-FEB-82 06:00
13-FEB-82 23:00    17-FEB-82 06:00
20-FEB-82 23:00    24-FEB-82 06:00
27-FEB-82 23:00    02-MAR-82 05:00
02-MAR-82 10:00    04-MAR-82 06:00
06-MAR-82 23:00    10-MAR-82 06:00
13-MAR-82 23:00    17-MAR-82 06:00
20-MAR-82 23:00    23-MAR-82 22:50
27-MAR-82 23:00    30-MAR-82 22:50
03-APR-82 23:00    06-APR-82 05:00
06-APR-82 10:00    09-APR-82 06:00
10-APR-82 23:00    13-APR-82 22:50
17-APR-82 23:00    20-APR-82 22:50
24-APR-82 23:00    27-APR-82 22:50
01-MAY-82 23:00    04-MAY-82 22:50
08-MAY-82 23:00    11-MAY-82 22:50
15-MAY-82 23:00    18-MAY-82 22:50
22-MAY-82 23:00    25-MAY-82 22:50
29-MAY-82 23:00    01-JUN-82 22:50
05-JUN-82 23:30    08-JUN-82 23:29
12-JUN-82 23:30    15-JUN-82 23:00
19-JUN-82 23:30    22-JUN-82 23:00
26-JUN-82 23:30    29-JUN-82 23:00
03-JUL-82 23:30    06-JUL-82 23:00
10-JUL-82 23:30    13-JUL-82 23:00
17-JUL-82 23:30    20-JUL-82 23:00
24-JUL-82 23:30    27-JUL-82 23:00
31-JUL-82 23:30    03-AUG-82 23:00
07-AUG-82 23:30    10-AUG-82 05:00
14-AUG-82 23:30    17-AUG-82 23:00
20-AUG-82 23:30    24-AUG-82 23:00
28-AUG-82 23:30    31-AUG-82 23:00
04-SEP-82 23:30    07-SEP-82 23:00
```

PVO maneuvers:

```
78/341/0536   DEC 07    m  alt. 200 km/ period trim
   341/0630   DEC 07    m  reor.
   342/0411   DEC 08    m  alt. 180 km
   345/0436   DEC 11    m  alt. 170 km
   347/0451   DEC 13    m  alt. 160 km
   347/1541   DEC 13    m  period trim
   349/0536   DEC 15    m  spin trim/alt. 155 km
   354/0103   DEC 20    m  reor.
   354/0537   DEC 20    m  alt. 150 km
   356/2343   DEC 22    m  reor.
   362/0040   DEC 28    m  reor. star map
   362/0240   DEC 28    m  reor. start Per. Eclipse #1
79/003/0736   JAN 03    m  alt. 148 km
   003/2145   JAN 03    m  period trim/spin trim
   011/0851   JAN 11    m  alt. 150 km/spin trim
   017/1026   JAN 17    m  alt. 147 km/spin trim
   024/0943   JAN 24    m  alt. 145 km/spin trim
   024/1113   JAN 24    m  reor.
   031/0928   JAN 31    m  alt. 142 km/spin trim
   033/0107   FEB 02    m  reor.
   038/0947   FEB 07    m  spin trim/alt. 142 km
   045/0932   FEB 14    m  spin trim/alt. 142 km
   048/0135   FEB 17    m  reor.
   052/0930   FEB 21    m  spin trim/alt. 142 km
   058/0855   FEB 27    m  spin trim/alt. 142 km
   062/0134   MAR 03    m  reor.
   072/1904   MAR 13    m  period trim/spin trim
   073/0037   MAR 14    m  reor.
   081 2328   MAR 22    m  reor.
   094/0058   APR 04    m  reor./spin trim
   115/0945   APR 25    m  reor.
   115/1145   APR 25    m  alt. 148 km/spin trim
   120/1125   APR 30    m  alt. 148 km/spin trim
   128/0930   MAY 08    m  reor./alt. reduct./spin trim
   136/0900   MAY 16    m  spin trim/alt. reduction
   144/1030   MAY 24    m  spin trim/reor./alt. reduct.
   149/1000   MAY 29    m  reor./spin trim/alt. reduct.
   151/2020   MAY 31    m  period trim/spin trim
   157/0830   JUN 06    m  reor./spin trim/alt. reduct.
   165/1130   JUN 14    m  reor./spin trim/alt. reduct.
   173/1120   JUN 22    m  reor./spin trim/alt. reduct.
   177/0327   JUN 26    m  reor.
   187/1740   JUL 06    m  reor.
   194/0325   JUL 13    m  reor.
   205/2220   JUL 24    m  period trim
   206/0425   JUL 25    m  spin trim/reor.
   216/0632   AUG 04    m  reor.
   255/0430   SEP 12    m  reor./spin trim/alt. reduct.
   256/2254   SEP 13    m  spin trim
   258/0455   SEP 15    m  reor.
   260/2312   SEP 17    m period trim/spin trim
   262/0350   SEP 19    m  reor./spin trim/alt. reduct.
   269/0440   SEP 26    m  reor./spin trim/alt. reduct.
   276/0550   OCT 03    m  reor.
   277/0134   OCT 04    m  spin trim/alt. reduction
   283/1354   OCT 10    m  reor.
```

```
   284/0223   OCT 11   m  spin trim/alt. reduction
   292/1650   OCT 19   m  reor.
   305/1705   NOV 01   m  reor.
   318/2001   NOV 14   m  reor.
   324/1957   NOV 20   m  reor.
   338/1948   DEC 04   m  reor./spin trim
   345/2123   DEC 11   m  alt. reduction/spin trim
   352/2120   DEC 18   m  alt. reduction/spin trim
   362/0016   DEC 28   m  reor.
   362/0744   DEC 28   m  alt. reduction/spin trim
80/003/2116   JAN 03   m  alt. reduct./spin trim/reor.
   008/2106   JAN 08   m  alt. reduction/spin trim
   009/0659   JAN 09   m period trim/spin trim
   015/2200   JAN 15   m  spin trim/alt. reduction
   022/2200   JAN 22   m  reor./spin trim/alt. reduct.
   029 2215   JAN 29   m  reor.
   030/0815   JAN 30   m  alt. reduction/spin trim
   037/0830   FEB 06   m  period trim
   038/1950   FEB 07   m  reor.
   039/0818   FEB 08   m  spin trim
   059/2215   FEB 28   m  reor.
   074/2345   MAR 14   m  reor./spin trim
   093/0150   APR 02   m  reor./alt. reduct./spin trim
   101/0035   APR 10   m  spin trim/alt. reduction
   107/0100   APR 16   m  alt. reduct./spin trim/reor.
   114/0050   APR 23   m  spin trim/alt. reduction
   121/0030   APR 30   m  reor./spin trim/alt. reduct.
   128/1007   MAY 07   m  spin trim/alt. reduction
   134/2330   MAY 13   m  reor./spin trim/alt. reduct.
   135/0932   MAY 14   m  period trim/spin trim
   141/0930   MAY 20   m  reor./spin trim/alt. reduct.
   142/0940   MAY 21   m  period trim/spin trim
   155/2120   JUN 03   m  reor. (special)
   164/2100   JUN 12   m  reor.
   171/2230   JUN 19   m  reor.
   179/0130   JUN 27   m  reor./spin trim
   192/1428   JUL 10   m  spin trim 6 sec
   195/1440   JUL 13   m  spin trim 12 sec
   210/2300   JUL 28   m  reor./spin trim
   269/1813   SEP 25   m  spin trim
   328/0115   NOV 23   m  reor.
   346/1722   DEC 11   m  reor.
81/006/0310   JAN 06   m  reor.
   036/0450   FEB 05   m  period trim
   036/1142   FEB 05   m  reor./spin trim
   037/0315   FEB 06   m  period trim
   037 1920   FEB 06   m  reor./spin trim
   057/0500   FEB 26   m  reor.
   076/0700   MAR 17   m  spin trim/reor.
   114/0100   APR 24   m  reor.
   127/0207   MAY 07   m  period trim
   127/2215   MAY 07   m   reor.
   141/0208   MAY 21   m  spin trim
   142/0208   MAY 22   m  spin trim
   148/0207   MAY 28   m  reor.
   168/0630   JUN 17   m  reor./spin trim
   203/0715   JUL 22   m  reor./spin trim (error)
   204/0745   JUL 23   m  reor./spin trim (correct)
```

```
    238/2015    AUG 26   m   reor./spin trim
    268/0830    SEP 25   m   reor./spin trim
    336/0152    DEC 02   m   spin trim
    363/0820    DEC 29   m   reor./spin trim
82/020/0630    JAN 20   m   reor.
    034/0645    FEB 03   m   reor./spin trim
    061/0740    MAR 02   m   reor.
    097/0144    APR 07   m   spin trim
    133/0730    MAY 13   m   spin trim/reor.
    176/0720    JUN 25   m   spin trim/reor.
    222/0700    AUG 10   m   reor.
    252/0620    SEP 09   m   reor.
    296/0125    OCT 23   m   reor.
    325/0630    NOV 21   m   reor.
    343/0615    DEC 09   m   reor.
```

Cycle 4 data arcs:

15-SEP-92 03:00	15-SEP-92 22:25	27-NOV-92 14:25	28-NOV-92 19:20
15-SEP-92 22:45	17-SEP-92 03:30	28-NOV-92 19:40	30-NOV-92 13:20
17-SEP-92 03:50	18-SEP-92 11:40	30-NOV-92 13:40	01-DEC-92 18:35
18-SEP-92 12:00	19-SEP-92 20:10	01-DEC-92 18:55	03-DEC-92 18:00
19-SEP-92 20:30	20-SEP-92 18:45	03-DEC-92 22:40	04-DEC-92 21:00
20-SEP-92 19:00	21-SEP-92 20:50	04-DEC-92 21:25	06-DEC-92 02:20
21-SEP-92 21:10	22-SEP-92 19:40	06-DEC-92 02:40	07-DEC-92 07:20
22-SEP-92 20:00	23-SEP-92 18:45	07-DEC-92 07:40	08-DEC-92 12:25
23-SEP-92 19:00	24-SEP-92 17:15	08-DEC-92 12:45	09-DEC-92 14:20
24-SEP-92 17:30	25-SEP-92 22:00	09-DEC-92 14:45	10-DEC-92 19:35
25-SEP-92 22:15	27-SEP-92 03:15	10-DEC-92 19:55	12-DEC-92 00:40
27-SEP-92 03:30	28-SEP-92 18:00	12-DEC-92 01:00	12-DEC-92 23:20
28-SEP-92 18:25	30-SEP-92 02:40	12-DEC-92 23:40	14-DEC-92 14:10
30-SEP-92 03:00	01-OCT-92 17:25	14-DEC-92 14:30	15-DEC-92 12:50
01-OCT-92 17:45	02-OCT-92 19:40	15-DEC-92 13:10	17-DEC-92 00:30
02-OCT-92 19:45	03-OCT-92 18:15	17-DEC-92 00:55	17-DEC-92 23:10
03-OCT-92 18:15	05-OCT-92 01:10	17-DEC-92 23:30	19-DEC-92 01:00
05-OCT-92 01:30	06-OCT-92 01:15	19-DEC-92 01:25	19-DEC-92 23:50
06-OCT-92 01:40	06-OCT-92 23:45	20-DEC-92 00:10	21-DEC-92 11:20
07-OCT-92 00:10	08-OCT-92 02:00	21-DEC-92 11:45	23-DEC-92 05:00
08-OCT-92 02:15	09-OCT-92 10:25	23-DEC-92 15:30	25-DEC-92 03:00
09-OCT-92 10:45	11-OCT-92 01:50	26-DEC-92 01:55	27-DEC-92 00:10
11-OCT-92 02:00	12-OCT-92 19:20	27-DEC-92 00:30	27-DEC-92 23:00
12-OCT-92 19:20	13-OCT-92 18:30	27-DEC-92 23:15	29-DEC-92 00:50
13-OCT-92 18:50	14-OCT-92 17:30	29-DEC-92 01:10	31-DEC-92 00:00
14-OCT-92 17:45	15-OCT-92 22:00	01-JAN-93 00:15	02-JAN-93 21:10
15-OCT-92 22:25	17-OCT-92 00:00	02-JAN-93 21:30	03-JAN-93 23:10
17-OCT-92 00:30	18-OCT-92 11:50	03-JAN-93 23:30	05-JAN-93 01:10
18-OCT-92 12:05	19-OCT-92 10:30	05-JAN-93 01:25	06-JAN-93 16:00
19-OCT-92 10:45	20-OCT-92 18:50	06-JAN-93 16:15	07-JAN-93 21:10
20-OCT-92 19:10	21-OCT-92 17:30	07-JAN-93 21:30	08-JAN-93 23:10
21-OCT-92 17:55	23-OCT-92 02:00	08-JAN-93 23:25	09-JAN-93 18:20
23-OCT-92 02:15	24-OCT-92 19:50	09-JAN-93 18:45	11-JAN-93 12:20
24-OCT-92 20:05	26-OCT-92 01:10	11-JAN-93 12:45	12-JAN-93 14:15
26-OCT-92 01:25	27-OCT-92 06:20	12-JAN-93 14:45	14-JAN-93 04:20
27-OCT-92 06:40	29-OCT-92 00:30	14-JAN-93 15:15	15-JAN-93 17:00
29-OCT-92 00:50	30-OCT-92 21:50	15-JAN-93 17:15	16-JAN-93 19:00
30-OCT-92 22:10	31-OCT-92 16:45	16-JAN-93 19:15	17-JAN-93 17:30
31-OCT-92 17:40	02-NOV-92 11:25	17-JAN-93 17:45	19-JAN-93 01:00
02-NOV-92 11:45	04-NOV-92 05:35	19-JAN-93 02:15	20-JAN-93 03:40
04-NOV-92 05:55	05-NOV-92 20:20	20-JAN-93 04:00	21-JAN-93 12:10
05-NOV-92 20:40	07-NOV-92 16:00	21-JAN-93 12:30	22-JAN-93 20:40
07-NOV-92 18:00	08-NOV-92 19:50	22-JAN-93 21:00	23-JAN-93 18:15
08-NOV-92 20:10	10-NOV-92 07:20	23-JAN-93 19:30	24-JAN-93 14:30
10-NOV-92 07:40	11-NOV-92 19:05	24-JAN-93 15:00	25-JAN-93 20:00
11-NOV-92 19:25	13-NOV-92 00:10	25-JAN-93 20:15	27-JAN-93 01:00
13-NOV-92 00:30	14-NOV-92 11:50	27-JAN-93 01:15	27-JAN-93 17:15
14-NOV-92 12:10	16-NOV-92 11:10	27-JAN-93 17:30	28-JAN-93 22:10
16-NOV-92 12:45	17-NOV-92 07:45	28-JAN-93 22:30	29-JAN-93 16:45
17-NOV-92 08:10	18-NOV-92 16:20	29-JAN-93 18:00	30-JAN-93 19:40
18-NOV-92 16:40	19-NOV-92 21:20	30-JAN-93 20:00	31-JAN-93 21:25
19-NOV-92 21:40	21-NOV-92 15:30	31-JAN-93 21:45	01-FEB-93 20:10
21-NOV-92 15:55	23-NOV-92 00:00	01-FEB-93 20:30	02-FEB-93 22:10
23-NOV-92 00:15	24-NOV-92 23:20	02-FEB-93 22:30	03-FEB-93 20:40
25-NOV-92 00:55	26-NOV-92 15:20	03-FEB-93 21:00	04-FEB-93 16:10
26-NOV-92 15:45	27-NOV-92 14:00	04-FEB-93 16:30	05-FEB-93 18:10
		05-FEB-93 18:30	06-FEB-93 16:40
		06-FEB-93 17:00	07-FEB-93 18:40

07-FEB-93 19:00	08-FEB-93 17:10	13-APR-93 13:15	14-APR-93 07:30
08-FEB-93 17:45	09-FEB-93 19:10	14-APR-93 08:45	15-APR-93 10:10
09-FEB-93 19:30	10-FEB-93 21:00	15-APR-93 10:30	16-APR-93 05:40
10-FEB-93 21:30	11-FEB-93 20:00	16-APR-93 06:00	17-APR-93 07:45
11-FEB-93 20:15	13-FEB-93 04:15	17-APR-93 08:00	18-APR-93 09:15
13-FEB-93 04:30	14-FEB-93 02:00	18-APR-93 09:45	19-APR-93 08:00
14-FEB-93 03:15	15-FEB-93 17:20	19-APR-93 08:30	20-APR-93 10:00
15-FEB-93 18:00	16-FEB-93 23:00	20-APR-93 10:15	21-APR-93 18:20
16-FEB-93 23:15	17-FEB-93 11:45	21-APR-93 18:45	22-APR-93 14:00
17-FEB-93 12:00	18-FEB-93 13:30	22-APR-93 14:15	23-APR-93 14:45
18-FEB-93 14:00	19-FEB-93 12:20	23-APR-93 16:00	24-APR-93 16:45
19-FEB-93 12:45	20-FEB-93 13:30	24-APR-93 18:00	25-APR-93 18:30
20-FEB-93 14:30	21-FEB-93 13:00	25-APR-93 19:45	26-APR-93 17:15
21-FEB-93 13:15	22-FEB-93 11:30	26-APR-93 18:30	27-APR-93 20:00
22-FEB-93 12:00	23-FEB-93 16:30	27-APR-93 20:30	29-APR-93 07:40
23-FEB-93 17:00	24-FEB-93 08:00	29-APR-93 08:00	30-APR-93 19:15
24-FEB-93 09:15	25-FEB-93 11:00	30-APR-93 19:45	01-MAY-93 17:45
25-FEB-93 11:15	26-FEB-93 12:30	01-MAY-93 18:15	02-MAY-93 20:00
26-FEB-93 13:00	27-FEB-93 18:00	02-MAY-93 20:15	03-MAY-93 22:00
27-FEB-93 18:15	28-FEB-93 13:15	03-MAY-93 22:00	04-MAY-93 20:15
28-FEB-93 13:45	01-MAR-93 11:45	04-MAY-93 20:45	05-MAY-93 18:30
01-MAR-93 12:15	02-MAR-93 09:45	05-MAY-93 19:30	06-MAY-93 20:00
02-MAR-93 11:00	03-MAR-93 12:30	06-MAY-93 21:15	07-MAY-93 19:30
03-MAR-93 13:00	04-MAR-93 11:00	07-MAY-93 20:00	09-MAY-93 07:10
04-MAR-93 11:30	05-MAR-93 16:15	09-MAY-93 07:30	10-MAY-93 05:00
05-MAR-93 16:45	06-MAR-93 11:30	10-MAY-93 06:15	11-MAY-93 11:00
06-MAR-93 12:00	07-MAR-93 12:45	11-MAY-93 11:15	12-MAY-93 05:30
07-MAR-93 14:00	08-MAR-93 11:45	12-MAY-93 06:45	13-MAY-93 08:10
08-MAR-93 12:45	09-MAR-93 14:15	13-MAY-93 08:30	14-MAY-93 16:30
09-MAR-93 14:30	11-MAR-93 05:15	14-MAY-93 17:00	15-MAY-93 14:15
11-MAR-93 05:30	12-MAR-93 16:15	15-MAY-93 15:30	16-MAY-93 16:15
12-MAR-93 17:00	13-MAR-93 15:20	16-MAY-93 17:30	18-MAY-93 04:45
13-MAR-93 15:45	15-MAR-93 09:20	18-MAY-93 05:00	19-MAY-93 06:30
15-MAR-93 09:45	17-MAR-93 00:15	19-MAY-93 07:00	20-MAY-93 12:20
17-MAR-93 00:45	17-MAR-93 16:20	20-MAY-93 12:40	21-MAY-93 16:45
17-MAR-93 17:00	18-MAR-93 18:15	21-MAY-93 17:15	22-MAY-93 15:15
18-MAR-93 18:45	20-MAR-93 16:00	22-MAY-93 15:45	23-MAY-93 10:00
20-MAR-93 16:15	22-MAR-93 00:10	23-MAY-93 11:15	24-MAY-93 06:15
22-MAR-93 00:30	23-MAR-93 08:40	24-MAY-93 06:45	24-MAY-93 21:30
23-MAR-93 09:00	24-MAR-93 13:15		
24-MAR-93 14:00	25-MAR-93 15:40		
25-MAR-93 16:00	27-MAR-93 03:10		
27-MAR-93 03:30	28-MAR-93 14:45		
28-MAR-93 15:15	29-MAR-93 13:15		
29-MAR-93 13:45	31-MAR-93 04:30		
31-MAR-93 04:45	01-APR-93 12:45		
01-APR-93 13:00	02-APR-93 14:40		
02-APR-93 15:00	03-APR-93 13:15		
03-APR-93 13:45	04-APR-93 15:10		
04-APR-93 15:30	05-APR-93 14:00		
05-APR-93 14:15	06-APR-93 12:20		
06-APR-93 12:45	07-APR-93 14:30		
07-APR-93 14:45	08-APR-93 13:15		
08-APR-93 13:30	09-APR-93 21:30		
09-APR-93 21:45	10-APR-93 13:40		
10-APR-93 14:00	11-APR-93 12:30		
11-APR-93 12:45	12-APR-93 07:30		
12-APR-93 08:00	13-APR-93 13:00		

Cycle 5 and 6 data arcs:

06-AUG-93 00:15	06-AUG-93 20:45
07-AUG-93 10:00	08-AUG-93 01:00
08-AUG-93 09:00	08-AUG-93 16:45
08-AUG-93 18:30	09-AUG-93 02:15
09-AUG-93 12:15	09-AUG-93 22:40
10-AUG-93 11:30	10-AUG-93 19:15
12-AUG-93 13:50	13-AUG-93 03:45
14-AUG-93 11:50	15-AUG-93 03:10
16-AUG-93 11:50	16-AUG-93 17:40
17-AUG-93 12:00	18-AUG-93 00:30
18-AUG-93 12:20	18-AUG-93 23:20
19-AUG-93 20:00	20-AUG-93 00:35
20-AUG-93 11:40	21-AUG-93 00:50
21-AUG-93 04:20	21-AUG-93 17:35
22-AUG-93 04:00	22-AUG-93 08:15
22-AUG-93 13:25	23-AUG-93 06:30
23-AUG-93 21:00	24-AUG-93 07:40
24-AUG-93 12:40	25-AUG-93 00:15
25-AUG-93 20:00	26-AUG-93 16:25
27-AUG-93 08:10	28-AUG-93 00:05
28-AUG-93 03:45	28-AUG-93 07:50
28-AUG-93 17:40	29-AUG-93 16:40
30-AUG-93 02:20	30-AUG-93 08:30
30-AUG-93 19:40	31-AUG-93 11:15
01-SEP-93 03:40	01-SEP-93 18:35
01-SEP-93 22:00	02-SEP-93 17:50
02-SEP-93 18:00	03-SEP-93 12:10
03-SEP-93 14:45	04-SEP-93 00:30
04-SEP-93 04:00	05-SEP-93 00:00
05-SEP-93 03:50	05-SEP-93 17:30
06-SEP-93 04:00	06-SEP-93 16:30
07-SEP-93 03:50	08-SEP-93 00:00
08-SEP-93 04:10	08-SEP-93 23:40
09-SEP-93 04:00	10-SEP-93 00:00
10-SEP-93 04:00	10-SEP-93 23:40
11-SEP-93 03:30	12-SEP-93 00:00
12-SEP-93 04:00	12-SEP-93 17:30
12-SEP-93 19:10	13-SEP-93 06:30
13-SEP-93 06:00	14-SEP-93 01:20
14-SEP-93 04:20	14-SEP-93 23:30
15-SEP-93 00:50	15-SEP-93 18:00
16-SEP-93 00:30	16-SEP-93 21:10
17-SEP-93 11:55	18-SEP-93 08:10
18-SEP-93 20:15	19-SEP-93 15:30
19-SEP-93 21:30	20-SEP-93 07:00
20-SEP-93 21:20	21-SEP-93 14:40
22-SEP-93 01:45	22-SEP-93 15:00
23-SEP-93 01:15	23-SEP-93 23:00
24-SEP-93 02:40	24-SEP-93 23:00
25-SEP-93 05:00	25-SEP-93 16:30
25-SEP-93 18:30	26-SEP-93 12:00
26-SEP-93 16:00	27-SEP-93 07:00
27-SEP-93 07:00	27-SEP-93 16:30
27-SEP-93 19:30	28-SEP-93 16:30

Periapse raise maneuver
Sept 28 17:18

28-SEP-93 19:30	29-SEP-93 16:30
29-SEP-93 19:30	30-SEP-93 16:00
01-OCT-93 05:00	01-OCT-93 22:00
02-OCT-93 04:00	02-OCT-93 22:30
03-OCT-93 00:45	03-OCT-93 16:30
03-OCT-93 19:30	04-OCT-93 07:00
04-OCT-93 07:00	04-OCT-93 17:45
04-OCT-93 19:40	05-OCT-93 16:10
05-OCT-93 20:45	06-OCT-93 12:45

Bistatic radar Oct 6, 12h to 22h

07-OCT-93 01:15	08-OCT-93 00:00
08-OCT-93 00:45	08-OCT-93 22:15
09-OCT-93 05:25	10-OCT-93 03:10
10-OCT-93 05:30	10-OCT-93 16:00
10-OCT-93 19:00	11-OCT-93 07:00
11-OCT-93 07:00	11-OCT-93 22:00
12-OCT-93 06:00	12-OCT-93 23:50
13-OCT-93 02:30	13-OCT-93 16:05
14-OCT-93 02:00	14-OCT-93 22:20
15-OCT-93 01:10	15-OCT-93 22:50
16-OCT-93 01:25	16-OCT-93 22:50
17-OCT-93 01:00	17-OCT-93 13:40
17-OCT-93 15:10	18-OCT-93 07:00
18-OCT-93 19:45	19-OCT-93 16:50
20-OCT-93 01:00	20-OCT-93 15:40
21-OCT-93 02:50	21-OCT-93 21:35
22-OCT-93 13:30	23-OCT-93 07:45
23-OCT-93 13:30	24-OCT-93 08:00
24-OCT-93 13:25	25-OCT-93 07:00

Began to lose lock 25-OCT 16:10
Lost lock from 25-OCT 18:33
to 26-OCT 01:00ish
Medium gain antenna, S-Band only

26-OCT-93 03:30	26-OCT-93 15:15
26-OCT-93 16:00	27-OCT-93 04:50
27-OCT-93 12:00	28-OCT-93 06:20
28-OCT-93 13:30	28-OCT-93 23:00

Return to high gain antenna
28-OCT 23:25

28-OCT-93 23:30	29-OCT-93 15:40
29-OCT-93 21:10	30-OCT-93 15:30
30-OCT-93 18:50	31-OCT-93 09:30
31-OCT-93 14:00	01-NOV-93 07:00
01-NOV-93 07:15	01-NOV-93 14:05
01-NOV-93 19:35	02-NOV-93 09:20
02-NOV-93 13:45	03-NOV-93 09:05
03-NOV-93 13:50	04-NOV-93 10:50
04-NOV-93 13:20	05-NOV-93 01:05
05-NOV-93 06:40	06-NOV-93 05:50
06-NOV-93 06:45	07-NOV-93 01:25
07-NOV-93 02:45	07-NOV-93 22:50
08-NOV-93 10:45	09-NOV-93 05:50

Bistatic radar
Nov 9, 8:35 to 11:44

09-NOV-93 13:35	10-NOV-93 03:30
10-NOV-93 10:05	11-NOV-93 00:20
11-NOV-93 00:25	11-NOV-93 15:40
11-NOV-93 20:20	12-NOV-93 17:40

12-NOV-93 18:30	13-NOV-93 15:50	05-JAN-94 21:30	06-JAN-94 08:20
13-NOV-93 16:40	14-NOV-93 15:25	27-JAN-94 10:30	28-JAN-94 00:00
14-NOV-93 19:40	15-NOV-93 08:50	28-JAN-94 15:00	29-JAN-94 14:00
15-NOV-93 16:20	16-NOV-93 13:15	30-JAN-94 01:25	30-JAN-94 15:20
16-NOV-93 14:00	17-NOV-93 10:55	31-JAN-94 09:00	31-JAN-94 23:00
17-NOV-93 18:50	18-NOV-93 09:15	01-FEB-94 10:05	02-FEB-94 06:30

No ramps beginning
Nov 18, 1993 at 18:55

18-NOV-93 14:25	19-NOV-93 00:35	02-FEB-94 11:15	03-FEB-94 00:50
19-NOV-93 08:25	19-NOV-93 16:00	03-FEB-94 13:10	04-FEB-94 08:10
19-NOV-93 20:50	20-NOV-93 15:00	04-FEB-94 10:30	05-FEB-94 09:50
20-NOV-93 19:25	21-NOV-93 08:40	05-FEB-94 11:45	06-FEB-94 08:00
21-NOV-93 14:00	22-NOV-93 01:10	06-FEB-94 22:30	07-FEB-94 12:20
22-NOV-93 07:15	22-NOV-93 22:50	07-FEB-94 14:10	08-FEB-94 10:17
23-NOV-93 07:25	24-NOV-93 04:15	08-FEB-94 12:10	09-FEB-94 11:30
24-NOV-93 05:05	24-NOV-93 23:05	09-FEB-94 22:50	10-FEB-94 12:45
25-NOV-93 08:00	25-NOV-93 21:10	10-FEB-94 23:55	11-FEB-94 20:30
26-NOV-93 01:20	26-NOV-93 15:15	11-FEB-94 22:10	12-FEB-94 18:50
26-NOV-93 20:10	27-NOV-93 09:20	12-FEB-94 20:15	13-FEB-94 16:25
27-NOV-93 13:35	27-NOV-93 22:50	13-FEB-94 21:30	14-FEB-94 17:50
28-NOV-93 07:45	28-NOV-93 21:45	14-FEB-94 22:30	15-FEB-94 12:30
29-NOV-93 05:05	29-NOV-93 12:15	15-FEB-94 23:50	16-FEB-94 23:05
29-NOV-93 20:40	30-NOV-93 08:40	17-FEB-94 11:05	18-FEB-94 06:50
30-NOV-93 12:20	01-DEC-93 00:15	18-FEB-94 21:00	19-FEB-94 11:35
01-DEC-93 07:40	02-DEC-93 04:50	19-FEB-94 15:40	20-FEB-94 06:20
02-DEC-93 08:30	03-DEC-93 02:45	20-FEB-94 16:45	21-FEB-94 07:00
03-DEC-93 03:30	03-DEC-93 20:00	21-FEB-94 18:20	22-FEB-94 14:30
03-DEC-93 22:25	04-DEC-93 16:30	22-FEB-94 16:30	23-FEB-94 05:40
04-DEC-93 17:20	05-DEC-93 04:50	23-FEB-94 17:40	24-FEB-94 04:30
05-DEC-93 09:00	06-DEC-93 06:20	24-FEB-94 09:20	25-FEB-94 01:40
06-DEC-93 08:00	07-DEC-93 00:00	25-FEB-94 10:30	26-FEB-94 00:30
07-DEC-93 06:20	07-DEC-93 23:30	26-FEB-94 11:40	27-FEB-94 05:00
08-DEC-93 04:30	09-DEC-93 00:20	27-FEB-94 13:00	28-FEB-94 00:00
09-DEC-93 09:00	10-DEC-93 07:35	28-FEB-94 10:50	01-MAR-94 04:15
10-DEC-93 12:10	11-DEC-93 06:15	01-MAR-94 12:05	02-MAR-94 11:00
11-DEC-93 08:35	12-DEC-93 07:20	02-MAR-94 14:00	03-MAR-94 13:10
12-DEC-93 12:30	13-DEC-93 07:25	03-MAR-94 14:15	04-MAR-94 11:00
13-DEC-93 13:00	14-DEC-93 05:20	04-MAR-94 12:20	05-MAR-94 02:50
14-DEC-93 08:50	14-DEC-93 22:30	05-MAR-94 14:15	06-MAR-94 04:00
15-DEC-93 09:00	16-DEC-93 02:20	06-MAR-94 11:40	07-MAR-94 02:00
16-DEC-93 08:30	17-DEC-93 05:50	07-MAR-94 09:45	08-MAR-94 08:30
17-DEC-93 08:30	18-DEC-93 07:05	08-MAR-94 17:20	09-MAR-94 07:30
18-DEC-93 12:25	19-DEC-93 08:00	09-MAR-94 16:00	10-MAR-94 16:20

Periapse lower maneuvers
March 10 16:10:18 and 17:44:35

19-DEC-93 13:25	20-DEC-93 06:20	10-MAR-94 19:50	11-MAR-94 06:45
20-DEC-93 14:20	21-DEC-93 10:45	11-MAR-94 14:20	12-MAR-94 08:00
21-DEC-93 19:45	22-DEC-93 15:15	12-MAR-94 13:00	13-MAR-94 08:30
22-DEC-93 17:00	23-DEC-93 13:10	13-MAR-94 16:40	14-MAR-94 04:00
23-DEC-93 18:30	24-DEC-93 06:00	14-MAR-94 21:20	15-MAR-94 20:50
25-DEC-93 17:30	26-DEC-93 14:00	15-MAR-94 21:30	16-MAR-94 18:50
26-DEC-93 15:30	27-DEC-93 12:00	17-MAR-94 07:10	18-MAR-94 03:00
27-DEC-93 13:40	28-DEC-93 07:00	18-MAR-94 06:45	19-MAR-94 06:40
28-DEC-93 18:00	29-DEC-93 14:20	19-MAR-94 07:20	20-MAR-94 04:30
29-DEC-93 16:00	30-DEC-93 06:15	20-MAR-94 08:40	21-MAR-94 08:40
30-DEC-93 14:15	31-DEC-93 10:30	21-MAR-94 10:00	22-MAR-94 08:15
01-JAN-94 07:10	02-JAN-94 03:40	22-MAR-94 17:00	23-MAR-94 08:00
02-JAN-94 05:15	03-JAN-94 01:30	24-MAR-94 01:20	24-MAR-94 12:00
03-JAN-94 09:45	04-JAN-94 08:15	24-MAR-94 19:00	25-MAR-94 07:00
04-JAN-94 10:50	05-JAN-94 07:00	25-MAR-94 14:00	26-MAR-94 08:10

```
26-MAR-94 12:00    27-MAR-94 08:30
27-MAR-94 10:00    28-MAR-94 02:45
28-MAR-94 11:10    29-MAR-94 02:00
29-MAR-94 15:30    30-MAR-94 12:30
30-MAR-94 13:30    31-MAR-94 10:40
31-MAR-94 15:35    01-APR-94 15:25
01-APR-94 17:15    02-APR-94 16:40
02-APR-94 17:15    03-APR-94 17:30
03-APR-94 18:30    04-APR-94 17:40
# Periapse raise on April 4,
# 17:31 and 20:40
04-APR-94 22:26    05-APR-94 13:50
# Apoapse lower on April 5,
# 14:48, 17:57, 21:06
05-APR-94 22:04    06-APR-94 19:40
06-APR-94 22:25    07-APR-94 19:00
07-APR-94 19:50    08-APR-94 19:40
08-APR-94 23:50    09-APR-94 18:00
09-APR-94 18:40    10-APR-94 16:00
10-APR-94 16:40    11-APR-94 15:00
# Apoapse lower on April 11,
# 14:58, 18:05, 21:12
11-APR-94 23:08    12-APR-94 15:50
# Apoapse lower on April 12,
# 15:53, 19:00, 22:06
13-APR-94 00:39    14-APR-94 00:45
14-APR-94 01:50    15-APR-94 03:00
15-APR-94 15:30    16-APR-94 07:50
16-APR-94 15:40    17-APR-94 08:00
17-APR-94 16:20    18-APR-94 07:10
18-APR-94 11:10    19-APR-94 11:00
19-APR-94 11:50    20-APR-94 11:40
20-APR-94 15:30    21-APR-94 09:30
21-APR-94 10:25    22-APR-94 07:45
22-APR-94 08:55    23-APR-94 07:50
23-APR-94 08:00    24-APR-94 07:45
24-APR-94 08:00    25-APR-94 07:05
25-APR-94 08:00    26-APR-94 08:30
26-APR-94 08:45    27-APR-94 07:50
27-APR-94 10:00    28-APR-94 08:00
28-APR-94 08:30    29-APR-94 07:50
29-APR-94 08:00    30-APR-94 07:45
30-APR-94 09:00    01-MAY-94 07:45
01-MAY-94 08:45    02-MAY-94 03:50
02-MAY-94 08:00    03-MAY-94 07:50
03-MAY-94 08:10    04-MAY-94 07:10
05-MAY-94 08:10    06-MAY-94 07:15
06-MAY-94 08:40    07-MAY-94 07:50
07-MAY-94 08:10    08-MAY-94 07:15
08-MAY-94 08:00    09-MAY-94 04:10
09-MAY-94 08:15    10-MAY-94 08:50
10-MAY-94 09:00    11-MAY-94 08:00
11-MAY-94 12:30    12-MAY-94 10:20
12-MAY-94 10:35    13-MAY-94 09:30
13-MAY-94 09:55    14-MAY-94 08:50
14-MAY-94 09:10    15-MAY-94 08:00
15-MAY-94 08:15    16-MAY-94 07:00
16-MAY-94 08:00    17-MAY-94 08:00
17-MAY-94 08:20    18-MAY-94 07:20
18-MAY-94 08:05    19-MAY-94 07:55
19-MAY-94 08:30    20-MAY-94 04:15
20-MAY-94 07:55    21-MAY-94 06:45
21-MAY-94 07:00    22-MAY-94 02:45
22-MAY-94 06:05    23-MAY-94 06:30
23-MAY-94 06:50    24-MAY-94 05:50
24-MAY-94 06:05    25-MAY-94 05:10
25-MAY-94 05:25    26-MAY-94 06:00
26-MAY-94 06:10    27-MAY-94 06:05
27-MAY-94 08:30    28-MAY-94 07:30
28-MAY-94 07:45    29-MAY-94 06:50
29-MAY-94 08:25    30-MAY-94 07:10
30-MAY-94 12:30    31-MAY-94 11:30
# Bistatic radar:
# May 31, 12:21 to 17:14 (SCET)
31-MAY-94 17:50    01-JUN-94 16:50
01-JUN-94 17:00    02-JUN-94 16:00
02-JUN-94 16:20    03-JUN-94 15:15
03-JUN-94 15:30    04-JUN-94 14:30
04-JUN-94 15:05    05-JUN-94 13:30
# Bistatic radar:
# Jun 5, 12:57 to 24:20 (SCET)
06-JUN-94 01:00    06-JUN-94 16:10
06-JUN-94 20:15    07-JUN-94 17:00
08-JUN-94 01:00    09-JUN-94 00:00
09-JUN-94 00:30    10-JUN-94 00:30
10-JUN-94 01:30    11-JUN-94 01:30
11-JUN-94 02:00    12-JUN-94 02:10
12-JUN-94 03:00    13-JUN-94 03:00
13-JUN-94 03:45    14-JUN-94 03:45
14-JUN-94 04:30    15-JUN-94 04:30
15-JUN-94 05:15    16-JUN-94 04:50
16-JUN-94 07:30    17-JUN-94 04:45
17-JUN-94 06:45    18-JUN-94 04:00
18-JUN-94 08:50    18-JUN-94 20:55
19-JUN-94 09:25    20-JUN-94 04:50
20-JUN-94 09:15    21-JUN-94 04:40
21-JUN-94 08:50    22-JUN-94 03:30
22-JUN-94 09:00    22-JUN-94 13:30
22-JUN-94 15:20    23-JUN-94 15:20
23-JUN-94 16:05    24-JUN-94 16:05
# Bistatic radar:
# June 24, 17:15 to 23:45
25-JUN-94 00:35    26-JUN-94 00:35
26-JUN-94 01:25    27-JUN-94 01:25
27-JUN-94 02:10    28-JUN-94 02:10
28-JUN-94 03:20    29-JUN-94 03:00
29-JUN-94 03:50    30-JUN-94 03:50
30-JUN-94 09:00    01-JUL-94 04:35
01-JUL-94 09:45    02-JUL-94 08:35
02-JUL-94 09:05    03-JUL-94 06:15
03-JUL-94 10:00    04-JUL-94 07:00
04-JUL-94 09:20    05-JUL-94 09:20
05-JUL-94 09:50    06-JUL-94 08:30
06-JUL-94 14:00    07-JUL-94 12:20
07-JUL-94 13:00    08-JUL-94 06:30
08-JUL-94 09:20    09-JUL-94 06:30
```

```
09-JUL-94 09:50   10-JUL-94 05:30
10-JUL-94 09:40   11-JUL-94 04:30
11-JUL-94 10:15   12-JUL-94 10:10
12-JUL-94 11:00   13-JUL-94 10:55
13-JUL-94 11:45   14-JUL-94 11:50
14-JUL-94 12:20   15-JUL-94 06:30
15-JUL-94 07:00   16-JUL-94 01:05
# Bistatic radar:
# July 16, 1:00-6
16-JUL-94 09:00   17-JUL-94 04:45
17-JUL-94 08:15   18-JUL-94 04:40
18-JUL-94 09:40   18-JUL-94 21:20
19-JUL-94 10:00   20-JUL-94 09:15
20-JUL-94 13:00   21-JUL-94 12:45
21-JUL-94 13:00   22-JUL-94 12:00
22-JUL-94 12:30   23-JUL-94 12:45
23-JUL-94 13:15   24-JUL-94 13:30
24-JUL-94 14:15   25-JUL-94 03:40
25-JUL-94 10:15   26-JUL-94 04:30
26-JUL-94 10:30   26-JUL-94 21:20
27-JUL-94 04:20   28-JUL-94 01:20
28-JUL-94 09:30   29-JUL-94 08:20
29-JUL-94 16:10   30-JUL-94 18:10
30-JUL-94 19:10   31-JUL-94 20:40
01-AUG-94 11:00   02-AUG-94 11:10
02-AUG-94 12:20   03-AUG-94 12:10
03-AUG-94 13:05   04-AUG-94 12:50
04-AUG-94 13:45   05-AUG-94 07:40
05-AUG-94 10:00   06-AUG-94 07:45
06-AUG-94 10:00   07-AUG-94 07:50
07-AUG-94 10:45   08-AUG-94 08:30
08-AUG-94 10:10   09-AUG-94 12:15
# Bistatic radar:
# Aug 9, 13:30-20:30
09-AUG-94 20:50   10-AUG-94 16:05
11-AUG-94 00:15   12-AUG-94 03:45
12-AUG-94 10:15   13-AUG-94 13:45
13-AUG-94 14:10   14-AUG-94 17:45
15-AUG-94 10:20   16-AUG-94 04:00
16-AUG-94 10:20   17-AUG-94 06:15
17-AUG-94 13:25   18-AUG-94 14:40
18-AUG-94 14:55   19-AUG-94 17:00
19-AUG-94 17:15   20-AUG-94 23:15
21-AUG-94 10:20   22-AUG-94 02:30
22-AUG-94 10:30   23-AUG-94 02:30
23-AUG-94 10:30   24-AUG-94 02:30
24-AUG-94 10:00   25-AUG-94 02:30
25-AUG-94 10:00   25-AUG-94 17:30
# Lower periapse for windmill
# Aug 25, 17:32, 19:05
25-AUG-94 19:15   26-AUG-94 08:00
26-AUG-94 10:00   27-AUG-94 06:30
27-AUG-94 10:20   28-AUG-94 13:45
28-AUG-94 13:50   29-AUG-94 14:30
29-AUG-94 14:35   30-AUG-94 15:10
# Windmill Aug 30 from
# 15:35-16:04,17:09-17:37 (SCET)
30-AUG-94 18:20   31-AUG-94 18:30
31-AUG-94 20:00   01-SEP-94 05:30
01-SEP-94 17:50   02-SEP-94 07:15
02-SEP-94 10:15   03-SEP-94 08:15
03-SEP-94 10:30   04-SEP-94 11:00
04-SEP-94 17:20   05-SEP-94 18:45
05-SEP-94 18:45   06-SEP-94 14:20
# Windmill Sep 06 from
# 14:26-14:55 (SCET)
06-SEP-94 15:00   07-SEP-94 15:05
# Windmill Sep 07 from
# 15:09-15:37 (SCET)
07-SEP-94 15:45   08-SEP-94 12:45
# Windmill Sep 08 from
# 12:46-13:15,14:19-14:48 (SCET)
08-SEP-94 14:55   09-SEP-94 13:25
# Windmill Sep 09 from
# 13:29-13:57 (SCET)
09-SEP-94 14:05   10-SEP-94 13:45
10-SEP-94 13:45   11-SEP-94 14:30
11-SEP-94 14:30   12-SEP-94 15:30
# Windmill Sep 12 from
# 15:36-16:05 (SCET)
12-SEP-94 16:10   13-SEP-94 13:05
# Windmill Sep 13 from
# 13:13-15:09 (SCET)
13-SEP-94 15:15   14-SEP-94 13:45
# Windmill Sep 14 from
# 13:51-17:30 (SCET)
14-SEP-94 17:35   15-SEP-94 16:30
15-SEP-94 17:00   16-SEP-94 17:10
16-SEP-94 17:45   17-SEP-94 17:55
17-SEP-94 18:25   18-SEP-94 18:25
18-SEP-94 19:00   19-SEP-94 22:15
19-SEP-94 23:00   21-SEP-94 00:20
21-SEP-94 11:50   22-SEP-94 15:10
22-SEP-94 15:45   23-SEP-94 15:50
23-SEP-94 16:10   24-SEP-94 16:30
24-SEP-94 16:50   25-SEP-94 17:10
25-SEP-94 17:50   26-SEP-94 17:30
26-SEP-94 19:15   27-SEP-94 18:40
27-SEP-94 19:10   28-SEP-94 16:50
# Lower periapse
# 28-SEP 16:49 and 18:21
28-SEP-94 19:50   29-SEP-94 19:50
29-SEP-94 20:30   30-SEP-94 20:30
30-SEP-94 21:00   01-OCT-94 21:15
01-OCT-94 21:40   02-OCT-94 21:45
02-OCT-94 22:15   03-OCT-94 22:45
04-OCT-94 11:20   05-OCT-94 09:00
05-OCT-94 13:15   06-OCT-94 15:00
06-OCT-94 15:30   07-OCT-94 17:40
07-OCT-94 19:15   08-OCT-94 21:45
08-OCT-94 23:00   09-OCT-94 22:30
10-OCT-94 12:00   11-OCT-94 13:00
# Termination maneuvers:
# 11-OCT-94
# 14:21,15:53,17:25,18:57,22:01
```

Magellan cycle 4 hide history.

First time is hide sequence epoch:
HA = Time past periapse of first hide, HH:MM:SS
LA = Length of first hide, HH:MM:SS
HB = Time past periapse of second hide, HH:MM:SS
LB = Length of second hide, HH:MM:SS

```
14-SEP-1992 04:55    HA=01:51:00 LA=00:20:00
25-SEP-1992 16:30    HA=01:41:55 LA=00:25:00
09-OCT-1992 17:31    HA=01:51:25 LA=00:40:00
23-OCT-1992 18:42    HA=01:07:00 LA=00:10:00    HB=02:21:31 LB=00:10:00
06-NOV-1992 19:43    HA=00:47:01 LA=00:40:00    HB=01:54:28 LB=00:40:00
20-NOV-1992 17:27    HA=00:41:00 LA=01:54:00
18-DEC-1992 16:31    HA=00:41:00 LA=00:49:00    HB=01:46:32 LB=00:49:00
15-JAN-1993 17:38    HA=00:41:31 LA=00:35:00    HB=02:00:51 LB=00:35:00
07-MAY-1993 17:06    HA=00:41:00 LA=00:49:00    HB=01:46:30 LB=00:49:00
```

Magellan cycle 5&6 hides.

First time is time of first hide on the given day.
L = Length of hide, HH:MM.
N = Time between beginning of hides, HH:MM:SS.

17-AUG-93	12:25	L=0:20	N=1:35:00
18-AUG-93	13:40	L=0:20	N=1:35:00
19-AUG-93	21:10	L=0:30	N=3:09:00
20-AUG-93	12:53	L=0:30	N=3:09:00
21-AUG-93	04:40	L=0:30	N=3:09:00
22-AUG-93	05:52	L=0:30	N=3:09:00
22-AUG-93	15:18	L=0:30	N=3:09:00
23-AUG-93	22:50	L=0:30	N=3:09:00
24-AUG-93	14:35	L=0:30	N=3:09:00
25-AUG-93	22:05	L=0:30	N=3:09:00
11-NOV-93	22:30	L=1:10	N=3:09:00
12-NOV-93	20:35	L=1:10	N=3:09:00
13-NOV-93	18:39	L=1:10	N=3:09:00
14-NOV-93	19:53	L=1:10	N=3:09:00
15-NOV-93	17:58	L=1:10	N=3:09:00
16-NOV-93	16:02	L=1:10	N=3:09:00
17-NOV-93	20:26	L=1:10	N=3:09:00
18-NOV-93	15:21	L=1:10	N=3:09:00
19-NOV-93	10:16	L=1:10	N=3:09:00
19-NOV-93	22:55	L=1:10	N=3:09:00
20-NOV-93	20:59	L=1:10	N=3:09:00
21-NOV-93	15:55	L=1:10	N=3:09:00
22-NOV-93	07:41	L=1:10	N=3:09:00
23-NOV-93	08:54	L=1:10	N=3:09:00
24-NOV-93	07:02	L=1:10	N=3:09:00
25-NOV-93	08:14	L=1:10	N=3:09:00
26-NOV-93	03:11	L=1:10	N=3:09:00
26-NOV-93	22:06	L=1:10	N=3:09:00
27-NOV-93	13:53	L=1:10	N=3:09:00
28-NOV-93	08:48	L=1:10	N=3:09:00
29-NOV-93	06:53	L=1:10	N=3:09:00
29-NOV-93	22:39	L=1:10	N=3:09:00
30-NOV-93	14:26	L=1:10	N=3:09:00
01-DEC-93	09:21	L=1:10	N=3:09:00
02-DEC-93	10:36	L=1:10	N=3:09:00
03-DEC-93	05:34	L=1:02	N=3:09:00
04-DEC-93	00:29	L=1:02	N=3:09:00
04-DEC-93	19:24	L=1:02	N=3:09:00
05-DEC-93	11:11	L=1:02	N=3:09:00
06-DEC-93	09:12	L=1:02	N=3:09:00
07-DEC-93	07:17	L=1:02	N=3:09:00
08-DEC-93	05:23	L=1:02	N=3:09:00
09-DEC-93	09:47	L=1:02	N=3:09:00
10-DEC-93	14:09	L=1:02	N=3:09:00
11-DEC-93	09:06	L=1:02	N=3:09:00
12-DEC-93	13:28	L=1:02	N=3:09:00
13-DEC-93	14:41	L=1:02	N=3:09:00
14-DEC-93	09:37	L=1:02	N=3:09:00
15-DEC-93	10:51	L=1:02	N=3:09:00
16-DEC-93	08:11	L=1:10	N=3:09:00
17-DEC-93	09:23	L=1:10	N=3:09:00
18-DEC-93	13:44	L=1:10	N=3:09:00
19-DEC-93	14:59	L=1:10	N=3:09:00
20-DEC-93	16:12	L=1:10	N=3:09:00
21-DEC-93	20:35	L=1:10	N=3:09:00
22-DEC-93	18:40	L=1:10	N=3:09:00
23-DEC-93	19:54	L=1:10	N=3:09:00
25-DEC-93	19:11	L=1:10	N=3:09:00
26-DEC-93	17:16	L=1:10	N=3:09:00
27-DEC-93	15:20	L=1:10	N=3:09:00
28-DEC-93	19:42	L=1:10	N=3:09:00
29-DEC-93	17:47	L=1:10	N=3:09:00
30-DEC-93	15:50	L=1:10	N=3:09:00
01-JAN-94	08:49	L=1:10	N=3:09:00
02-JAN-94	06:54	L=1:10	N=3:09:00
03-JAN-94	11:16	L=1:10	N=3:09:00
04-JAN-94	12:27	L=1:10	N=3:09:00
05-JAN-94	23:10	L=1:10	N=3:09:00
27-JAN-94	12:01	L=0:58	N=3:09:00
28-JAN-94	16:22	L=0:58	N=3:09:00
30-JAN-94	03:01	L=0:58	N=3:09:00
31-JAN-94	10:31	L=0:58	N=3:09:00
01-FEB-94	11:43	L=0:58	N=3:09:00
02-FEB-94	12:56	L=0:58	N=3:09:00
03-FEB-94	14:06	L=0:58	N=3:09:00
04-FEB-94	12:09	L=0:58	N=3:09:00
05-FEB-94	13:21	L=0:58	N=3:09:00
07-FEB-94	00:00	L=0:58	N=3:09:00
07-FEB-94	15:45	L=0:58	N=3:09:00
08-FEB-94	13:48	L=0:58	N=3:09:00
10-FEB-94	00:27	L=0:58	N=3:09:00
11-FEB-94	01:39	L=0:58	N=3:09:00
11-FEB-94	23:42	L=0:58	N=3:09:00
12-FEB-94	21:46	L=0:58	N=3:09:00
13-FEB-94	22:59	L=0:58	N=3:09:00
15-FEB-94	00:11	L=0:58	N=3:09:00
16-FEB-94	01:24	L=0:58	N=3:09:00
17-FEB-94	12:04	L=0:58	N=3:09:00
18-FEB-94	22:43	L=0:58	N=3:09:00
19-FEB-94	17:37	L=0:58	N=3:09:00
20-FEB-94	18:49	L=0:58	N=3:09:00
21-FEB-94	20:00	L=0:58	N=3:09:00
22-FEB-94	18:03	L=0:58	N=3:09:00
23-FEB-94	19:15	L=0:58	N=3:09:00
24-FEB-94	11:00	L=0:58	N=3:09:00
25-FEB-94	12:12	L=0:58	N=3:09:00
26-FEB-94	13:23	L=0:58	N=3:09:00
27-FEB-94	14:35	L=0:58	N=3:09:00
28-FEB-94	12:38	L=0:58	N=3:09:00
01-MAR-94	13:50	L=0:58	N=3:09:00
02-MAR-94	15:02	L=0:58	N=3:09:00
03-MAR-94	16:13	L=0:58	N=3:09:00
04-MAR-94	14:17	L=0:58	N=3:09:00
05-MAR-94	15:28	L=0:58	N=3:09:00
06-MAR-94	13:30	L=0:58	N=3:09:00
07-MAR-94	11:33	L=0:58	N=3:09:00
08-MAR-94	19:03	L=0:58	N=3:09:00

```
09-MAR-94 17:06    L=0:58    N=3:09:00        20-JUN-94 12:00    L=0:25    N=3:06:00
10-MAR-94 21:14    L=0:58    N=3:08:30        21-JUN-94 09:42    L=0:25    N=3:06:00
11-MAR-94 16:04    L=0:58    N=3:08:30        22-JUN-94 10:30    L=0:25    N=3:06:00
12-MAR-94 14:03    L=0:58    N=3:08:30        22-JUN-94 17:24    L=1:05    N=3:06:00
13-MAR-94 18:20    L=0:58    N=3:08:30        23-JUN-94 18:12    L=1:05    N=3:06:00
14-MAR-94 22:37    L=0:58    N=3:08:30        25-JUN-94 02:44    L=1:05    N=3:06:00
15-MAR-94 23:45    L=0:58    N=3:08:30        26-JUN-94 03:31    L=1:05    N=3:06:00
17-MAR-94 10:16    L=0:58    N=3:08:30        27-JUN-94 04:19    L=1:05    N=3:06:00
18-MAR-94 08:16    L=0:58    N=3:08:30        28-JUN-94 05:05    L=1:05    N=3:06:00
19-MAR-94 09:24    L=0:58    N=3:08:30        29-JUN-94 05:52    L=1:05    N=3:06:00
20-MAR-94 10:32    L=0:58    N=3:08:30        30-JUN-94 09:46    L=1:05    N=3:06:00
21-MAR-94 11:40    L=0:58    N=3:08:30        01-JUL-94 10:34    L=1:05    N=3:06:00
22-MAR-94 19:05    L=0:58    N=3:08:30        02-JUL-94 11:20    L=1:05    N=3:06:00
24-MAR-94 02:28    L=0:58    N=3:08:30        03-JUL-94 12:08    L=1:05    N=3:06:00
24-MAR-94 11:40    L=0:58    N=3:08:30        04-JUL-94 09:47    L=1:05    N=3:06:00
25-MAR-94 16:10    L=0:58    N=3:08:30        05-JUL-94 10:35    L=1:05    N=3:06:00
26-MAR-94 14:09    L=0:58    N=3:08:30        06-JUL-94 14:28    L=1:05    N=3:06:00
27-MAR-94 12:06    L=0:58    N=3:08:30        07-JUL-94 15:16    L=1:05    N=3:06:00
28-MAR-94 13:14    L=0:58    N=3:08:30        08-JUL-94 09:51    L=1:05    N=3:06:00
29-MAR-94 17:30    L=0:58    N=3:08:30        09-JUL-94 10:38    L=1:05    N=3:06:00
30-MAR-94 15:30    L=0:58    N=3:08:30        10-JUL-94 11:25    L=1:05    N=3:06:00
31-MAR-94 17:15    L=0:50    N=3:08:30        11-JUL-94 12:13    L=1:05    N=3:06:00
01-APR-94 18:23    L=0:50    N=3:08:30        12-JUL-94 13:01    L=1:05    N=3:06:00
02-APR-94 19:27    L=0:50    N=3:08:30        13-JUL-94 13:48    L=1:05    N=3:06:00
03-APR-94 20:38    L=0:50    N=3:08:30        14-JUL-94 14:46    L=0:55    N=3:06:00
05-APR-94 00:55    L=0:50    N=3:08:30        15-JUL-94 09:19    L=0:55    N=3:06:00
05-APR-94 22:56    L=0:50    N=3:08:30        16-JUL-94 10:06    L=0:55    N=3:06:00
07-APR-94 00:04    L=0:50    N=3:08:30        17-JUL-94 10:53    L=0:55    N=3:06:00
07-APR-94 22:01    L=0:50    N=3:08:30        18-JUL-94 11:40    L=0:55    N=3:06:00
09-APR-94 02:17    L=0:50    N=3:08:30        19-JUL-94 12:27    L=0:55    N=3:06:00
09-APR-94 21:05    L=0:50    N=3:08:30        20-JUL-94 13:15    L=0:55    N=3:06:00
10-APR-94 18:57    L=0:50    N=3:08:30        21-JUL-94 14:02    L=0:55    N=3:06:00
12-APR-94 02:17    L=0:50    N=3:06:00        22-JUL-94 14:49    L=0:55    N=3:06:00
13-APR-94 03:10    L=0:50    N=3:06:00        23-JUL-94 15:36    L=0:55    N=3:06:00
14-APR-94 03:58    L=0:50    N=3:06:00        24-JUL-94 16:23    L=0:55    N=3:06:00
15-APR-94 17:08    L=0:50    N=3:06:00        25-JUL-94 10:58    L=0:55    N=3:06:00
16-APR-94 17:52    L=0:50    N=3:06:00        26-JUL-94 11:46    L=0:55    N=3:06:00
17-APR-94 18:40    L=0:50    N=3:06:00        27-JUL-94 06:22    L=0:55    N=3:06:00
18-APR-94 13:16    L=0:50    N=3:06:00        28-JUL-94 10:15    L=0:55    N=3:06:00
19-APR-94 14:04    L=0:50    N=3:06:00        29-JUL-94 17:14    L=0:55    N=3:06:00
20-APR-94 17:58    L=0:50    N=3:06:00        30-JUL-94 21:07    L=0:55    N=3:06:00
02-JUN-94 19:30    L=0:25    N=3:06:00        01-AUG-94 13:23    L=0:55    N=3:06:00
03-JUN-94 17:10    L=0:25    N=3:06:00        02-AUG-94 14:10    L=0:55    N=3:06:00
04-JUN-94 17:59    L=0:25    N=3:06:00        03-AUG-94 14:57    L=0:55    N=3:06:00
06-JUN-94 04:04    L=0:25    N=3:06:00        15-SEP-94 19:20    L=0:50    N=3:05:00
06-JUN-94 22:40    L=0:25    N=3:06:00        16-SEP-94 20:02    L=0:50    N=3:05:00
08-JUN-94 02:33    L=0:25    N=3:06:00        17-SEP-94 20:44    L=0:50    N=3:05:00
09-JUN-94 03:20    L=0:25    N=3:06:00        18-SEP-94 21:26    L=0:50    N=3:05:00
10-JUN-94 04:07    L=0:25    N=3:06:00        20-SEP-94 01:13    L=0:50    N=3:05:00
11-JUN-94 04:55    L=0:25    N=3:06:00        21-SEP-94 14:15    L=0:50    N=3:05:00
12-JUN-94 05:42    L=0:25    N=3:06:00        22-SEP-94 18:02    L=0:50    N=3:05:00
13-JUN-94 06:30    L=0:25    N=3:06:00        23-SEP-94 18:44    L=0:50    N=3:05:00
14-JUN-94 07:16    L=0:25    N=3:06:00        24-SEP-94 19:26    L=0:50    N=3:05:00
15-JUN-94 08:04    L=0:25    N=3:06:00        25-SEP-94 20:07    L=0:50    N=3:05:00
16-JUN-94 08:51    L=0:25    N=3:06:00        26-SEP-94 20:49    L=0:50    N=3:05:00
17-JUN-94 09:38    L=0:25    N=3:06:00        27-SEP-94 21:30    L=0:50    N=3:05:00
18-JUN-94 10:26    L=0:25    N=3:06:00        28-SEP-94 22:11    L=0:50    N=3:05:00
19-JUN-94 11:12    L=0:25    N=3:06:00        29-SEP-94 22:49    L=0:50    N=3:05:00
```

```
30-SEP-94 23:26   L=0:50   N=3:05:00
02-OCT-94 00:03   L=0:50   N=3:05:00
03-OCT-94 00:42   L=0:50   N=3:05:00
04-OCT-94 13:40   L=0:50   N=3:05:00
05-OCT-94 14:18   L=0:50   N=3:05:00
06-OCT-94 18:00   L=0:50   N=3:05:00
07-OCT-94 21:43   L=0:50   N=3:05:00
09-OCT-94 01:26   L=0:50   N=3:05:00
10-OCT-94 14:23   L=0:50   N=3:05:00
```

Appendix G

MGNP90LSAAP Gravity Coefficients to Degree and Order 40

The columns contain the following information (in order):
1. Degree n
2. Order m
3. Normalized $C_{nm} \times 10^{-10}$ ($C_{20} = -J_2$)
4. Normalized $S_{nm} \times 10^{-10}$
5. Uncertainty of normalized $C_{nm} \times 10^{-10}$
6. Uncertainty of normalized $S_{nm} \times 10^{-10}$

The reference radius to use with this gravity field is 6051 km.
The gravitational mass is 324858.601 km^3/s^2.

n	m	C_{nm}	S_{nm}	σC	σS	n	m	C_{nm}	S_{nm}	σC	σS
2	0	-19716.2	.0	7.1	.0	9	2	-2417.0	-1734.8	3.0	2.8
2	1	290.3	142.8	4.7	4.6	9	3	2586.1	139.5	2.7	2.4
2	2	8546.5	-998.9	9.4	9.0	9	4	-1578.9	604.1	2.2	2.1
3	0	7989.9	.0	3.6	.0	9	5	561.5	264.0	1.7	1.7
3	1	23479.0	5393.3	3.6	2.9	9	6	-355.2	-1913.7	1.7	1.7
3	2	-95.0	8095.7	5.1	5.5	9	7	-1915.5	1457.0	1.5	1.5
3	3	-1848.8	2126.4	8.6	8.0	9	8	947.8	1354.5	1.8	1.8
4	0	7152.1	.0	2.7	.0	9	9	2525.7	-34.2	2.7	2.7
4	1	-4587.7	4911.9	2.9	2.6	10	0	-2439.1	.0	3.9	.0
4	2	1264.4	4839.8	2.2	2.1	10	1	-1014.2	772.6	4.0	3.3
4	3	-1744.9	-1173.6	3.1	3.0	10	2	-291.7	-1359.5	3.3	3.2
4	4	1769.2	13762.3	6.5	6.3	10	3	-572.7	677.4	2.9	2.7
5	0	-1419.7	.0	2.8	.0	10	4	3410.5	-593.9	2.5	2.4
5	1	1738.0	4422.0	2.6	2.3	10	5	-2931.5	235.3	2.0	2.0
5	2	786.3	-8674.4	2.4	2.2	10	6	-2042.4	1730.4	1.7	1.7
5	3	5093.4	5547.1	1.9	2.0	10	7	-1620.2	-720.3	1.5	1.5
5	4	3925.1	2382.2	3.4	3.2	10	8	1400.5	1221.9	1.6	1.6
5	5	2549.8	-3702.2	4.6	5.0	10	9	1727.0	-826.9	1.6	1.6
6	0	322.6	.0	2.8	.0	10	10	-431.4	161.0	2.3	2.4
6	1	3939.2	-3380.9	2.6	2.4	11	0	495.4	.0	4.2	.0
6	2	1865.3	2604.9	2.3	2.1	11	1	-2304.6	-717.9	4.4	3.7
6	3	7429.5	-2139.7	2.1	1.9	11	2	1086.7	397.9	3.8	3.5
6	4	-1989.2	2597.1	2.0	1.9	11	3	1097.9	-1224.5	3.4	3.0
6	5	2560.6	-591.0	2.3	2.3	11	4	734.5	1742.7	2.8	2.8
6	6	-282.4	2158.7	4.3	4.1	11	5	-45.3	199.6	2.4	2.4
7	0	-721.5	.0	3.0	.0	11	6	-68.0	1437.2	2.0	2.0
7	1	3059.4	-2777.1	2.9	2.5	11	7	-46.4	-1133.4	1.6	1.6
7	2	1518.7	347.8	2.3	2.2	11	8	399.1	292.6	1.5	1.6
7	3	1669.2	-4501.0	2.1	2.0	11	9	67.9	827.3	1.4	1.5
7	4	-200.3	-876.9	2.1	2.0	11	10	374.7	215.3	1.5	1.5
7	5	316.6	-400.3	1.6	1.7	11	11	1285.9	-1229.9	2.2	2.2
7	6	-66.0	-2794.5	2.3	2.2	12	0	-442.5	.0	4.7	.0
7	7	-1948.6	-4835.3	3.5	3.3	12	1	79.7	-880.8	5.0	4.0
8	0	-4251.3	.0	3.3	.0	12	2	1314.0	-199.1	4.2	3.9
8	1	-203.7	-2618.0	3.2	2.8	12	3	-721.5	-586.7	3.8	3.4
8	2	1627.4	-2328.1	2.7	2.5	12	4	661.1	262.9	3.1	3.1
8	3	3060.7	-53.1	2.2	2.0	12	5	1780.7	-1943.5	2.7	2.8
8	4	-1095.8	825.8	1.9	1.8	12	6	1305.4	848.9	2.3	2.4
8	5	-356.3	-176.5	1.7	1.7	12	7	1075.0	-424.4	1.9	2.0
8	6	-602.3	-2973.3	1.8	1.7	12	8	-1062.3	167.6	1.6	1.7
8	7	-317.6	-854.7	1.9	1.9	12	9	-584.0	23.2	1.4	1.5
8	8	964.4	2957.7	3.1	3.1	12	10	1805.4	-366.0	1.4	1.5
9	0	-2353.1	.0	3.6	.0	12	11	816.9	-2326.3	1.4	1.5
9	1	-978.0	227.7	3.5	3.1	12	12	-1611.1	-470.2	1.9	1.9

13	0	-1645.3	.0	5.2	.0		17	10	108.3	785.5	2.8	2.7
13	1	-186.7	-535.9	5.6	4.5		17	11	-484.4	566.3	2.3	2.2
13	2	402.4	-1096.7	4.8	4.4		17	12	-642.6	2.3	2.0	1.9
13	3	1054.5	-582.2	4.4	3.9		17	13	419.9	515.9	1.6	1.6
13	4	-1205.7	-1091.1	3.6	3.6		17	14	561.7	-233.0	1.4	1.4
13	5	1030.6	-1227.8	3.1	3.2		17	15	466.9	-1602.1	1.3	1.3
13	6	-474.0	482.9	2.7	2.8		17	16	259.6	-262.0	1.2	1.2
13	7	-27.0	1113.6	2.2	2.3		17	17	-789.6	11.1	1.4	1.4
13	8	-17.2	383.0	1.9	2.0		18	0	-286.2	.0	8.9	.0
13	9	148.5	1742.1	1.6	1.6		18	1	162.8	44.0	10.0	7.2
13	10	326.3	-433.4	1.4	1.4		18	2	-51.4	25.3	9.0	7.6
13	11	-1122.6	-542.4	1.4	1.4		18	3	20.9	-137.7	8.4	7.1
13	12	-683.4	-255.3	1.4	1.4		18	4	364.3	-236.5	7.0	7.4
13	13	864.9	345.1	1.9	1.9		18	5	-12.2	235.2	6.6	6.7
14	0	-330.7	.0	5.8	.0		18	6	721.3	342.1	5.7	6.3
14	1	-615.2	-1314.3	6.3	4.9		18	7	80.6	722.0	5.1	5.7
14	2	588.9	-285.2	5.5	4.9		18	8	81.2	562.7	4.5	4.8
14	3	1492.3	-751.6	5.0	4.4		18	9	-344.3	406.0	3.9	4.1
14	4	283.9	-173.9	4.2	4.2		18	10	-93.8	-450.5	3.4	3.3
14	5	-152.0	24.5	3.6	3.8		18	11	247.1	190.9	2.9	2.8
14	6	140.5	-677.1	3.1	3.2		18	12	72.5	803.6	2.3	2.3
14	7	-318.8	-549.8	2.6	2.8		18	13	-117.1	369.4	2.0	2.0
14	8	60.8	-789.2	2.2	2.4		18	14	68.2	-699.3	1.6	1.6
14	9	-223.7	-684.3	1.9	2.0		18	15	130.5	-233.0	1.4	1.4
14	10	-71.8	217.3	1.6	1.6		18	16	-260.7	-141.1	1.2	1.2
14	11	-1191.1	-676.8	1.5	1.4		18	17	-70.3	-100.5	1.2	1.2
14	12	-913.3	490.2	1.4	1.3		18	18	194.0	-203.8	1.3	1.2
14	13	-426.2	-408.3	1.3	1.3		19	0	122.1	.0	9.9	.0
14	14	-426.9	-1284.0	1.6	1.6		19	1	502.4	-19.3	11.2	7.9
15	0	-712.9	.0	6.5	.0		19	2	-112.4	-472.0	10.1	8.4
15	1	245.5	652.0	7.0	5.4		19	3	-327.9	-78.3	9.5	8.0
15	2	459.9	-815.6	6.2	5.5		19	4	-502.6	-323.3	8.0	8.4
15	3	-112.2	567.9	5.7	5.0		19	5	389.1	277.5	7.4	7.8
15	4	473.0	-765.4	4.8	4.8		19	6	264.2	561.3	6.8	7.2
15	5	120.1	756.0	4.2	4.4		19	7	-185.1	-306.1	5.9	6.7
15	6	-295.3	505.6	3.6	3.9		19	8	315.2	-196.6	5.5	5.7
15	7	383.1	-174.8	3.1	3.3		19	9	-317.5	438.3	4.7	4.8
15	8	625.8	-22.3	2.6	2.8		19	10	128.7	-362.0	4.1	4.1
15	9	-690.0	225.1	2.2	2.3		19	11	610.9	-170.0	3.6	3.3
15	10	-446.1	701.0	1.9	1.9		19	12	171.8	-121.8	2.9	2.8
15	11	118.9	-38.5	1.6	1.6		19	13	304.2	-596.0	2.4	2.3
15	12	5.1	-37.5	1.4	1.4		19	14	-184.4	-148.0	2.0	2.0
15	13	485.6	-187.1	1.3	1.3		19	15	231.6	251.5	1.6	1.6
15	14	68.3	113.2	1.3	1.2		19	16	-246.0	-420.9	1.4	1.4
15	15	55.3	-70.5	1.6	1.6		19	17	-296.7	178.6	1.2	1.2
16	0	-16.6	.0	7.2	.0		19	18	394.1	189.3	1.2	1.1
16	1	-466.6	1067.5	7.9	5.9		19	19	544.1	-96.2	1.3	1.3
16	2	-319.2	797.4	7.1	6.1		20	0	264.8	.0	11.0	.0
16	3	359.5	271.1	6.5	5.7		20	1	175.8	274.3	12.7	8.7
16	4	551.1	719.7	5.5	5.6		20	2	-38.7	-242.4	11.4	9.3
16	5	-957.4	291.6	4.9	5.1		20	3	-189.2	-58.5	10.7	9.0
16	6	234.6	525.9	4.2	4.6		20	4	-92.6	-158.5	9.1	9.4
16	7	206.8	980.8	3.7	3.9		20	5	-299.6	-2.5	8.4	9.0
16	8	2.2	-144.6	3.1	3.4		20	6	22.6	294.6	7.8	8.2
16	9	-229.1	-846.2	2.7	2.8		20	7	-25.5	-222.1	7.0	7.8
16	10	103.5	144.3	2.3	2.2		20	8	-134.1	454.0	6.3	6.8
16	11	327.6	51.5	2.0	2.0		20	9	26.4	422.2	5.7	5.8
16	12	519.5	-305.5	1.6	1.6		20	10	-374.6	-140.0	4.9	4.9
16	13	845.6	690.8	1.4	1.4		20	11	-92.5	-37.6	4.3	4.1
16	14	903.4	102.0	1.3	1.3		20	12	-37.0	-34.5	3.6	3.4
16	15	115.8	-421.4	1.3	1.2		20	13	374.8	62.9	3.0	2.9
16	16	-307.0	-1174.2	1.4	1.4		20	14	-176.4	-5.1	2.4	2.3
17	0	-287.1	.0	8.0	.0		20	15	244.7	135.0	2.1	2.1
17	1	-633.4	-672.4	8.9	6.5		20	16	-39.0	44.1	1.6	1.6
17	2	146.0	796.9	8.0	6.8		20	17	-357.8	458.5	1.4	1.4
17	3	-582.5	248.4	7.4	6.3		20	18	285.7	888.3	1.2	1.1
17	4	-305.7	-345.6	6.2	6.5		20	19	106.8	-401.0	1.2	1.1
17	5	469.6	-694.8	5.7	5.8		20	20	23.3	-230.6	1.1	1.1
17	6	-25.4	607.4	4.9	5.5		21	0	270.0	.0	12.2	.0
17	7	186.0	573.9	4.4	4.7		21	1	195.9	114.6	14.1	9.5
17	8	-425.8	-10.9	3.8	4.0		21	2	118.7	-206.3	12.8	10.3
17	9	542.6	180.1	3.2	3.4		21	3	79.0	130.1	12.0	10.1

21	4	-206.8	7.3	10.3	10.6		24	7	31.3	61.1	11.9	13.0
21	5	198.9	169.9	9.5	10.2		24	8	-9.1	183.7	11.4	11.8
21	6	-119.8	534.4	8.9	9.5		24	9	-95.1	-15.3	10.6	11.1
21	7	-218.2	107.0	8.2	8.8		24	10	318.4	-58.3	10.0	9.4
21	8	144.3	134.5	7.5	8.2		24	11	194.7	117.7	8.9	8.3
21	9	539.5	102.6	6.8	6.8		24	12	-326.1	-56.5	7.6	7.2
21	10	147.2	114.6	6.0	5.9		24	13	91.8	-148.5	6.6	6.3
21	11	-244.3	107.8	5.2	4.9		24	14	.4	-159.4	5.6	5.4
21	12	5.9	2.2	4.4	4.2		24	15	-4.3	-183.7	4.6	4.6
21	13	9.5	161.0	3.7	3.6		24	16	-197.3	.7	3.8	3.9
21	14	-549.0	181.4	3.0	2.9		24	17	122.0	-80.3	3.2	3.1
21	15	-219.4	-189.9	2.4	2.4		24	18	245.3	186.2	2.6	2.4
21	16	174.0	-63.7	2.0	2.0		24	19	-113.9	-328.0	2.2	2.1
21	17	-576.0	279.1	1.7	1.7		24	20	43.0	-173.0	1.7	1.6
21	18	163.5	81.0	1.4	1.4		24	21	-438.3	195.9	1.5	1.4
21	19	674.3	-60.5	1.2	1.2		24	22	157.1	173.2	1.1	1.1
21	20	109.9	-51.0	1.1	1.1		24	23	129.5	-191.7	1.1	1.1
21	21	325.6	-345.7	1.2	1.2		24	24	-143.5	153.2	1.0	1.0
22	0	74.2	.0	13.5	.0		25	0	253.0	.0	17.8	.0
22	1	-48.3	-209.3	15.7	10.3		25	1	-136.4	-120.4	21.1	13.3
22	2	118.6	-134.6	14.3	11.4		25	2	-162.0	-257.9	19.3	14.9
22	3	-8.5	428.4	13.4	11.1		25	3	-90.8	31.8	18.2	15.0
22	4	-52.4	-208.7	11.6	11.9		25	4	-18.5	20.5	15.6	16.2
22	5	-90.9	263.5	10.9	11.3		25	5	-205.1	-81.1	15.2	15.3
22	6	-5.0	171.8	9.9	10.9		25	6	-141.9	-183.3	14.2	14.9
22	7	-290.2	-45.8	9.5	10.0		25	7	-40.6	-41.0	13.1	14.7
22	8	577.9	263.3	8.5	9.4		25	8	34.4	-46.9	12.8	13.2
22	9	-208.9	-345.6	8.1	8.1		25	9	395.4	156.4	12.0	12.5
22	10	91.7	-249.0	7.1	6.8		25	10	229.7	-101.4	11.4	11.1
22	11	-462.5	84.0	6.2	5.9		25	11	44.2	-48.2	10.5	9.5
22	12	190.5	189.9	5.4	5.0		25	12	-31.1	-36.1	9.1	8.6
22	13	63.9	364.5	4.5	4.3		25	13	172.9	270.7	7.8	7.4
22	14	-156.0	377.6	3.7	3.6		25	14	231.3	-172.6	6.7	6.5
22	15	-16.5	301.5	3.0	3.0		25	15	-257.6	-243.3	5.7	5.6
22	16	-81.4	83.6	2.4	2.4		25	16	217.8	-141.8	4.7	4.7
22	17	-302.6	-131.9	2.1	2.1		25	17	-87.8	167.7	4.0	3.9
22	18	317.7	.3	1.7	1.6		25	18	-147.6	-262.2	3.3	3.1
22	19	570.9	-422.1	1.5	1.4		25	19	-134.4	-198.0	2.6	2.6
22	20	-26.3	-362.2	1.1	1.1		25	20	-145.1	269.6	2.2	2.1
22	21	81.4	-155.2	1.1	1.1		25	21	15.4	-23.1	1.8	1.7
22	22	2.9	637.5	1.1	1.1		25	22	292.8	117.7	1.4	1.4
23	0	-157.1	.0	14.9	.0		25	23	123.4	-117.3	1.1	1.2
23	1	82.3	92.4	17.4	11.2		25	24	-251.6	-1.3	1.1	1.1
23	2	72.5	49.3	15.9	12.4		25	25	121.5	-102.6	1.0	1.0
23	3	16.0	-133.9	14.9	12.4		26	0	-17.4	.0	19.4	.0
23	4	-124.9	287.3	12.8	13.3		26	1	-264.4	242.9	23.0	14.3
23	5	-110.4	165.2	12.4	12.5		26	2	104.8	-157.4	21.1	16.2
23	6	-349.0	58.9	11.2	12.4		26	3	-.7	8.8	19.9	16.2
23	7	268.1	-144.2	10.7	11.3		26	4	-19.2	-89.7	17.1	17.8
23	8	371.9	278.9	9.9	10.7		26	5	-359.9	-403.5	16.6	16.8
23	9	-227.3	327.8	9.2	9.5		26	6	-128.1	3.2	15.7	16.2
23	10	119.2	14.6	8.6	8.1		26	7	-90.8	-43.3	14.5	16.2
23	11	163.3	128.9	7.4	7.0		26	8	342.8	59.3	14.2	14.9
23	12	-143.8	79.1	6.5	6.1		26	9	142.5	-54.3	13.5	13.7
23	13	-351.3	-113.0	5.5	5.3		26	10	-3.2	-138.0	12.7	12.7
23	14	102.6	108.7	4.5	4.5		26	11	-190.4	-267.9	12.1	11.1
23	15	231.7	-24.7	3.8	3.8		26	12	324.9	161.8	10.5	9.9
23	16	227.8	-117.7	3.1	3.1		26	13	-38.5	11.0	9.2	8.9
23	17	-93.0	24.8	2.5	2.5		26	14	25.5	-134.5	7.9	7.6
23	18	27.1	222.7	2.1	2.0		26	15	68.7	16.9	6.8	6.8
23	19	95.8	-257.6	1.7	1.7		26	16	-23.0	177.3	5.8	5.8
23	20	-34.2	-600.2	1.4	1.4		26	17	133.6	-461.4	4.8	4.8
23	21	114.0	163.0	1.2	1.2		26	18	-251.8	-134.3	4.1	3.9
23	22	-123.6	240.8	1.1	1.1		26	19	238.3	-95.1	3.4	3.2
23	23	339.1	104.0	1.1	1.1		26	20	-222.6	-120.6	2.6	2.6
24	0	33.1	.0	16.3	.0		26	21	169.2	-70.5	2.2	2.2
24	1	-214.5	-196.8	19.3	12.2		26	22	15.6	-27.8	1.7	1.7
24	2	-236.4	-240.9	17.6	13.6		26	23	308.7	356.0	1.5	1.5
24	3	51.5	-225.0	16.5	13.6		26	24	-93.4	-237.7	1.1	1.1
24	4	-168.0	231.4	14.2	14.8		26	25	-20.9	-44.8	1.1	1.1
24	5	-321.1	386.5	13.7	13.8		26	26	-167.4	92.6	1.0	1.0
24	6	-71.0	-268.6	12.6	13.7		27	0	7.8	.0	20.9	.0

27	1	-116.0	149.1	24.9	15.5	29	16	16.2	-19.7	9.5	9.5
27	2	-88.0	-21.6	22.8	17.6	29	17	-29.4	166.4	8.3	8.2
27	3	-250.0	32.1	21.6	17.7	29	18	-85.8	-52.8	7.2	7.1
27	4	102.9	-46.3	18.6	19.2	29	19	161.3	-124.2	6.2	6.1
27	5	-341.9	86.1	18.1	18.3	29	20	-204.8	144.7	5.1	5.0
27	6	-247.9	90.1	17.0	17.7	29	21	27.0	213.5	4.3	4.3
27	7	139.8	-119.7	16.0	17.5	29	22	-65.9	-77.8	3.4	3.5
27	8	128.1	-110.2	15.4	16.7	29	23	71.5	85.9	2.8	2.9
27	9	-14.3	-202.6	15.1	15.0	29	24	-13.4	115.7	2.3	2.3
27	10	-165.7	57.5	14.1	14.2	29	25	21.9	-83.3	1.8	1.9
27	11	210.5	208.1	13.6	12.6	29	26	186.1	79.9	1.5	1.5
27	12	-41.3	140.7	12.3	11.4	29	27	175.2	114.6	1.2	1.2
27	13	-54.8	-31.1	10.6	10.2	29	28	-179.0	-17.3	1.1	1.1
27	14	-149.2	-88.9	9.4	9.0	29	29	202.2	-71.8	.9	.9
27	15	221.0	148.0	8.0	7.9	30	0	85.7	.0	25.4	.0
27	16	42.1	169.0	6.9	6.9	30	1	5.9	34.5	29.9	19.3
27	17	57.7	-14.1	5.9	5.9	30	2	-97.6	-158.3	27.4	21.9
27	18	174.1	83.9	5.0	4.8	30	3	58.5	109.0	26.2	21.7
27	19	346.9	80.9	4.2	4.1	30	4	-53.3	-72.0	22.6	23.9
27	20	-109.7	-166.5	3.3	3.3	30	5	156.5	-133.2	22.4	21.8
27	21	150.5	59.9	2.7	2.7	30	6	10.7	-108.1	21.0	21.5
27	22	23.2	-8.0	2.2	2.2	30	7	-260.5	-.8	19.6	21.2
27	23	206.2	115.9	1.8	1.8	30	8	1.6	12.4	19.0	20.2
27	24	199.3	-191.2	1.5	1.5	30	9	-84.9	-65.3	18.3	19.8
27	25	-383.1	112.4	1.1	1.2	30	10	-27.6	-30.2	18.5	17.9
27	26	91.7	-78.3	1.1	1.1	30	11	-253.5	-112.5	17.5	16.8
27	27	-81.1	-102.8	1.0	1.0	30	12	56.8	51.1	16.5	15.8
28	0	157.9	.0	22.4	.0	30	13	-6.6	87.0	15.4	14.5
28	1	62.1	-51.8	26.8	16.7	30	14	50.5	-1.5	13.8	13.5
28	2	-21.8	-32.3	24.4	18.9	30	15	77.4	-188.4	12.5	12.1
28	3	-57.7	-92.9	23.2	19.2	30	16	-86.7	127.3	10.8	10.8
28	4	-144.2	-206.8	20.0	20.8	30	17	112.6	76.5	9.7	9.6
28	5	-22.6	26.2	19.7	19.5	30	18	-113.7	-154.4	8.4	8.3
28	6	-96.1	88.4	18.4	19.2	30	19	71.9	-69.9	7.3	7.2
28	7	76.8	-153.4	17.4	18.6	30	20	-25.0	158.8	6.2	6.2
28	8	64.8	-40.3	16.5	18.3	30	21	-64.7	124.3	5.2	5.2
28	9	-206.6	-226.2	16.5	16.6	30	22	-106.9	-28.4	4.3	4.4
28	10	154.1	166.9	15.7	15.4	30	23	-25.7	86.7	3.5	3.6
28	11	178.3	245.1	14.8	14.3	30	24	39.0	38.0	2.8	2.9
28	12	-3.9	13.4	13.9	12.8	30	25	192.2	51.5	2.3	2.4
28	13	48.0	44.7	12.3	11.8	30	26	97.5	144.9	1.8	1.8
28	14	21.1	257.5	10.8	10.4	30	27	75.1	-176.5	1.6	1.6
28	15	-81.1	-73.3	9.5	9.3	30	28	88.1	17.5	1.1	1.1
28	16	-45.7	-92.1	8.1	8.1	30	29	68.8	119.6	1.1	1.1
28	17	-119.8	-48.8	7.0	7.0	30	30	67.1	-159.6	.9	.9
28	18	7.6	40.9	6.1	5.9	31	0	57.5	.0	26.6	.0
28	19	-17.2	-140.6	5.1	4.9	31	1	160.2	-57.0	31.1	20.7
28	20	162.1	-122.7	4.2	4.1	31	2	-31.5	-45.7	28.6	23.3
28	21	-34.7	-134.3	3.4	3.4	31	3	-14.8	-10.9	27.4	23.0
28	22	84.5	54.4	2.7	2.8	31	4	39.0	-20.2	23.6	25.2
28	23	-13.3	-23.2	2.3	2.3	31	5	15.8	-119.8	23.7	22.9
28	24	-11.0	29.8	1.7	1.8	31	6	-139.3	-89.8	22.0	22.3
28	25	-360.6	-50.7	1.5	1.5	31	7	-166.8	-107.7	20.6	22.4
28	26	20.7	271.3	1.1	1.1	31	8	-41.4	-18.9	20.0	21.0
28	27	186.8	-206.6	1.1	1.1	31	9	-42.8	-92.2	19.2	20.6
28	28	-34.0	54.9	.9	.9	31	10	-113.1	-40.1	19.3	19.3
29	0	183.4	.0	23.9	.0	31	11	100.0	-50.8	18.8	17.7
29	1	95.2	80.6	28.5	18.0	31	12	15.6	9.3	17.4	17.0
29	2	-231.6	-200.9	26.0	20.4	31	13	-80.0	-5.6	16.6	15.7
29	3	-13.3	151.5	24.8	20.5	31	14	-56.2	-147.8	15.2	14.8
29	4	-142.9	-54.2	21.4	22.3	31	15	-178.2	225.2	13.9	13.5
29	5	16.6	40.2	21.1	20.7	31	16	-150.5	-42.3	12.4	12.3
29	6	-96.5	24.1	19.8	20.5	31	17	-15.8	-69.1	10.9	10.9
29	7	-116.5	-14.0	18.6	19.9	31	18	-25.1	4.1	9.8	9.7
29	8	-29.7	-9.0	17.8	19.4	31	19	-168.6	15.7	8.5	8.4
29	9	-86.6	204.6	17.5	18.3	31	20	46.1	-69.4	7.4	7.2
29	10	180.2	-62.3	17.3	16.6	31	21	-26.3	111.4	6.3	6.3
29	11	-93.4	-52.2	16.1	15.7	31	22	-144.2	118.9	5.2	5.3
29	12	77.6	-127.5	15.4	14.3	31	23	9.7	4.2	4.4	4.5
29	13	72.4	185.4	13.9	13.2	31	24	101.3	-87.8	3.5	3.6
29	14	82.5	69.0	12.4	12.0	31	25	85.7	13.8	2.9	3.0
29	15	16.2	49.0	10.9	10.6	31	26	50.2	-34.8	2.4	2.4

31	27	-8.5	-76.8	1.9	1.9
31	28	71.9	-105.6	1.6	1.5
31	29	101.6	-31.4	1.2	1.2
31	30	-146.7	152.9	1.1	1.1
31	31	75.3	-166.2	.9	.9
32	0	-135.8	.0	27.8	.0
32	1	144.7	94.3	32.2	22.1
32	2	-1.5	10.9	29.4	24.7
32	3	-92.9	-67.4	28.4	24.3
32	4	81.9	105.2	24.4	26.3
32	5	48.4	-31.9	24.8	23.8
32	6	75.9	-43.3	23.0	23.0
32	7	-54.8	5.7	21.2	23.0
32	8	-98.0	-163.6	20.8	22.0
32	9	-87.4	-86.9	20.0	21.1
32	10	-37.9	-87.3	19.6	20.3
32	11	-62.1	183.5	19.8	18.7
32	12	40.1	-63.6	18.5	17.9
32	13	93.4	-14.7	17.5	16.9
32	14	82.0	-43.5	16.4	15.8
32	15	32.6	140.4	15.1	14.8
32	16	-34.3	28.1	13.8	13.5
32	17	102.4	-26.8	12.3	12.3
32	18	-3.3	146.8	10.9	10.8
32	19	-27.7	130.5	9.8	9.7
32	20	25.4	42.5	8.5	8.4
32	21	156.0	65.8	7.4	7.4
32	22	-126.0	17.3	6.3	6.4
32	23	54.6	7.3	5.3	5.3
32	24	110.3	4.0	4.4	4.5
32	25	-158.4	-59.8	3.7	3.7
32	26	61.0	78.2	3.0	2.9
32	27	85.4	-82.8	2.5	2.4
32	28	-146.1	-59.0	1.9	1.8
32	29	-75.7	23.8	1.6	1.6
32	30	-95.4	119.6	1.2	1.2
32	31	170.4	224.8	1.1	1.1
32	32	-289.3	-7.3	.9	.9
33	0	10.5	.0	28.7	.0
33	1	139.2	86.0	32.9	23.4
33	2	26.7	10.5	29.8	26.2
33	3	-49.2	52.3	29.3	25.2
33	4	-32.0	25.8	24.9	27.3
33	5	144.0	38.8	25.6	24.4
33	6	114.0	-7.5	23.9	23.6
33	7	-103.0	-55.0	21.7	23.4
33	8	-127.6	-20.6	21.2	22.6
33	9	-68.9	-65.6	20.6	21.5
33	10	49.3	-186.7	20.0	20.8
33	11	5.6	-14.7	20.1	19.5
33	12	-45.0	-74.7	19.4	18.5
33	13	52.9	35.6	18.1	17.8
33	14	28.5	162.6	17.4	16.6
33	15	164.6	9.6	16.1	15.8
33	16	113.4	-12.4	14.9	14.6
33	17	-125.1	6.8	13.5	13.5
33	18	-114.5	35.4	12.3	12.2
33	19	49.0	5.3	10.9	10.8
33	20	-85.2	-14.0	9.8	9.7
33	21	-111.4	173.6	8.6	8.6
33	22	-4.6	-69.5	7.4	7.5
33	23	57.3	-107.8	6.5	6.5
33	24	36.9	31.6	5.4	5.3
33	25	-134.2	-29.4	4.6	4.5
33	26	-55.1	-17.3	3.7	3.6
33	27	144.4	-159.3	3.1	2.9
33	28	-152.1	-145.9	2.5	2.4
33	29	-53.5	200.5	2.0	1.9
33	30	71.9	105.6	1.6	1.6
33	31	34.6	229.3	1.2	1.2
33	32	-45.4	-10.8	1.1	1.1
33	33	20.0	116.3	.9	.9
34	0	-132.9	.0	29.6	.0
34	1	32.0	-7.2	33.3	24.8
34	2	98.5	4.3	30.0	27.7
34	3	-38.6	10.9	29.9	26.1
34	4	-30.4	92.0	25.3	28.3
34	5	54.8	112.8	26.2	24.8
34	6	8.9	-1.3	24.6	23.9
34	7	-3.5	23.2	22.1	23.8
34	8	-23.9	-15.3	21.5	22.8
34	9	-26.0	150.7	20.9	21.9
34	10	84.8	-40.2	20.4	20.9
34	11	93.2	-145.1	20.1	20.0
34	12	-66.5	-6.7	19.9	19.0
34	13	-46.5	104.3	18.7	18.4
34	14	-62.6	119.1	18.0	17.3
34	15	34.8	-108.6	16.9	16.4
34	16	21.4	-56.2	15.8	15.6
34	17	-12.8	40.6	14.7	14.5
34	18	63.9	-32.3	13.4	13.4
34	19	4.8	-55.4	12.3	12.2
34	20	-68.1	82.9	10.9	10.9
34	21	-16.6	-85.6	9.8	9.8
34	22	102.8	-125.7	8.5	8.7
34	23	-143.7	51.3	7.5	7.6
34	24	-33.8	-34.6	6.5	6.4
34	25	-31.1	41.2	5.5	5.4
34	26	-130.8	151.1	4.7	4.5
34	27	-145.6	23.0	3.9	3.6
34	28	-82.6	-22.1	3.1	2.9
34	29	-45.0	69.0	2.6	2.5
34	30	127.0	-53.8	2.0	1.9
34	31	63.7	127.6	1.7	1.7
34	32	-91.2	88.6	1.2	1.2
34	33	-9.3	106.9	1.1	1.1
34	34	32.3	117.1	.9	.9
35	0	-129.3	.0	30.2	.0
35	1	-24.4	82.5	33.5	26.2
35	2	61.1	6.4	30.1	29.1
35	3	-20.4	79.0	30.3	26.9
35	4	-57.1	-56.7	25.6	29.1
35	5	-.9	36.7	26.8	25.1
35	6	131.0	-43.7	25.0	24.1
35	7	21.5	24.7	22.4	24.1
35	8	47.5	136.9	21.6	22.7
35	9	-42.7	79.1	20.9	22.1
35	10	43.3	-31.5	20.7	20.9
35	11	12.3	-88.1	20.1	20.1
35	12	132.8	-58.3	19.9	19.3
35	13	-26.0	23.9	19.0	18.7
35	14	149.2	-110.3	18.3	17.9
35	15	114.5	-1.9	17.6	16.9
35	16	-82.1	90.3	16.4	16.2
35	17	-51.5	-70.4	15.5	15.3
35	18	72.2	-52.4	14.4	14.4
35	19	73.1	14.5	13.3	13.2
35	20	-59.3	28.3	12.1	12.1
35	21	-13.8	-84.7	10.8	10.9
35	22	-72.4	65.3	9.7	9.9
35	23	-53.7	-63.8	8.6	8.8
35	24	62.9	-5.5	7.6	7.6
35	25	-27.8	10.7	6.6	6.6
35	26	-111.5	122.2	5.7	5.5
35	27	-162.3	-60.3	4.9	4.6
35	28	-46.8	-38.8	4.0	3.8
35	29	64.8	128.8	3.2	3.2
35	30	162.0	-96.6	2.6	2.6
35	31	103.2	-91.9	2.0	2.0
35	32	33.9	33.9	1.7	1.7
35	33	-13.1	38.1	1.2	1.2
35	34	-110.7	-133.3	1.1	1.1
35	35	21.7	42.7	.9	.9
36	0	44.6	.0	31.0	.0

36	1	57.2	-13.8	33.4	27.7	37	36	114.5	-26.0	1.2	1.1
36	2	57.7	11.9	30.0	30.3	37	37	-17.2	-20.1	.9	.9
36	3	-68.8	-167.1	30.6	27.7	38	0	-49.0	.0	32.1	.0
36	4	-103.1	-45.3	25.8	29.7	38	1	-69.8	39.5	33.3	30.1
36	5	43.2	-81.6	27.3	25.3	38	2	45.9	58.3	29.5	32.7
36	6	19.2	21.2	25.4	24.2	38	3	8.0	23.4	31.3	28.7
36	7	16.0	20.6	22.5	24.3	38	4	-82.4	-62.9	26.1	30.9
36	8	99.6	130.2	21.7	22.6	38	5	-24.1	-70.6	28.1	25.8
36	9	-74.4	34.9	20.8	22.0	38	6	6.6	15.3	26.0	24.4
36	10	12.5	-9.9	20.7	20.9	38	7	34.5	-14.6	22.6	24.7
36	11	-61.2	76.1	20.1	19.9	38	8	-75.2	-30.0	22.1	22.4
36	12	44.5	-110.4	19.6	19.4	38	9	-79.1	3.4	20.8	21.5
36	13	27.5	-51.5	19.1	18.8	38	10	11.8	14.1	20.3	20.5
36	14	73.7	-140.3	18.4	18.1	38	11	42.8	-59.9	20.1	19.5
36	15	-25.9	-47.3	17.9	17.2	38	12	-8.7	53.5	19.1	19.2
36	16	1.5	56.3	17.0	16.6	38	13	49.4	89.5	18.6	18.7
36	17	54.1	-65.2	16.0	15.8	38	14	-44.0	159.5	18.3	18.1
36	18	17.2	59.6	15.2	15.1	38	15	8.1	77.9	18.0	17.3
36	19	-75.0	-30.1	14.1	14.2	38	16	-13.2	20.4	17.4	16.9
36	20	38.3	6.7	13.1	13.1	38	17	1.7	104.6	16.6	16.4
36	21	69.0	9.9	12.0	12.1	38	18	8.3	2.7	16.0	15.9
36	22	-122.2	38.6	10.7	10.9	38	19	-7.4	-.9	15.2	15.3
36	23	-15.3	41.8	9.8	9.9	38	20	78.7	-28.2	14.6	14.6
36	24	59.1	.4	8.7	8.8	38	21	-114.1	-23.4	13.7	13.8
36	25	11.3	63.8	7.7	7.8	38	22	-6.7	45.5	12.7	12.9
36	26	-27.6	4.2	6.8	6.6	38	23	7.0	-123.6	11.9	11.9
36	27	8.7	86.6	5.9	5.5	38	24	29.0	1.1	10.8	10.8
36	28	44.7	19.5	5.0	4.7	38	25	-13.0	46.3	9.8	10.0
36	29	48.2	60.5	4.1	4.0	38	26	27.7	71.7	8.9	8.8
36	30	37.2	77.8	3.3	3.2	38	27	15.2	-45.5	8.0	7.7
36	31	2.1	-33.7	2.7	2.7	38	28	.5	75.8	7.0	6.8
36	32	50.2	6.9	2.0	2.1	38	29	161.7	-40.5	6.0	5.9
36	33	95.5	15.7	1.7	1.8	38	30	-88.3	-22.5	5.1	5.1
36	34	-47.3	-172.2	1.2	1.3	38	31	-71.4	-3.7	4.2	4.2
36	35	-1.3	133.8	1.1	1.2	38	32	22.9	-74.7	3.3	3.5
36	36	-85.6	-11.5	.9	.9	38	33	-159.6	32.5	2.8	2.9
37	0	-97.5	.0	31.5	.0	38	34	-7.5	119.6	2.1	2.2
37	1	3.2	12.6	33.4	29.0	38	35	16.0	-48.1	1.8	1.8
37	2	5.8	-73.1	29.7	31.6	38	36	15.0	-16.4	1.3	1.3
37	3	61.1	-15.2	30.9	28.2	38	37	-37.8	93.0	1.2	1.2
37	4	-143.5	-18.3	26.0	30.4	38	38	-80.5	91.5	.9	.9
37	5	-17.7	46.7	27.7	25.4	39	0	-32.2	.0	32.5	.0
37	6	45.6	8.2	25.7	24.3	39	1	-69.0	-11.8	33.3	31.0
37	7	95.1	27.2	22.6	24.4	39	2	74.2	-21.7	29.3	33.7
37	8	-51.4	1.9	21.9	22.6	39	3	16.9	12.1	31.7	29.2
37	9	-37.9	-24.3	20.8	21.7	39	4	74.6	-70.4	26.5	31.3
37	10	19.2	59.2	20.4	20.7	39	5	-7.6	-56.5	28.4	26.1
37	11	-46.5	-84.3	20.2	19.7	39	6	-12.7	-107.0	26.3	24.6
37	12	23.1	77.3	19.3	19.3	39	7	126.9	-43.8	22.7	25.0
37	13	46.0	-40.0	18.9	18.8	39	8	-11.3	106.4	22.4	22.4
37	14	54.2	37.0	18.4	18.2	39	9	-5.7	121.3	20.8	21.5
37	15	-2.8	58.5	18.0	17.4	39	10	-94.6	63.3	20.3	20.4
37	16	34.9	-56.3	17.3	16.8	39	11	.0	71.8	20.0	19.4
37	17	17.7	63.9	16.4	16.2	39	12	12.9	56.3	19.1	19.1
37	18	-144.6	23.3	15.6	15.5	39	13	-57.3	85.6	18.4	18.7
37	19	28.6	-24.4	14.8	14.9	39	14	-27.5	-37.4	18.2	18.0
37	20	15.4	39.2	14.0	13.9	39	15	-8.2	87.9	17.9	17.3
37	21	-36.1	-.4	12.9	13.0	39	16	91.0	93.9	17.4	16.8
37	22	43.8	-60.3	11.9	12.1	39	17	-2.3	70.8	16.7	16.6
37	23	-.3	33.0	10.8	10.9	39	18	72.5	-9.5	16.1	16.1
37	24	107.9	-83.3	9.8	9.9	39	19	-58.9	-24.3	15.5	15.6
37	25	-55.8	48.9	8.8	8.9	39	20	-26.6	-36.1	14.9	15.0
37	26	77.3	-23.1	7.8	7.7	39	21	1.3	-34.3	14.3	14.4
37	27	121.5	65.7	7.0	6.7	39	22	-13.2	47.4	13.4	13.7
37	28	50.5	-8.3	5.9	5.7	39	23	-79.7	-48.1	12.7	12.7
37	29	18.1	-154.8	5.0	5.0	39	24	-63.5	31.1	11.8	11.8
37	30	89.1	-52.6	4.1	4.1	39	25	-22.2	18.2	10.8	10.9
37	31	-4.0	86.6	3.4	3.4	39	26	17.7	-33.3	9.9	9.9
37	32	-17.1	39.4	2.7	2.8	39	27	-91.4	-119.1	9.1	8.8
37	33	43.0	-.9	2.1	2.2	39	28	49.0	-12.7	8.0	7.8
37	34	-91.1	-2.4	1.7	1.8	39	29	69.5	25.7	7.0	7.0
37	35	-131.9	64.7	1.3	1.3	39	30	-64.5	-4.9	6.0	6.0

39	31	-27.3	47.2	5.2	5.2
39	32	-39.4	-39.0	4.2	4.3
39	33	-30.2	18.6	3.5	3.6
39	34	74.9	68.1	2.8	2.9
39	35	67.5	.4	2.2	2.3
39	36	-12.6	55.2	1.8	1.8
39	37	72.1	53.1	1.3	1.3
39	38	69.8	-70.9	1.2	1.2
39	39	-28.6	76.2	.9	.9
40	0	-28.9	.0	32.9	.0
40	1	-1.8	-43.9	33.5	31.7
40	2	-51.1	-36.2	29.2	34.5
40	3	30.4	-6.3	32.0	29.6
40	4	134.9	-.8	27.0	31.5
40	5	72.0	57.7	28.9	26.4
40	6	43.6	-21.0	26.7	25.0
40	7	82.9	-5.6	22.9	25.4
40	8	66.3	15.9	22.7	22.5
40	9	5.6	65.2	21.2	21.5
40	10	-64.0	21.6	20.3	20.4
40	11	-18.1	-21.8	19.9	19.4
40	12	-43.9	-10.4	19.2	19.1
40	13	26.2	33.8	18.3	18.7
40	14	8.2	18.5	18.1	18.0
40	15	90.0	85.2	17.9	17.4
40	16	72.4	13.0	17.4	16.7
40	17	76.2	-79.3	16.8	16.6
40	18	84.7	-17.5	16.2	16.3
40	19	6.7	-52.0	15.7	15.7
40	20	-4.8	-28.1	15.3	15.3
40	21	-52.2	-46.2	14.7	14.7
40	22	-36.9	-18.5	14.0	14.3
40	23	-47.3	55.3	13.4	13.3
40	24	17.0	38.0	12.6	12.5
40	25	63.2	131.0	11.7	11.8
40	26	-90.9	-56.9	10.8	10.8
40	27	-43.2	-22.3	10.1	9.8
40	28	-43.3	-39.6	9.1	8.9
40	29	-30.6	17.5	8.0	8.0
40	30	-56.9	-12.2	7.1	7.2
40	31	23.4	40.8	6.1	6.2
40	32	55.5	96.3	5.2	5.4
40	33	-39.2	44.2	4.3	4.5
40	34	-30.3	-65.7	3.5	3.6
40	35	57.1	-11.7	2.9	3.0
40	36	-34.6	-78.8	2.3	2.2
40	37	75.7	30.8	1.9	1.9
40	38	54.8	-59.6	1.3	1.3
40	39	-50.9	28.3	1.2	1.2
40	40	37.5	51.0	.9	.9

Appendix H
Correlations Between Estimated Parameters

The correlations between the nongravity global parameters and the first 5th degree and order coefficients are given in this appendix.

The following names are for the orientation of Venus:

ZACPL2 = Venus pole right ascension in Earth-Mean-Equator of J2000
ZDEPL2 = Venus pole declination in Earth-Mean-Equator of J2000
WDP2 = Venus rotation rate

The following are the Venus (2) and Earth-Moon barycenter (B) Set III parameters from Brouwer and Clemence (1969):

DMW = Δ longitude + Δ rotation about z-axis (ecliptic north, heliocentric)
EDW = eccentricity * Δ rotaton about z-axis
DA = Δ semi-major axis / semi-major axis
DE = Δ eccentricity
DP = Δ rotation about x-axis (p and q give the inclination)
DQ = Δ rotation about y-axis

And the rest of the parameters are:

2K2_2 = Love number of Venus (k_2)
GM2 = GM of Venus
J20n = Zonal coefficient of degree n
C20n0m = C_{nm}
S20n0m = S_{nm}

	ZACPL2	ZDEPL2	WDF2	DW2	EDW2	DA2	DE2	DP2	DQ2	DWB	EDWB
ZACPL2	1.0										
ZDEPL2	1.0E-01	1.0									
WDF2	-3.8E-02	7.5E-02	1.0								
DW2	6.1E-03	-2.9E-03	7.2E-04	1.0							
EDW2	1.7E-04	-6.1E-03	-8.0E-03	-2.2E-01	1.0						
DA2	-8.3E-04	-4.5E-03	2.9E-03	8.0E-03	-3.8E-01	1.0					
DE2	2.0E-03	-4.4E-03	-3.2E-03	8.6E-02	-1.1E-01	1.4E-01	1.0				
DP2	-8.0E-03	-2.4E-03	8.4E-04	-2.3E-01	-1.5E-01	6.2E-02	-1.3E-01	1.0			
DQ2	3.0E-02	1.0E-02	5.2E-03	-6.3E-02	-3.5E-02	-6.5E-03	-8.5E-03	8.8E-02	1.0		
DWB	6.8E-03	-2.9E-03	1.1E-03	1.0	-2.4E-01	8.1E-01	9.2E-02	-1.7E-01	-2.7E-02	1.0	
EDWB	1.0E-02	-2.2E-03	-9.5E-03	-6.6E-02	6.4E-01	-3.8E-01	5.2E-01	-3.8E-01	2.1E-02	-8.2E-02	1.0

	DAB	DEB	DPB	DQB	2K2_2	GM2	J202	C20201	S20201	C20202	S20202
ZACPL2	-9.3E-04	6.6E-04	-2.3E-02	2.2E-02	-3.9E-02	1.1E-02	-1.4E-02	-3.3E-03	1.1E-02	6.5E-02	3.5E-02
ZDEPL2	-4.7E-03	3.6E-03	-7.0E-03	7.3E-03	4.5E-02	-6.3E-03	-3.4E-02	7.2E-02	4.7E-03	-1.1E-02	4.4E-03
WDF2	3.0E-03	2.0E-03	-1.4E-03	5.4E-03	3.2E-02	1.8E-02	-1.8E-02	7.8E-02	1.6E-02	-2.2E-02	1.6E-02
DW2	8.1E-01	2.5E-01	-1.7E-01	-1.6E-01	1.9E-02	-5.9E-03	-7.5E-03	-1.6E-02	2.9E-02	-2.0E-02	1.5E-02
EDW2	-3.9E-01	-8.2E-01	-1.6E-01	-1.3E-01	-1.4E-02	2.9E-03	1.7E-02	-4.1E-01	-1.1E-02	1.5E-02	-1.3E-02
DA2	1.0	2.6E-01	1.1E-01	-1.7E-02	4.1E-03	4.6E-03	-4.7E-04	3.2E-03	4.9E-03	-1.4E-02	1.1E-02
DE2	1.5E-01	5.3E-01	-1.2E-01	-1.2E-02	1.0E-02	-5.8E-03	5.3E-03	1.0E-03	-5.5E-02	-2.5E-02	-1.3E-02
DP2	6.9E-02	-1.4E-01	8.6E-01	5.5E-01	-3.7E-03	1.0E-02	2.6E-03	-5.5E-03	-9.1E-03	-6.7E-03	1.0E-02
DQ2	-7.3E-02	5.3E-02	-4.3E-01	8.8E-01	2.1E-02	-3.6E-04	6.6E-03	1.5E-02	-2.6E-02	-2.0E-02	-5.8E-03
DWB	8.2E-01	2.5E-01	-1.3E-01	-9.4E-02	2.0E-02	-2.4E-03	-6.6E-03	1.3E-04	2.8E-02	-2.1E-02	1.6E-02
EDWB	-3.7E-01	-2.1E-01	-3.9E-01	-1.4E-01	1.4E-02	-5.9E-03	1.4E-05	4.8E-03	3.4E-02	-3.1E-02	-7.6E-03
DAB	1.0	2.7E-01	1.2E-01	-2.0E-02	4.4E-03	-5.7E-03	-1.0E-02	4.6E-02	5.4E-03	-1.4E-02	1.2E-02
DEB		1.0	-1.3E-01	1.9E-02	2.0E-02	-2.1E-02	-2.0E-02	-3.6E-02	-3.9E-02	-2.9E-02	-1.8E-02
DPB			1.0	4.9E-02	-1.5E-02	6.3E-03	6.7E-03	1.3E-02	6.2E-03	-4.4E-03	1.2E-02
DQB				1.0	1.7E-02	-2.1E-03	5.9E-02	2.3E-01	2.6E-02	-4.1E-02	-1.3E-03
2K2_2					1.0	1.9E-02	5.9E-02	-6.2E-04	7.4E-02	-2.7E-01	7.4E-02
GM2						1.0	1.0	1.7E-01	-1.2E-02	-3.2E-03	-3.4E-02
J202								1.0	1.6E-02	-4.7E-02	1.8E-01
C20201									2.5E-01	-8.3E-02	4.5E-01
S20201									1.0		2.0E-01
C20202										1.0	-1.5E-01
S20202											1.0

	J203	C20301	S20301	C20302	S20302	C20303	S20303	J204	C20401	S20401	C20402
ZACPL2	4.4E-03	6.8E-03	-1.8E-02	6.9E-02	3.3E-02	1.9E-02	-2.3E-03	-4.2E-03	8.9E-03	4.8E-02	5.1E-02
ZDEPL2	-4.9E-03	-5.9E-02	6.3E-05	2.3E-02	4.3E-03	4.2E-03	-8.1E-03	2.0E-02	-1.3E-02	1.8E-02	6.4E-03
WDF2	-5.3E-02	-5.1E-02	1.2E-01	-5.8E-02	-3.6E-02	-3.6E-02	-3.3E-02	-5.7E-02	-1.5E-02	9.5E-03	-1.1E-01
DM2	1.6E-02	3.0E-03	3.0E-03	1.1E-02	-1.4E-02	1.6E-02	9.9E-03	-1.2E-02	2.6E-03	5.9E-03	-2.9E-03
EDM2	-1.3E-02	1.2E-02	-4.1E-03	-1.4E-02	1.9E-02	-1.6E-02	7.3E-03	1.5E-02	2.3E-04	-9.0E-04	1.0E-02
DA2	2.3E-03	7.2E-03	5.9E-03	7.2E-03	-3.2E-03	6.3E-03	3.0E-02	-3.7E-03	-1.0E-02	-4.6E-03	-8.1E-04
DE2	-3.6E-03	-6.3E-03	-6.7E-03	8.6E-03	2.0E-02	1.6E-03	1.2E-02	1.4E-02	-6.1E-03	-2.3E-03	2.5E-03
DP2	1.0E-02	2.5E-03	9.3E-03	4.9E-03	-2.9E-03	-3.1E-04	-1.3E-02	2.3E-03	5.6E-03	2.9E-03	-7.6E-03
DQ2	-2.6E-03	3.6E-03	-8.2E-03	-2.6E-02	-2.9E-03	6.8E-03	2.5E-02	4.8E-02	8.0E-03	-5.8E-03	2.7E-03
DWB	1.6E-02	3.2E-02	3.4E-03	1.1E-02	-1.4E-02	1.7E-02	1.0E-02	-1.2E-02	2.8E-03	5.8E-03	-2.9E-03
EDWB	1.2E-02	1.2E-02	-5.4E-03	-1.3E-02	6.4E-03	-8.8E-03	1.3E-02	2.3E-03	1.5E-02	1.2E-02	9.2E-03
DAB	1.3E-03	7.0E-03	6.2E-03	6.9E-03	-3.2E-03	6.5E-03	3.0E-03	-3.2E-03	-1.1E-02	-4.7E-03	-6.3E-04
DEB	6.2E-03	-1.9E-02	-6.5E-03	1.4E-02	2.2E-03	2.0E-02	6.9E-03	1.3E-02	-5.9E-03	5.3E-04	-7.8E-03
DPB	1.2E-02	-7.7E-05	1.2E-02	7.4E-03	-6.8E-04	-4.5E-03	-2.6E-02	2.4E-03	-2.4E-03	5.3E-03	-8.5E-03
DQB	2.2E-03	4.2E-03	-3.2E-03	-1.6E-03	-1.6E-03	4.9E-03	1.6E-02	4.7E-03	9.2E-03	-2.1E-03	-5.8E-04
2K2_2	4.0E-02	8.6E-02	2.2E-03	1.7E-01	-5.6E-03	1.3E-01	-9.8E-02	-4.9E-02	1.4E-01	3.1E-02	2.0E-01
GM2	-7.6E-03	4.0E-03	-9.2E-03	5.2E-03	-6.2E-03	-1.6E-02	-6.2E-03	5.2E-02	-5.5E-02	3.1E-01	9.3E-03
J202	-2.3E-01	3.3E-01	1.7E-01	7.7E-02	4.6E-02	7.3E-02	-3.1E-01	3.2E-01	-7.1E-02	1.3E-01	6.3E-02
C20201	-4.4E-01	5.8E-02	-4.7E-02	-1.9E-01	-2.3E-01	-1.4E-01	6.3E-02	1.9E-02	-1.2E-01	-1.5E-01	-1.9E-02
S20201	1.4E-01	1.4E-01	4.8E-02	9.2E-02	-4.2E-01	-1.4E-01	-2.1E-02	-6.5E-02	-7.3E-02	-1.8E-01	-1.6E-01
C20202	-3.7E-03	2.1E-01	-8.9E-02	-3.3E-01	-1.7E-01	-3.5E-01	2.0E-01	-3.4E-02	-9.8E-03	-1.2E-01	6.7E-02
S20202	-3.9E-02	1.5E-01	3.1E-01	1.2E-01	-1.7E-01	2.5E-01	-2.7E-01	-2.4E-02	8.1E-02	-5.6E-02	1.3E-01
J203	1.0	-4.3E-02	1.4E-01	1.6E-01	2.0E-01	6.1E-01	-2.0E-02	-1.7E-01	6.3E-01	4.2E-01	6.9E-02
C20301		1.0	2.8E-01	-5.3E-02	-9.8E-02	-9.3E-02	4.0E-02	-3.0E-02	4.1E-02	-5.2E-02	-1.9E-01
S20301			1.0	6.4E-02	5.2E-02	1.4E-01	1.5E-04	-1.1E-02	9.3E-02	-1.4E-02	1.2E-01
C20302				1.0	-2.6E-02	1.3E-01	-2.5E-01	4.6E-02	1.5E-01	7.8E-02	-1.4E-02
S20302					1.0	4.1E-01	6.7E-02	6.0E-02	5.0E-02	3.3E-01	1.8E-01
C20303						1.0	2.7E-02	4.2E-02	1.1E-02	1.4E-01	1.5E-01
S20303							1.0	-1.3E-02	-3.0E-02	-1.9E-02	-4.1E-02
J204								1.0	-2.0E-01	2.7E-02	1.3E-01
C20401									1.0	3.5E-01	1.2E-01
S20401										1.0	5.5E-02
C20402											1.0

	S20402	C20403	S20403	C20404	S20404	J205	C20501	S20501	C20502	S20502	C20503
ZACFL2	-4.3E-02	-5.2E-03	-6.5E-04	1.2E-01	2.5E-02	-2.1E-02	6.5E-03	-3.9E-02	-3.3E-02	-4.2E-02	1.3E-01
ZDEFL2	-4.7E-02	3.4E-02	-1.2E-04	-1.7E-02	7.8E-03	3.4E-02	-7.7E-03	-4.9E-02	3.1E-02	-3.3E-02	7.5E-02
WDF2	-5.1E-02	1.0E-02	-4.5E-02	-8.9E-02	5.4E-02	-3.0E-02	-4.5E-02	-9.7E-02	9.0E-02	-4.3E-02	-7.4E-03
DW2	4.3E-03	-8.4E-03	1.7E-02	-9.5E-03	2.2E-02	1.4E-02	-2.1E-03	2.5E-03	1.0E-02	-8.2E-03	1.3E-02
EDW2	-6.6E-03	-6.0E-03	8.8E-03	4.2E-03	-3.7E-03	-1.4E-02	1.3E-02	-5.4E-03	-1.5E-02	1.3E-02	-1.9E-02
DA2	-6.9E-04	-2.1E-03	3.8E-03	-6.2E-03	4.4E-03	1.3E-02	2.4E-03	1.1E-03	1.4E-02	1.9E-04	2.1E-04
DE2	-2.1E-03	-7.2E-03	-3.4E-02	-4.7E-03	-6.5E-03	6.4E-03	3.3E-03	-8.2E-04	1.5E-02	9.1E-03	-1.2E-02
DP2	-9.2E-04	-4.1E-03	1.2E-02	-4.8E-04	-7.2E-03	-2.5E-02	1.9E-03	3.2E-03	2.6E-03	-4.7E-03	-3.5E-02
DQ2	-6.9E-03	-3.8E-04	-3.3E-03	-1.1E-02	1.8E-02	-1.2E-02	2.3E-03	-1.2E-02	-5.4E-03	-1.9E-04	5.2E-03
DWB	4.1E-03	-7.8E-03	1.6E-02	-9.7E-03	2.2E-02	1.4E-02	-1.6E-03	2.7E-03	1.0E-02	-8.4E-03	1.3E-02
EDWB	-3.6E-03	-2.5E-02	-5.1E-03	1.5E-03	1.8E-02	-2.8E-03	9.5E-03	-2.9E-03	-1.0E-02	-2.4E-03	-1.2E-02
DAB	-5.6E-04	6.1E-03	3.1E-03	-5.9E-03	4.0E-03	1.3E-02	2.7E-03	1.7E-03	1.4E-02	4.7E-04	-5.0E-05
DEB	8.3E-03	7.2E-03	-1.5E-02	-1.3E-02	-1.7E-03	8.0E-03	-8.1E-04	8.1E-04	2.6E-02	-1.0E-03	1.4E-02
DPB	2.6E-03	-8.0E-03	2.7E-03	4.7E-03	-1.5E-03	5.8E-03	-6.4E-04	8.4E-03	4.0E-03	-4.9E-03	-5.4E-03
DQB	-6.1E-03	1.6E-02	-5.1E-03	-8.3E-03	1.0E-02	-1.0E-02	3.5E-03	-8.3E-03	-2.8E-03	-2.8E-03	1.7E-03
ZK2_2	7.9E-04	-1.5E-01	-1.3E-01	-2.3E-01	3.1E-01	-3.7E-02	-3.4E-02	-5.9E-02	-2.1E-02	-6.5E-02	-9.2E-02
QM2	4.6E-03	-2.2E-02	-8.2E-03	1.8E-02	-2.0E-02	1.2E-02	1.7E-02	1.0E-02	7.2E-03	-9.7E-03	-8.9E-03
J202	-2.2E-01	1.4E-02	-1.3E-01	2.0E-03	-1.3E-02	6.4E-02	2.9E-01	-2.2E-02	-7.0E-02	4.1E-04	-1.3E-01
C20201	5.9E-02	1.8E-01	2.7E-02	1.2E-01	1.6E-02	-1.0E-01	-1.7E-02	-3.8E-02	-1.1E-01	5.6E-02	-8.8E-02
S20201	1.1E-01	-9.3E-02	1.3E-01	9.4E-03	7.3E-02	-3.8E-02	1.6E-02	7.0E-03	-2.0E-02	-5.0E-02	-3.8E-02
C20202	-9.9E-02	2.0E-01	2.4E-01	1.9E-01	-7.6E-02	-2.9E-02	1.2E-01	-6.0E-02	-2.1E-01	-1.6E-02	2.7E-02
S20202	3.6E-02	-1.9E-02	-9.3E-02	3.3E-02	8.1E-02	-2.7E-02	6.0E-02	1.1E-02	-5.9E-02	-1.2E-01	-6.1E-02
J203	-8.6E-02	-4.2E-02	1.1E-02	-8.1E-02	1.1E-02	-1.2E-01	-4.7E-02	3.8E-02	3.2E-02	-3.1E-02	4.8E-02
C20301	-2.6E-01	-1.1E-02	1.3E-02	2.7E-02	9.0E-02	-7.5E-02	3.0E-01	-1.6E-02	-1.2E-01	-8.4E-02	-6.9E-02
S20301	-4.4E-01	4.5E-02	-8.5E-02	-1.8E-03	3.1E-02	-9.5E-03	1.6E-01	1.9E-01	9.9E-03	6.7E-02	-1.5E-01
C20302	-1.5E-01	-4.5E-01	-1.7E-01	-7.9E-02	1.1E-02	1.2E-01	-3.2E-02	4.1E-02	2.6E-01	-1.8E-01	1.3E-01
S20302	-9.4E-02	1.4E-01	-4.1E-01	-1.5E-01	-1.7E-01	1.6E-02	5.5E-02	-6.6E-02	1.0E-01	2.0E-01	1.5E-02
C20303	6.7E-03	-8.1E-02	-4.5E-01	-3.9E-01	-3.9E-02	3.0E-02	2.7E-04	9.6E-03	9.2E-02	3.0E-02	2.9E-01
S20303	-2.5E-02	3.5E-02	3.7E-02	-9.9E-02	-2.4E-01	-2.4E-02	8.8E-02	-2.9E-02	-5.7E-02	7.6E-02	6.8E-02
J204	1.3E-01	2.2E-02	-6.5E-02	1.6E-02	-1.1E-01	1.3E-01	5.1E-02	3.5E-02	1.1E-02	5.4E-02	-6.4E-02
C20401	-8.8E-02	-5.6E-02	1.3E-02	-3.4E-01	9.9E-02	-5.4E-01	-1.5E-02	-6.6E-02	-4.4E-01	-3.7E-01	-2.5E-02
S20401	2.5E-02	1.0E-02	-1.1E-02	8.2E-02	-2.5E-02	-2.3E-01	4.2E-02	-2.3E-01	1.6E-01	-5.3E-01	8.3E-02
C20402	-8.6E-02	5.5E-02	1.3E-01	-8.2E-02	6.2E-02	6.9E-02	7.0E-02	1.1E-02	-2.7E-01	1.7E-02	-3.6E-02
S20402	1.0	1.8E-02	-9.3E-03	-2.1E-02	5.2E-02	-8.0E-02	-9.9E-02	1.1E-01	1.4E-02	1.8E-01	-1.4E-02
C20403		1.0	7.6E-03	-5.3E-03	-2.5E-01	-4.5E-02	8.6E-02	-1.4E-02	-3.0E-02	2.5E-02	-1.2E-01
S20403			1.0	3.4E-01	9.1E-02	-3.5E-02	-2.5E-02	3.5E-02	-1.2E-01	6.9E-02	3.1E-01
C20404				1.0	-6.0E-02	1.2E-02	1.1E-02	1.1E-01	-7.7E-02	6.9E-02	-2.7E-03
S20404					1.0	-3.6E-03	-3.0E-02	2.7E-02	-6.5E-02	-9.2E-02	-1.4E-02
J205						1.0	8.0E-02	1.1E-01	2.3E-01	3.3E-01	2.2E-02
C20501							1.0	2.7E-01	1.1E-01	1.8E-02	-9.7E-02
S20501								1.0	3.6E-02	1.5E-01	-4.8E-02
C20502									1.0	1.9E-02	1.7E-01
S20502										1.0	1.4E-02
C20503											1.0

	S20503	C20504	S20504	C20505	S20505
ZACFL2	-3.5E-02	2.0E-02	-3.1E-02	-2.8E-02	-5.8E-02
ZDEFL2	-4.1E-02	-6.3E-03	1.6E-02	1.7E-02	-3.8E-03
WDF2	9.3E-02	-5.0E-02	-3.8E-03	3.4E-02	-2.1E-02
DW2	8.8E-03	1.2E-02	-9.2E-03	-9.1E-03	-9.5E-04
EW2	3.4E-03	-1.3E-02	1.8E-02	9.2E-04	4.0E-03
DX2	5.9E-03	3.3E-04	-7.4E-04	-5.3E-03	-5.4E-04
DE2	-5.3E-03	-1.3E-02	4.0E-02	1.3E-02	1.3E-02
DP2	-2.0E-03	2.1E-03	-3.4E-03	3.0E-03	1.5E-04
DQ2	7.4E-03	9.6E-03	1.8E-02	-1.1E-02	1.6E-02
DWB	9.0E-03	1.2E-02	-8.5E-03	-9.2E-03	1.4E-04
EWB	5.3E-03	-5.4E-03	2.6E-02	2.3E-03	1.7E-03
DAB	5.8E-03	-1.3E-04	-6.9E-04	-5.1E-03	1.9E-05
DEB	-5.9E-03	4.5E-03	7.7E-03	4.2E-03	9.8E-03
DPB	-5.8E-03	-2.6E-03	-1.3E-02	8.5E-03	-9.0E-03
DQB	5.5E-03	7.2E-03	1.6E-02	-7.6E-03	1.4E-02
ZK2_2	-8.8E-02	1.9E-02	9.4E-02	1.4E-01	-3.3E-03
GM2	-2.3E-02	3.2E-02	5.7E-02	2.0E-02	-6.0E-02
J202	1.0E-01	-2.3E-02	6.3E-02	6.8E-02	7.7E-02
C20201	-7.8E-03	-2.3E-02	-4.6E-02	-2.8E-02	4.6E-02
S20201	6.7E-02	2.1E-02	1.6E-02	-1.9E-02	8.5E-02
C20202	1.6E-01	-8.7E-02	-1.8E-01	-2.0E-01	-4.7E-02
S20202	-6.5E-02	4.2E-02	7.1E-02	7.7E-02	8.4E-02
J203	1.3E-02	8.4E-02	-1.3E-02	4.8E-04	-7.3E-02
C20301	2.7E-01	-1.2E-02	-4.4E-02	-1.9E-03	3.0E-02
S20301	1.3E-02	-2.1E-02	-4.5E-02	1.2E-01	4.9E-02
C20302	-9.9E-03	2.5E-01	6.7E-02	9.0E-02	-1.1E-01
S20302	-3.3E-02	-1.1E-02	1.6E-01	1.3E-01	-5.5E-02
C20303	-1.5E-02	5.2E-02	3.5E-01	2.8E-01	1.3E-02
S20303	4.2E-01	-2.8E-01	4.4E-02	-2.8E-01	3.3E-01
J204	-2.1E-02	-6.5E-02	6.3E-02	2.8E-02	5.6E-02
C20401	-7.1E-02	4.8E-02	-3.9E-02	2.3E-02	-6.6E-03
S20401	3.5E-02	4.4E-02	9.5E-02	5.5E-02	-1.1E-02
C20402	-1.1E-01	1.7E-02	7.8E-02	8.3E-02	-2.5E-02
S20402	-4.1E-01	-3.6E-02	5.1E-02	-2.3E-02	2.3E-02
C20403	3.2E-03	-5.0E-01	-2.7E-02	-1.5E-01	2.0E-01
S20403	-4.3E-02	-6.3E-02	-4.7E-01	-3.3E-01	-4.8E-02
C20404	-5.8E-02	-1.0E-01	-4.4E-01	-3.1E-01	-1.6E-01
S20404	-1.1E-01	4.0E-01	-1.0E-01	-2.8E-01	-3.3E-01
J205	4.0E-03	5.5E-02	-1.2E-02	2.1E-02	-4.5E-02
C20501	1.3E-01	-7.8E-02	1.4E-02	-4.3E-03	4.2E-02
S20501	-9.8E-02	3.9E-02	-7.5E-02	2.0E-02	-1.9E-02
C20502	1.6E-02	4.2E-02	1.3E-01	3.6E-02	-4.9E-02
S20502	1.2E-02	-1.3E-01	1.5E-03	3.3E-03	-1.3E-02
C20503	2.0E-02	1.2E-02	2.9E-02	4.2E-02	-1.2E-01
S20503	1.0	-4.3E-02	7.0E-02	-1.1E-01	2.0E-01
C20504		1.0	1.2E-03	1.6E-01	-3.7E-01
S20504			1.0	3.0E-01	1.4E-01
C20505				1.0	-7.2E-02
S20505					1.0

Appendix I
Regional Gravity Maps for MGNP90LSAAP

The following vertical gravity plots in milligal contours are included in this appendix:

1. Eistla Regio
2. Ishtar Terra
3. Bell Regio
4. Beta Regio
5. Atla Regio
6. Aphrodite Terra
7. Atalanta Planitia

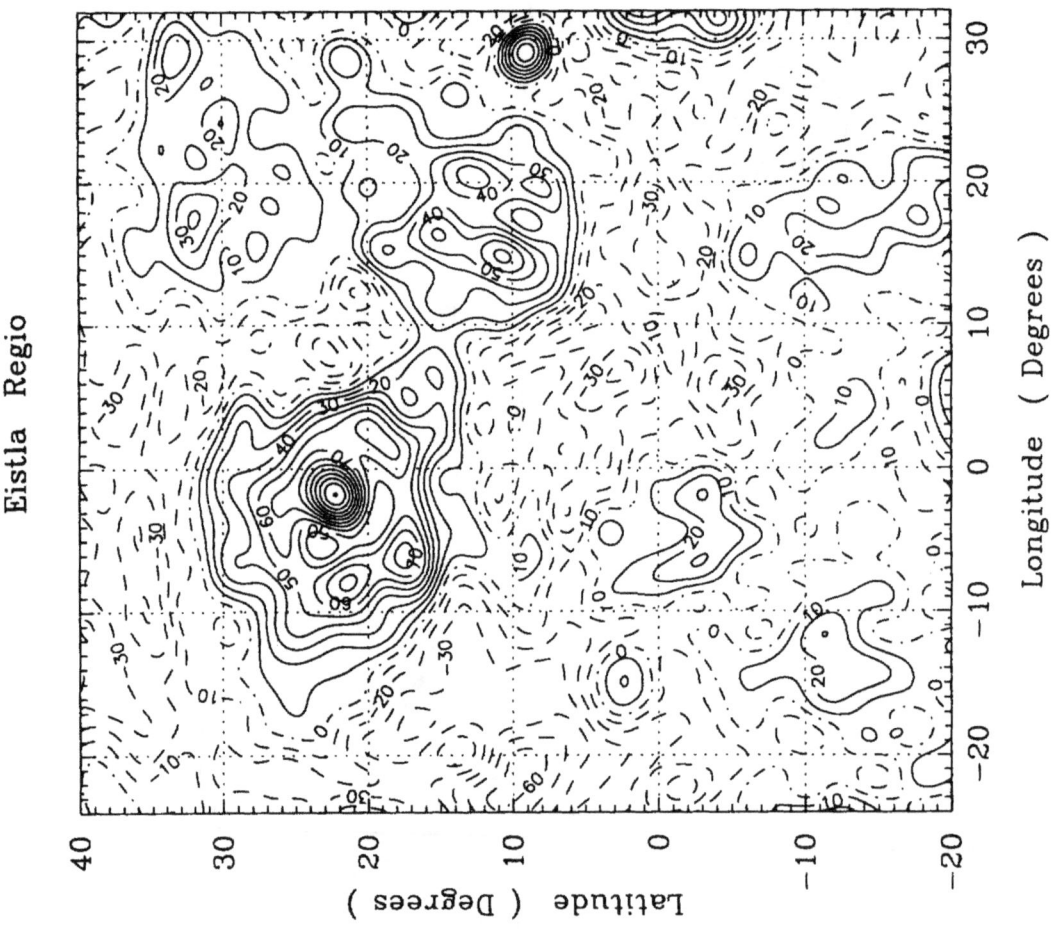

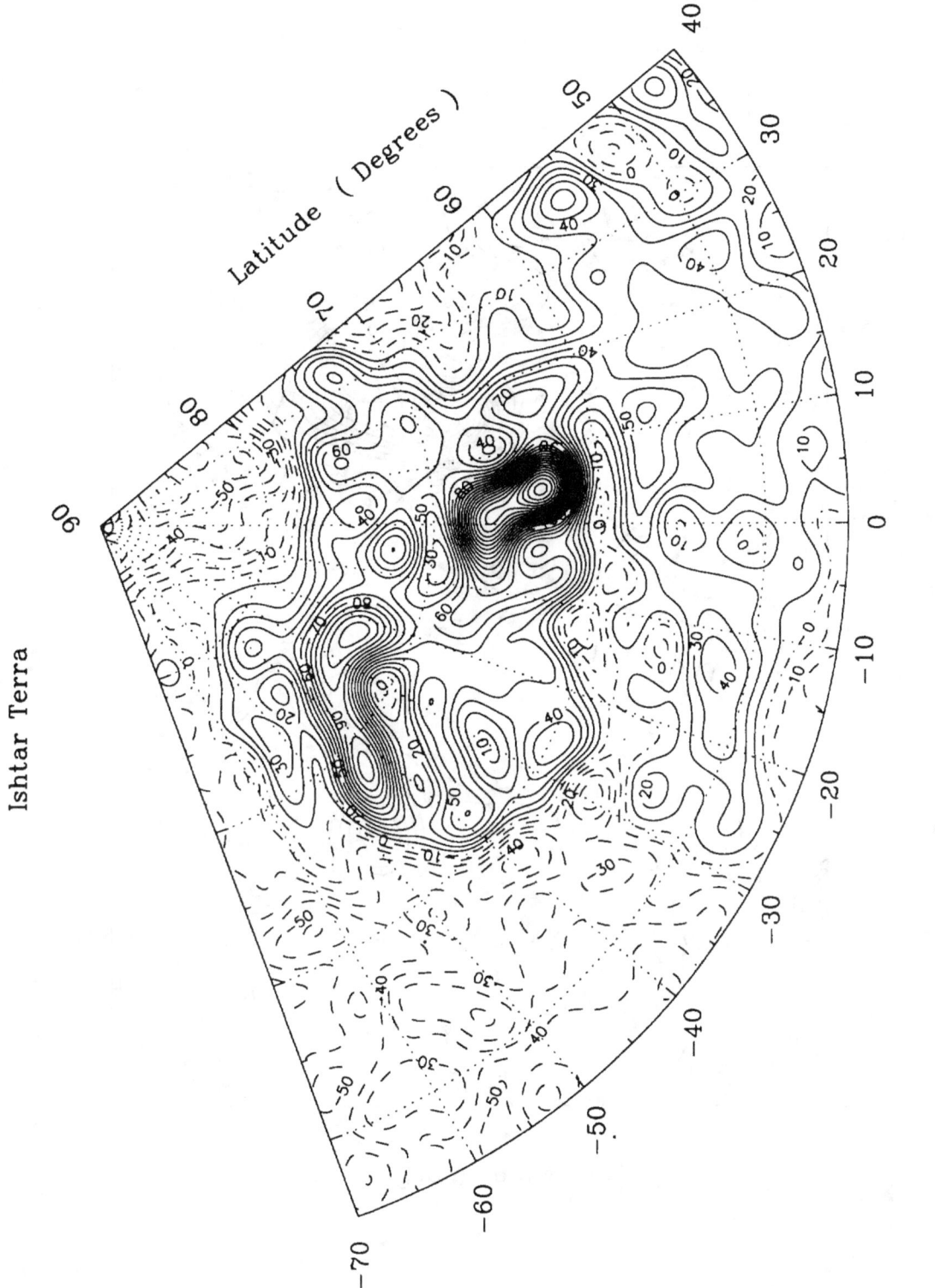

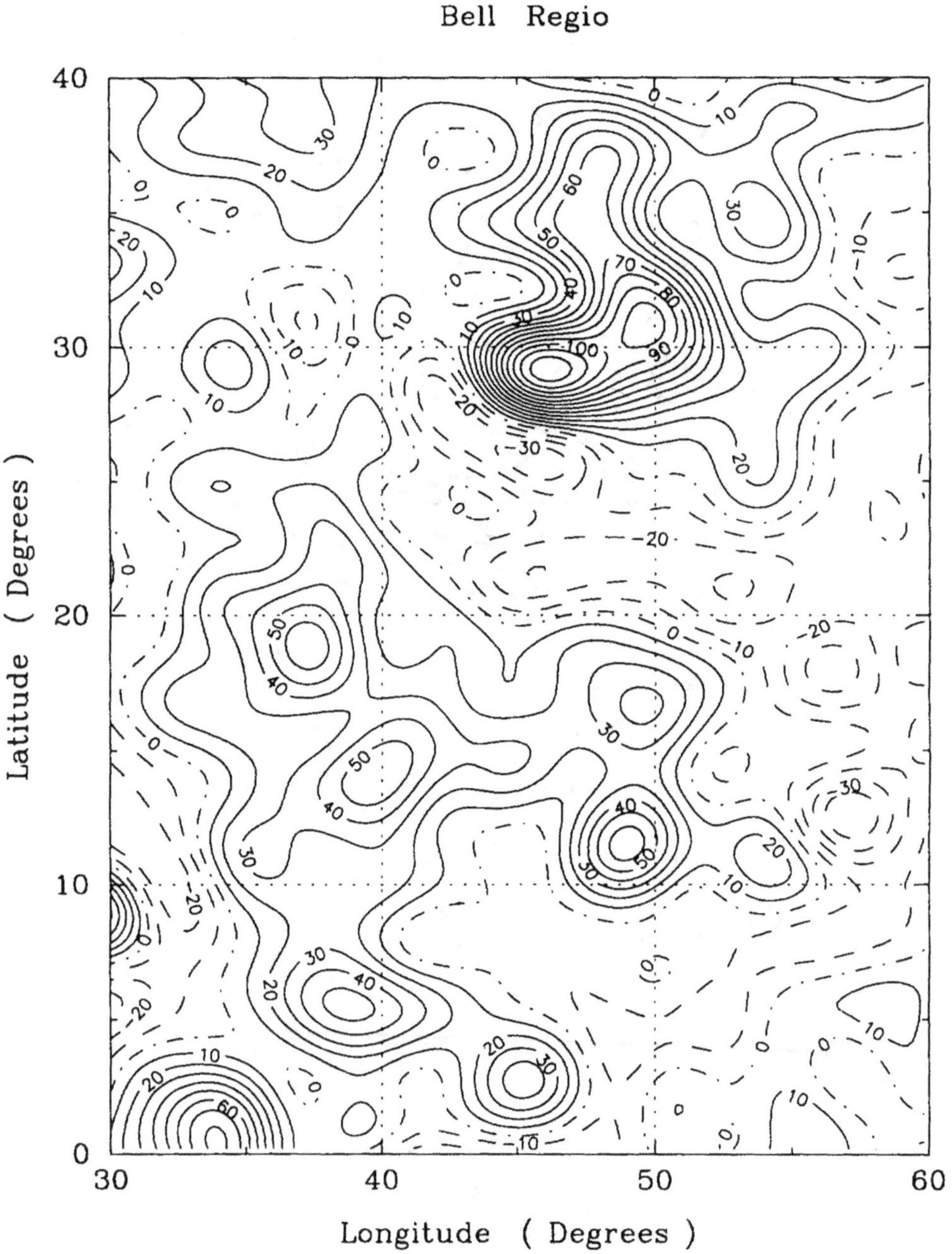

Bell Regio
Latitude (Degrees)
Longitude (Degrees)
0
10
20
30
40
30
40
50
60

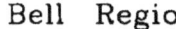

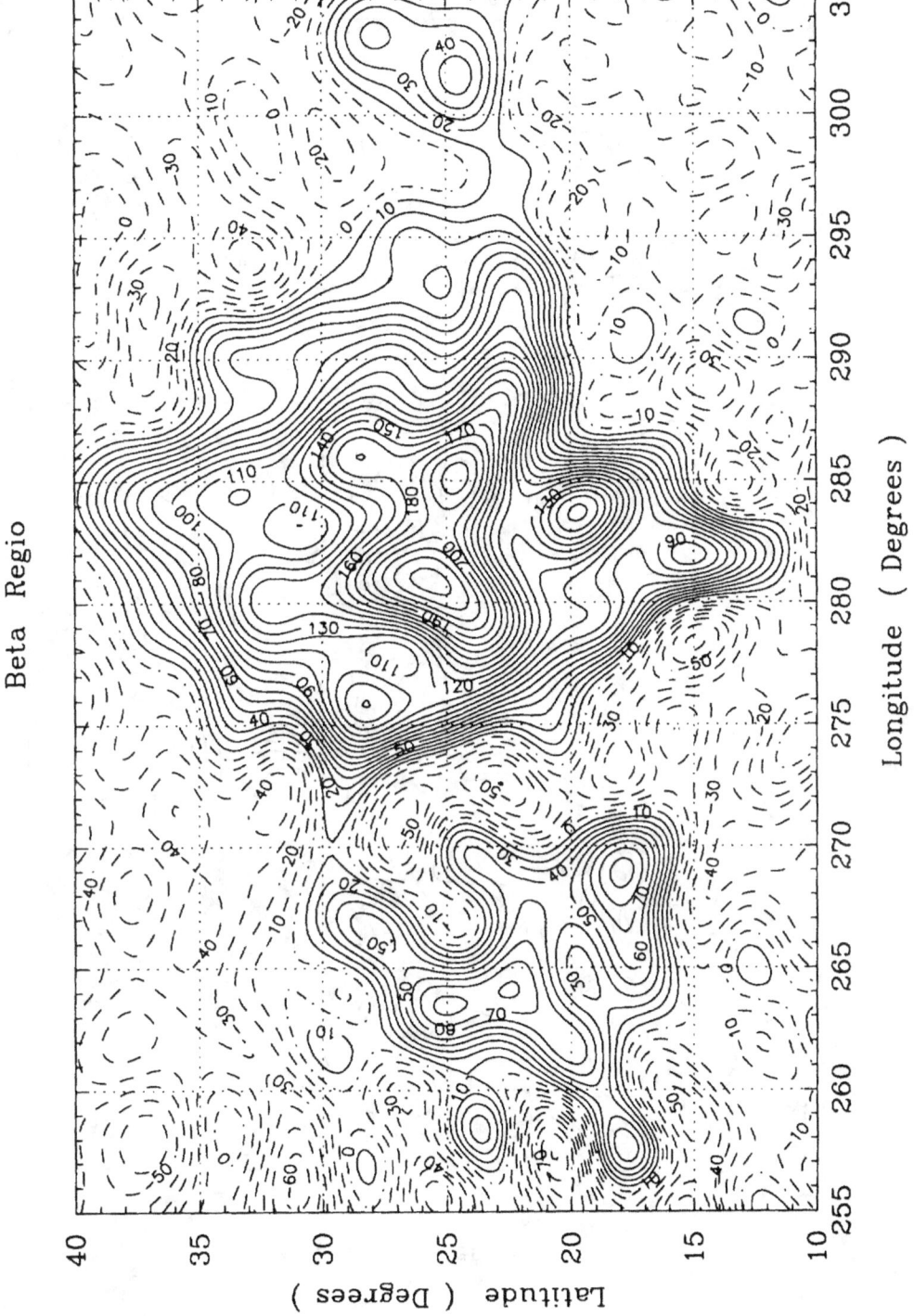

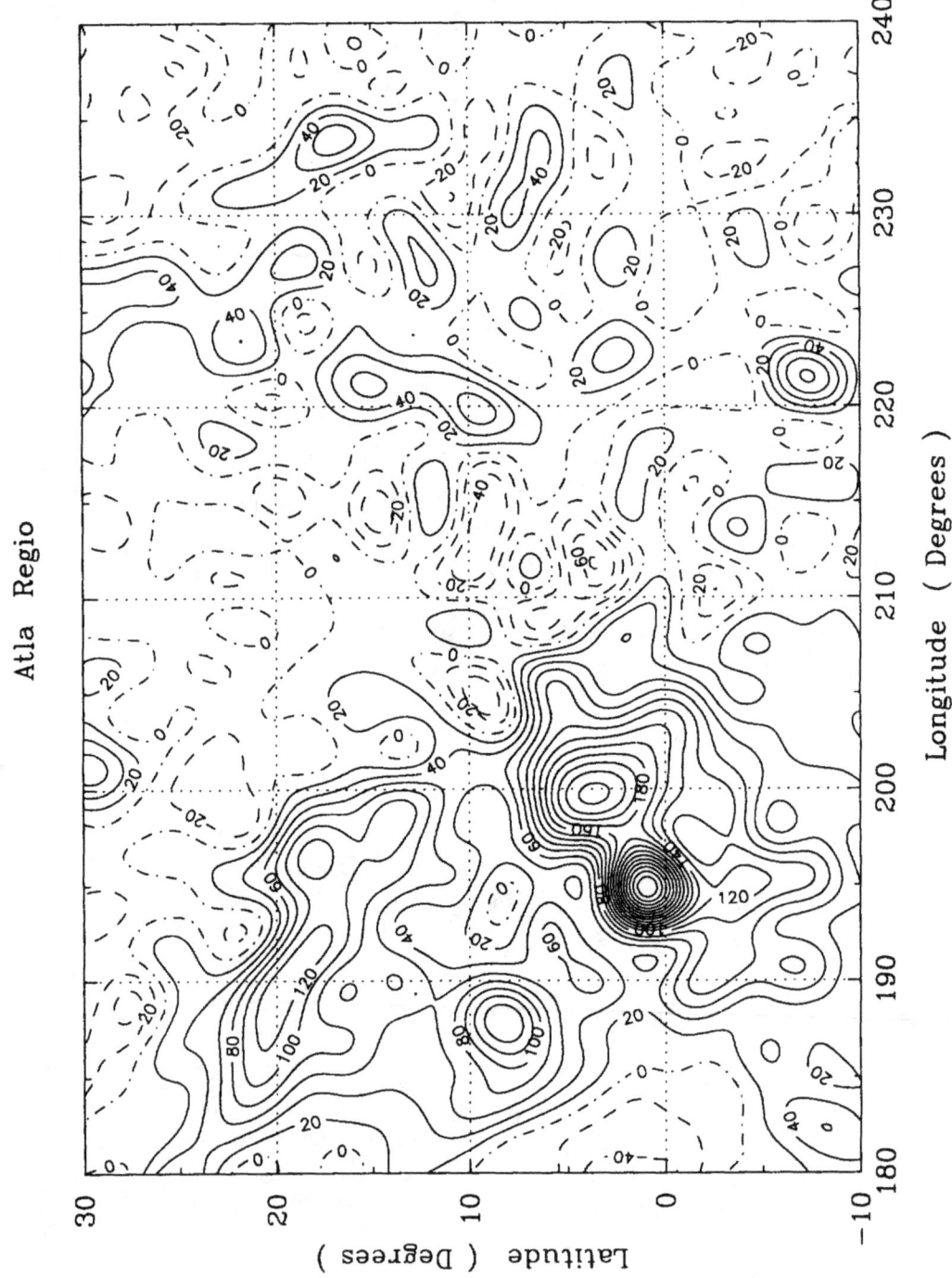

Atla Regio

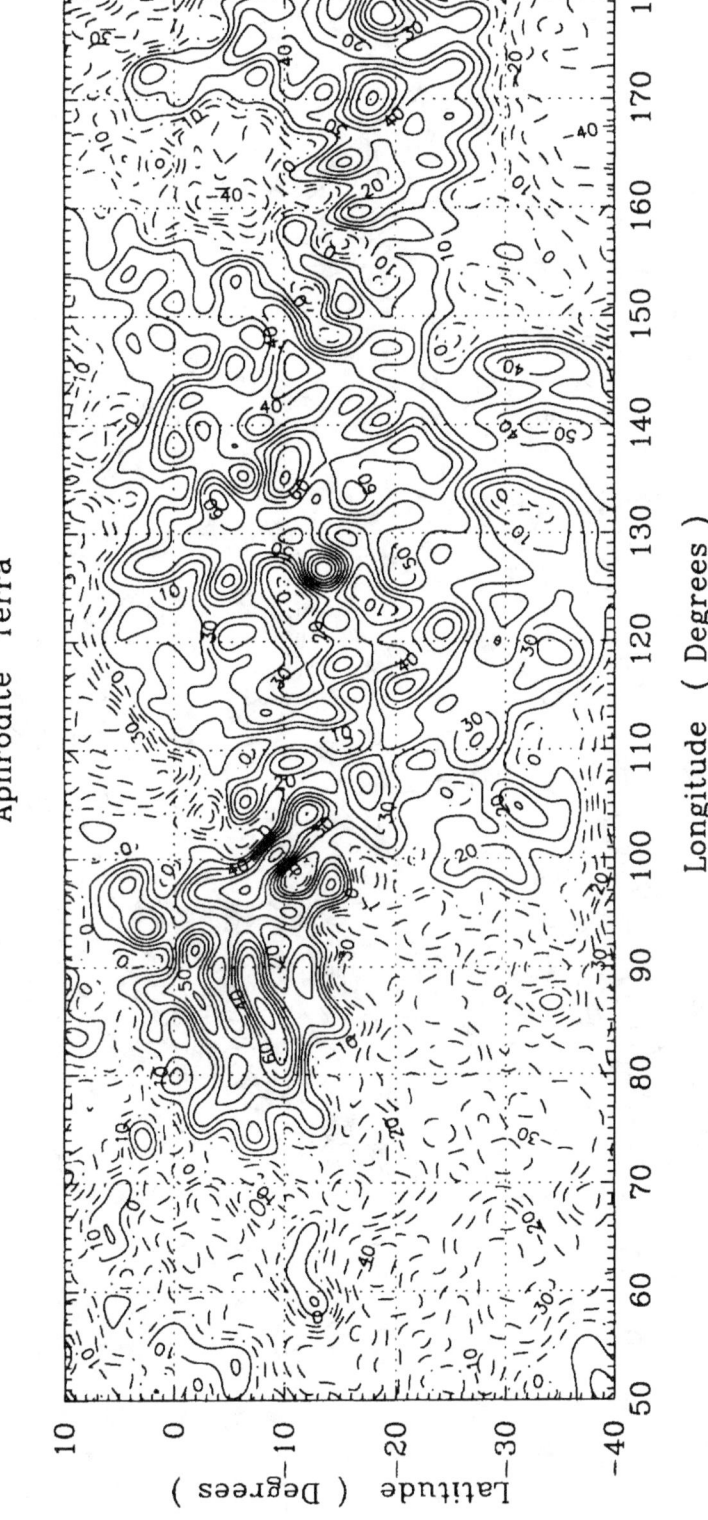

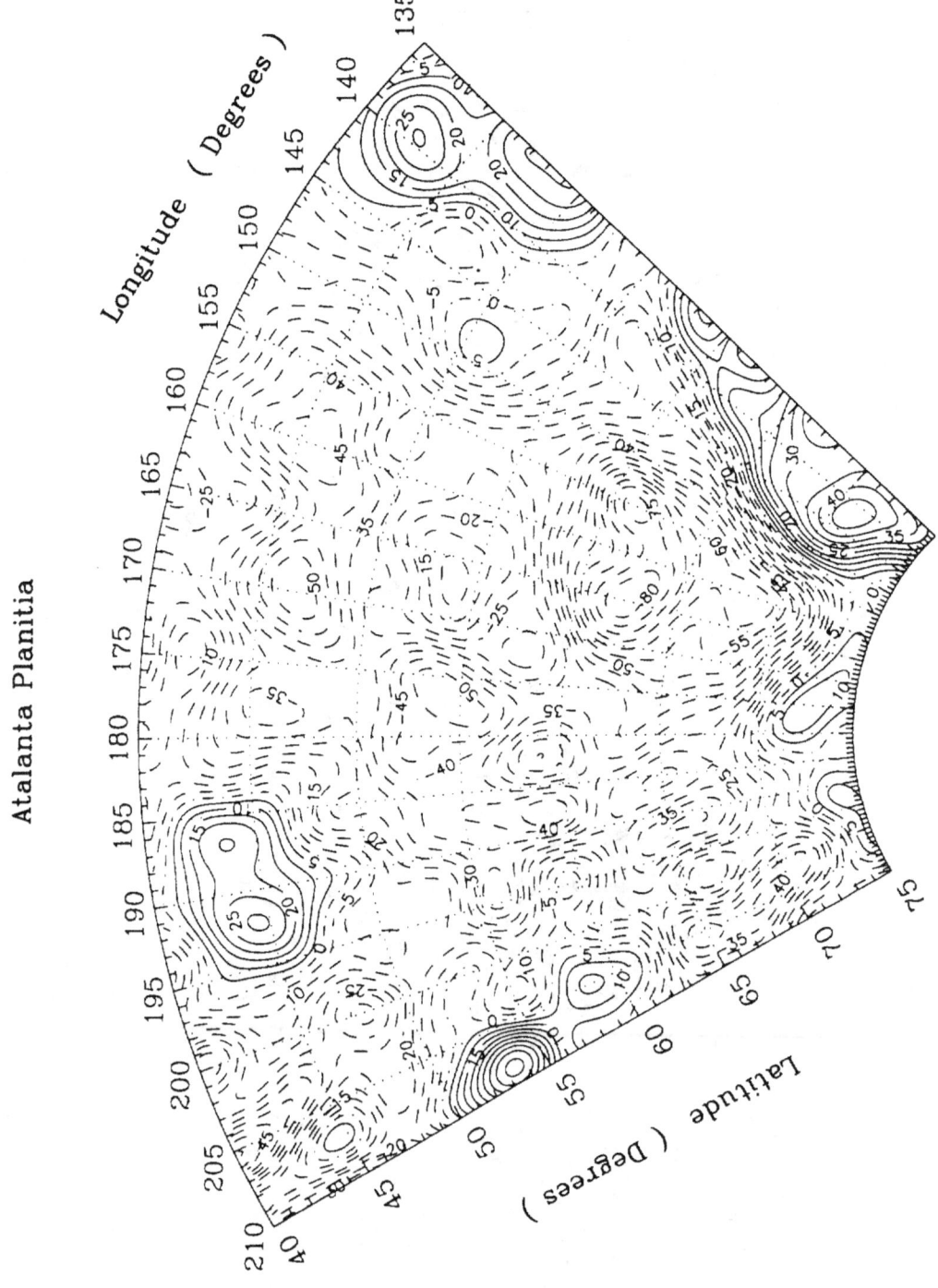

Atalanta Planitia

www.ingramcontent.com/pod-product-compliance
Lightning Source LLC
Chambersburg PA
CBHW081724170526
45167CB00009B/3685